Cooperative Extension   NRAES–130

Managing Nutrients and Pathogens from Animal Agriculture

Proceedings of a Conference for Nutrient Management Consultants, Extension Educators, and Producer Advisors
Camp Hill, Pennsylvania
March 28–30, 2000

Natural Resource, Agriculture, and Engineering Service (NRAES)
Cooperative Extension
152 Riley-Robb Hall
Ithaca, New York 14853-5701

NRAES-130

March 2000

ISBN 0-935817-54-9

Library of Congress Cataloging-in-Publication Data (TO COME)

NRAES—Natural Resource, Agriculture, and Engineering Service
Cooperative Extension, 152 Riley-Robb Hall
Ithaca, New York 14853-5701
Phone: (607) 255-7654 • Fax: (607) 254-8770 • E-mail: NRAES@CORNELL.EDU
Web site: WWW.NRAES.ORG

Table of Contents

● ●

Table of Contents

· ·

Table of Contents

Table of Contents

●●●

Table of Contents

· ·

Session 1

Nutrients and Water Quality

Nutrients and Water Quality

Hank Zygmunt
United States Environmental Protection Agency

This conference sponsored by the Natural Resource, Agriculture, and Engineering Service with its fourteen cooperative extension agencies from many excellent land grant universities provides an excellent forum to continue to discuss and learn more about the continued challenge of improving the nation's water quality and public health. Local, state, and federal governments, academic educators, the agricultural community, environmental organizations, corporate leaders, and in particular the youth of tomorrow are all dedicated to becoming better educated to make needed improvements in our environment as we immerse ourselves in the next decade of the new millennium.

The nation has made great strides in dramatically improving the health of rivers, lakes, and coastal waters. By implementing the Clean Water Act, it has prevented billions of pounds of pollution from adversely impacting waterways throughout the nation and in so doing has doubled the number of waterways to be safe for fishing and swimming. Also, as many rivers, lakes, and coasts being part of the traditional natural heritage have been restored these improvements have helped the economic viability for healthy communities of tomorrow.

Despite the tremendous progress that has been achieved in reducing water pollution, almost 40 % of the Nation's waters that have been assessed by States do not meet water quality goals. Based upon water quality monitoring information submitted to EPA by State Water Quality Agencies, about 15,000 waterbodies are impacted by siltation, nutrient, bacteria, oxygen-depleting substances, metals, habitat alteration, pesticides, and organic toxic chemicals. With much of the pollution associated from factories and sewage treatment plants being dramatically reduced, polluted runoff from city streets, rural areas, and other sources continue to degrade the environment and place drinking water at risk. As many of you have know, EPA's Water Quality Inventory report, based upon data furnished by States, have reported that agriculture is the leading source of impairment in the Nation's rivers and lakes, and a major source of impairment in estuaries. The impairment comes from nutrients associated with agricultural operations. These farming activities are not the only source of nutrient pollution. Other loadings come from wastewater treatment plants, industrial plants, and septic tanks. Atmospheric deposition of pollutants is yet another source of nitrogen, an area of research receiving much needed attention. Amongst all of these sources of pollution it should be recognized that investments of responsibilities have been and will continue to be made by the conservation movement with commensurate levels of support from state and federal governments for both technical and financial resources. In consideration of the conservation ethic and land stewardship that the agricultural community has and will continue to demonstrate one of the critical components of future success is the development and implementation of comprehensive nutrient management plans (CNMPs). These CNMPs represent the cornerstone of both voluntary and regulatory activites and as such will manage nutrients and quantities of manure on our nation's animal feeding operations resulting in improvements to water quality and protection of public health. It is clear that as the agricultural community is led by the farm community, their conservation efforts should directly correlate to improvements in the environment. Without clean water and air that makes up a part of their environment, continued reliance upon these precious natural resources could provide situations that are difficult to support sustainability and impact future generations.

NITROGEN AND PHOSPHORUS

EPA in 1988 reported that nitrogen and phosphorus from agriculture accelerate production in receiving waters resulting in a variety of problems including clogged pipelines, fish kills, and reduced recreational opportunities. Nitrate is also a potential human health threat.

The Economic Research Service describes the existence of regional problems based upon a variety of pollutants impacting water quality in the Great Lakes, the Chesapeake Bay, and the Gulf of Mexico. In the Great Lakes most of the shoreline is polluted from toxic chemicals, primarily polychlorinated biphenyls, mercury, pesticides, and dioxins found in fish samples. Atmospheric deposition of toxics, point sources, and contaminated sediment are the leading sources of water quality impairment.

The Chesapeake Bay has been impacted primarily for elevated concentrations of nitrogen and phosphorus leaving the Bay overenriched. Much progress has been made in implementing both voluntary and regulatory programs yet additional reductions in nutrients that promote algae growth and have led to poor water quality which has dramatically reduced shellfish harvest needs to occur.

The Gulf of Mexico has experienced a doubling in the size of an oxygen-deficient "dead" zone to a 7,000 square mile area. The primary cause is believed to be increased levels of nitrates carried to the Gulf by the Mississippi and Atchafalaya Rivers. A major source of nitrates is fertilizers from the Upper Mississippi Basin.

GROUND WATER

Unlike the nations comprehensive surface water quality program, a comprehensive groundwater program presently does not exist. However due to the concerns about the impacts on surface waters and the relationship between both surface and ground waters, State agencies do report on the quality of ground water resources in State Water Quality reports under Section 305(b) of the Clean Water Act. Generally, States report that sources of ground water contamination is localized with over 45 States reporting that pesticides and fertilizer applications were sources of contamination. Other indications of ground water quality are derived from EPA's National Survey of Pesticide in Drinking Water Wells. This survey provided the first national estimate of the frequency and concentration of nitrates and pesticides in community water system wells and rural domestic drinking water wells.

HUMAN HEALTH CONCERNS

Based upon a report completed in December,1997 by the United States Senate Committee on Agriculture, Nutrition, and Forestry, the report indicates that much attention is focused on the direct impact of animal waste on the aquatic environment but there are human health concerns associated with animal waste pollution that need to be further studied.

Animal manure contain pathogens that can be harmful to humans. Some of these include Salmonella, Cryptosporidium, and Giardia which have polluted drinking water supplies and adversely impacting the health of humans. Microorganisms in livestock can cause several diseases through direct contact with contaminated water, consumption of contaminated drinking water or consumption of contaminated shellfish. Fortunately proper animal management practices and water treatment can minimize the risk to human health posed by these pathogens.

THE USDA/EPA UNIFIED ANIMAL FEEDING OPERATION STRATEGY

Nutrients from manure is an increasing concern given the recent trend toward larger, more specialized beef, swine, and poultry operations. As many of you know the nature of the animal feeding industry has changed dramatically over the past decades. Advances in technologies for raising and feeding animals, decreases in transportation costs, and organizational changes in agricultural businesses and corporations have transformed the industry. The data overwhelmingly shows a shift in the industry from smaller to much larger operations. The total number of feeding operations has declined in every livestock sector. Yet over the last several years the total number of animals in each livestock group has increased.

As a result of the nation's increasing awareness of waterbodies being adversely impacted by polluted runoff and in part the recognition of the number of animal feeding operations and the trends in agriculture towards greater consolidation of these operations, EPA and the U.S. Department of Agriculture developed the Unified Strategy for Animal Feeding Operations announced on March 9, 1999. The Strategy outlines a flexible, common-sense approach to minimize the water quality and public health impacts of animal feeding operations while ensuring the long term sustainability of livestock production in the nation.

Farmers were among the first stewards of our nation's natural resources and farmers have consistently recognize the value of protecting water quality and the environment. Conservation translates to improving water and air quality and enhancing our natural resources. In consideration of the Strategy a national performance expectation has been established that all AFO owners and operators should develop and implement technically sound, economically feasible, and site specific comprehensive nutrient management plans for properly managing the animal wastes produced at their facilities.

While the vast majority of the estimated 450,000 AFOs nationwide are encouraged to develop CNMPs on a voluntary basis, between 15,000-20,000 large AFOs, generally those with 1,000 Animal Units or more, will be required to implement CNMPs as part of their NPDES discharge permit, under the authority of the Clean Water Act. Currently, EPA estimates that several thousand CAFOs now have NPDES permits under the Clean Water Act.

FUTURE EFFORTS

As we prepare for the next ten years, EPA will continue to implement the provisions of the AFO Strategy which call for a number of innovative opportunities to be developed in continuing to address water quality and public health impacts from animal feeding operations.

Seven major strategic issues are included in the AFO Strategy. They are:

1. Building Capacity for CNMP Development and Implementation;

2. Accelerating Voluntary, Incentive-Based Programs;

3. Implementing and Improving the Existing Regulatory Program;

4. Coordinated Research, Technical Innovation, Compliance Assistance, and Technology Transfer;

5. Encouraging Industry Leadership;

6. Data Coordination; and

7. Performance Measures and Accountability.

Also described in the Strategy is a two phased regulatory approach that will improve upon the existing Clean Water Act's permit program as the National Pollutant Discharge Elimination System is administered by State Water Quality Agencies and EPA.

Throughout 2000 EPA and States will issue permits to CAFOs under the existing CWA regulations. To further explain how the EPA NPDES program for CAFOs needs to be implemented, a Guidance document is planned to be released by EPA this March. This document has been developed based upon extensive public comment review and should reflect a well balanced approach of applying a common sense approach to permitting CAFOs.

In December of 2000, EPA is planning on releasing draft CAFO regulations that will represent an update to existing regulations based upon revised effluent guidelines for the livestock industry. These draft regulations will convey to the general public the most up to date cost effective technology based pollution control abatement efforts that will protect water quality and public health. Final regulations, based upon completing the effluent guideline process, is planned to go into effect in 2005.

Throughout the development of both the voluntary program and regulatory program for animal feeding operations the technical underpinning for success is associated with the fundamental design and implementation of the comprehensive nutrient management plan. It is clear as both USDA and EPA continue to implement the key actions of the AFO Strategy that many professionals both within the public and private sectors will need to play a vital role in developing levels of expertise as CNMPs are developed for certification purposes. Some of these levels of expertise will be associated with technical advancements in the area of phosphorus as a limiting nutrient that will need to be incorporated into the development of CNMPs and the resulting potential need for using quantities of manure in alternative ways.

MANAGING NUTRIENTS

USDA's Natural Resources Conservation Service has adopted a new policy for nutrient management. Fundamental to the new policy and the Nutrient Management Conservation Practice Standard (#590) released in May 1999 are both the production and environmental protection considerations of nutrient management.

Presently, State NRCS offices are developing state specific nutrient management programs in response to the 590 Standard which are due to be completed in the Summer of 2001. One of the basic requirements of the new policy is the development of nutrient management plans that will include: Field and Soil Maps, Crop Sequence, Realistic Yields, Soil Test and Other Analysis, Sources of Nutrients, Nutrient Budget, Rates, Methods and Timing of Nutrient Application, Identification of Sensitive Areas, and Guidance for Implementation, Operation and Maintenance for Recordkeeping. Of particular importance as the nation manages the application to land of both commercial fertilizers and animal manures will be the heightened awareness of certified specialists, both public and private, in determining the appropriate application rates of nutrients, both nitrogen, phosphorus, and potassium in producing crop yields as well as for protecting the environment.

PHOSPHORUS

The role of phosphorus in agriculture and water quality is receiving significant attention form agricultural organizations, environmental groups, government , and the news media. Since phosphorus can contribute to water quality impairment, there is a growing awareness that effective management of phosphorus in traditional conservation approaches is undergoing a national review. In particular, as animal manure continues to be widely used as an excellent source of plant nutrients, typical application rates determined based upon the nitrogen requirement of the crop produce an increase in the amount of phosphorus in the soil. Over time the accumulation of phosphorus in the soil can become significant and overall contribute to eutrophication in many impaired waterbodies throughout the nation. The fundamental goal to reduce phosphorus losses from agriculture should be to balance off-farm inputs of it in feed and fertilizer with outputs in product, and to manage soils in ways that retain crop nutrients resources and do not result in excessive accumulations in the soils. As we continue to further understand how best to manage phosphorus as comprehensive nutrient management plans are developed and implemented, it will be of utmost importance to minimize soil phosphorus buildup in excess of crop requirements, reduce surface runoff and erosion and improve our capability to identify fields that are major sources of phosphorus loss to waterways.

FUTURE MARKETS

The agricultural community is recognizing that in certain on farm situations there are opportunities for third parties to use value added manure for a variety of uses that not only benefit both the farmer and the third party but potentially the environment. As phosphorus considerations become incorporated into nutrient management plans where appropriate and as called for by conservation offices and are also required by law, we will see a call on the part of not only the farm community but a variety of entrepreneurs who claim to have the " silver bullet" to help solve the nations "excess" manure situation. We have already established third parties who are benefitting from the use of value added manure products but we are only in the early stage of development. As state conservation offices finalize their responses to NRCS #590 nutrient standard by the summer of 2000, progressive communities and counties will plan accordingly to fully utilize quantities of manure generated from livestock operations for resource generating opportunities that benefit those involved in the production, marketing and use sector. The opening of value driven, environmental markets are being evaluated today by a number of companies that understand that changes are potentially being discussed particularly in geographic areas throughout the nation that are generating more manure than the adjacent land areas can absorb in any environmentally sound manner. The role of the agricultural community in this developing science driven equation is to understand the basis for the concern and the relevant opportunities that will present itself to essentially transport organic materials away from generation areas to benefit other agricultural or other uses.

FEDERAL PROGRAMS THAT ADDRESS AGRICULTURAL NONPOINT SOURCE POLLUTION

The United States Environmental Protection Agency is primarily responsible for policies and programs that address water quality under the authority of the Clean Water Act. In mentioning a few of these national programs the National Pollutant Discharge Elimination System (NPDES) establishes limits on individual point source discharges is the program that will be continued to be used to permit Concentrated Animal Feeding Operations (CAFOs) as the National USDA/EPA Animal Feeding Operation Strategy is implemented for those facilities that require a discharge permit. As mentioned above the design of a comprehensive nutrient management plan is the technical underpinning of the permit.

When technology based controls are not sufficient for waters to meet State established water quality standards, Section 303(d) of the Clean Water Act requires States to identify those waters and to develop total maximum daily loads. The TMDL process is flexible and is a locally driven process although both EPA and State Water Quality Agencies are involved in the approval process.

Overall management of nonpoint sources primarily is associated with Section 319, the Nonpoint Source Program, Section 314, the Clean Lakes Program , and the National Estuary Program under Section 320. Section 319 currently is providing $200 million in grants to States to promote the implementation of NPS management programs. Section 319 funds also are available to assist reduce nonpoint source pollution for the nation's Clean Lakes program. The National Estuary Program assist States develop and implement comprehensive basin wide programs to conserve and manage precious estuary resources.

The Safe Drinking Water Act (SDWA) requires EPA to establish standards for drinking water quality and requirements for water treatment by public water systems. The SDWA authorized the Wellhead Protection Program to protect supplies of ground water used as public drinking water from contamination by chemicals, pesticides, nutrients, and other agricultural chemicals.

USDA also administer a number of excellent programs that assist the agricultural community better manage nutrients and thus help protect both surface and ground water. These programs rely upon providing technical assistance, education, research, and financial assistance to help achieve water quality and other environmental objectives. These programs include: the Environmental Quality Incentives Program, the Water Quality Program, Conservation Technical Assistance, Conservation Compliance, Conservation Reserve, the Wetlands Reserve Program, and the Small Watersheds Program.

CONCLUSION

Much of the progress in reducing water pollution has been associated with implementing controls as part of effective national programs over discharges from sewage and industrial wastes. As we continue to address these significant pollution sources, a major water pollution challenge associated with nonpoint source runoff pollution from our urban areas, construction sites, forest harvesting operations and agricultural activities is taking center stage.

As the nation directs needed attention to a variety of sources of pollution that stem from nonpoint source runoff, a greater level of responsibility of those that are part of the rural and urban areas need to continue to further understand how their operations are impacting the environment and to further understand the solutions that are being offered from a voluntary incentive based approaches or common sense regulatory programs. In consideration of the amount of land that is private in our nation, approximately 70 percent, national attention for those that own these valuable lands in the future will become an increasing national priority. Priorities should be based in part upon the strong reliance upon smart conservation that translates into environmental benefits in achieving clean air and water. Other approaches that are regulatory in nature will draw upon common sense values that provide a flexible system for local governments to best address solutions. Innovative and creative approaches are needed as the nation makes significant strides in reducing pollution abatement from all sources of pollution.

This conference does provide a superb gathering of important and relevant topics and professional speakers that can deliver a strong a precise message. The Planning group is to be strongly commended and the Natural Resource, Agriculture, and Engineering Service with its fourteen universities equally commended for having such a timely, thought provoking, science based conference.

Sources of Nutrients
in the Nation's Watersheds

Richard A. Smith and Richard B. Alexander
Hydrologists
U.S. Geological Survey
Reston, Virginia

Biographies for most speakers are in alphabetical order after the last paper.

Introduction

Animal agriculture is a common source of nutrients in watersheds, but it is never the only source. Indeed, the diverse and ubiquitous nature of nitrogen and phosphorus forms in the environment introduces significant complexity to the increasingly important task of managing nutrients in watersheds. Thus, it is appropriate near the outset of this conference to attempt a systematic quantification of nutrient sources in surface waters as a means of exploring the relative importance of animal agriculture's influence on the nutrient balance in aquatic ecosystems under different conditions.

In this paper, we present estimates of the percentage contribution of five categories of nutrient sources to the total nitrogen and total phosphorus flux from watersheds in the major water resources regions of the conterminous United States. It is noteworthy that our estimates pertain to "in-stream" conditions rather than "input-level" contributions from each of the source categories. The latter represent a simpler way to quantify nutrient source contributions (see for example Puckett, 1995; Jaworski *et al.* 1992) but do not account for the effects of landscape and stream processing of nutrient material, and thus may give a very different picture of the importance of a particular source on water quality conditions. For example, agricultural fertilizer inputs to watersheds may be estimated from state- or county-level sales data or from estimated usage rates and cropland acreage. But fertilizer inputs generally exceed stream nutrient yields (mass per area per time) by a factor of two or more (Howarth *et al.* 1996; Carpenter *et al.* 1998) due to crop uptake and removal. In input terms, therefore, agricultural fertilizers appear to be a larger contributor to watershed nutrients than when they are expressed in in-stream terms.

Our estimates of in-stream source contributions are obtained through application of SPARROW (SPAtially Referenced Regressions On Watersheds; Smith, *et al*, 1997), a recently developed technique for interpreting water quality monitoring data in relation to watershed sources and characteristics. We begin with a brief discussion of methods for relating in-stream nutrient flux to source inputs and develop the rationale for spatial referencing of model terms. Next we provide a brief overview of the SPARROW model followed by a description of the data sources used here. The results pertaining to nutrient sources in general are presented in Tables 1 and 2. The results for animal agriculture are presented in map form in Figure 1. A brief discussion of the results and conclusions completes the report.

Quantifying watershed nutrient contributions by source category

A variety of deterministic and statistical methods have been used to develop estimates of nutrient contributions to watersheds from human and natural sources. The simplest deterministic approaches consist of a simple accounting of the inputs and outputs of nutrients. A mass balance is achieved by comparing major source inputs (*e.g.*, fertilizer application, livestock waste, atmospheric deposition, and point sources) with outputs (*e.g.*, river export, crop removal) and by assuming that total losses to volatilization, soil adsorption, sedimentation, groundwater storage and denitrification equal the difference between the total inputs and outputs. Such simple models must assume that loss processes operate equally on all sources and that the relative contributions of sources to watershed export are proportional to the inputs. More complex deterministic models of nutrients in watersheds describe transport and loss processes in more detail and incorporate terms for spatial and temporal variations in sources and sinks. A major limitation on the applicability of such models at the regional or national scale is the problem of obtaining the necessary data for process description, especially if processes are treated dynamically.

Statistical approaches to modeling nutrients in watersheds have their origins in simple correlations of stream nutrient measurements with watershed sources and landscape properties. Recent examples include regressions of coastal total nitrogen flux on population density, net anthropogenic sources, and atmospheric deposition (Caraco and Cole, 1999; Howarth *et al*, 1996). A noteworthy advantage of the statistical approach is the ability to quantify errors in model parameters and predictions. Simple correlative models consider sources and sinks to be homogeneously distributed in space, do not separate terrestrial from in-stream loss processes, and rarely account for the interactions between sources and watershed processes. By contrast, more complex empirical approaches (Smith *et al*. 1997; Preston and Brakebill, 1999; Alexander *et al*. 2000; Alexander *et al*. *in press*; Johnes, 1996; Johnes and Heathwaite, 1997) indicate that knowledge of spatial variations in watershed properties that influence nutrient attenuation can significantly improve the accuracy of estimates of export and source contributions at larger watershed and regional scales.

SPARROW (Smith *et al*. 1997), a hybrid statistical/deterministic approach, expands on previous methods by using a mechanistic regression equation to correlate measured stream nutrient flux with spatial data on sources, landscape characteristics (*e.g.*, soil permeability, temperature), and stream properties (*e.g.*, flow, water time of travel). The model separately estimates the quantities of nutrients delivered to streams and the outlets of watersheds from point and diffuse sources. Spatial referencing of land-based and water-based variables is accomplished via superposition of a set of contiguous land-surface polygons on a digitized network of stream reaches that define surface-water flow paths for the region of interest. Water-quality measurements are available from monitoring stations located in a subset of the stream reaches. Water-quality predictors in the model are developed as functions of both reach and land-surface attributes and include quantities describing nutrient sources (point and nonpoint) as well as factors associated with rates of material transport through the watershed (such as soil permeability and stream velocity). Predictor formulae describe the land-to-water transport of nutrient mass from specific sources in the watershed surrounding each reach, and in-stream transport from reach to reach in downstream order. Loss of nutrient mass occurs during both land-to-water and in-stream transport. In calibrating the model, measured rates of contaminant transport are regressed on the set of predictor formulae evaluated at the locations of the monitoring stations, giving rise to a set of estimated linear and nonlinear coefficients from the predictor formulae. Once calibrated, the model can be used to estimate contaminant transport (and concentration) in all stream reaches. In addition, because the nutrient contribution from each source is tracked separately in the model, the percent contribution from each source category (*e.g.*, fertilizer, animal agriculture, etc) can also be computed for each reach. A study of model reliability is given in Alexander *et al.* (*in press*).

SPARROW has been applied nationally in the United States (Smith *et al.* 1997) with separate studies of nitrogen flux in the Chesapeake Bay watershed (Preston and Brakebill, 1999), the Mississippi River and its tributaries (Alexander *et al.* 2000), the watersheds of major U.S. estuaries (Alexander *et al. in press*), and watersheds of New Zealand (Alexander, R.B., U.S. Geological Survey, written comm., 1999).

Data sources and methods

Detailed descriptions of the data sources and calibration results for the SPARROW total nitrogen (TN) and total phosphorus (TP) models used in this study are given in Smith *et al.* (1997). Observations of in-stream nutrient transport (i.e., the dependent variables in model calibrations) were based on U.S. Geological Survey (USGS) long-term stream monitoring records of TN and TP for the period 1974 to 1989 for 414 (TN) and 381 (TP) sites in the conterminous United States. Data for nutrient sources were developed for five major source categories: 1) municipal and industrial point sources, 2) commercial fertilizer, 3) animal agriculture, 4) nonagricultural runoff, and 5) atmospheric deposition (TN model only). Watershed inputs of nutrients for the source category fertilizer are based on fertilizer sales data. Nitrogen contributions from leguminous crop fixation are assumed to be reflected in the estimated coefficient for the fertilizer source category. Nutrient inputs for the source category animal agriculture are based on federal surveys of animal populations and literature data on animal waste production and the nutrient content of animal wastes. Atmospheric deposition sources are based on measured inputs of wet nitrate deposition, which are scaled by the model to reflect additional atmospheric contributions from such sources as wet deposition of ammonium and organic nitrogen and dry deposition of inorganic nitrogen. The source category nonagricultural runoff is scaled according to nonagricultural land area, and includes remaining nutrient sources unaccounted for by the other categories. This source may include surface and subsurface runoff from wetlands and urban, forested, and barren lands.

Data on the source inputs and terrestrial characteristics, available for nearly 20,000 land-surface polygons, were referenced to approximately 60,000 stream reaches in a digital stream network using conventional spatial disaggregating methods in a geographic information system [see Smith *et al.* 1997]. The surface water flow paths, defined according to a 1:500,000 scale digital network of rivers for the conterminous United States, cover nearly one million kilometers of channel, and are obtained from the U.S. Environmental Protection Agency River Reach File 1 (RF1). The river reach network provides the spatial framework in the model for relating in-stream measurements of flux at monitoring stations to landscape and stream channel properties in the watersheds above these stations. The median watershed size of the reaches is 82 km^2 with an interquartile range from 40 to 150 km^2. Stream attributes of the digital network include estimates of mean streamflow and velocity from which water time of travel is computed as the quotient of stream length and mean water velocity.

Model predictions of nutrient export were summarized for the 2,057 nontidal watersheds (hydrologic cataloging units; see Seaber *et al.* 1987) comprising the major water-resources regions of the conterminous United States (see Smith *et al.* 1997; model output is available at http://water.usgs.gov/nawqa/sparrow/wrr97/results.html). These watersheds are a logical choice for national level water-quality characterization because they represent a systematically developed and widely recognized delineation of U.S. watersheds, and provide a spatially representative view of water-quality conditions (Smith *et al.* 1997; Seaber *et al.* 1987).

Results and discussion

A statistical summary of TN and TP export from the watersheds of the major water resources regions of the conterminous United States is given in Tables 1 and 2. The tables give median total export and the median and quartile percent contributions to export from each of the five source categories. According to Table 1, for example, the estimated median TN export from the watersheds in the Mid Atlantic region is 9.0 kg ha^{-1} yr^{-1}. The median contribution to TN export from animal agriculture in the same region is 15.5 percent with quartile (i.e., 25th and 75th percentile) values of 8.2 and 23.0 percent.

Median export of TN and TP varies among the regions by more than an order of magnitude, with the highest rates for both elements occurring in the Upper Mississippi and Ohio regions and the lowest occurring in the Great Basin and Rio Grande regions. Recognizing that these figures refer to the median rate in each region, it is clear that the total range of variation in nutrient export rates among all watersheds is much larger.

From Tables 1 and 2, it is also clear that the relative importance of the different categories of nutrient sources in watersheds vary greatly from one region to another. Not surprisingly, point sources, which generally represent the smallest contributors to nitrogen and phosphorus export from watersheds, are seen to reach their highest importance in the densely populated Northeast, Mid Atlantic, and Great Lakes regions. In the Northeast region, in fact, point sources contribute more than half of the total phosphorus export in at least twenty-five percent (i.e., upper quartile) of the watersheds. Fertilizer is a large contributor to nutrient loads in the high-export Upper Mississippi and Ohio regions, but makes its highest contribution in percentage terms (median TN=75 percent; median TP=64 percent) in the Red-Rainy basin in the northern plains where total export is low. In the southwestern regions, where total export is also low, nonagricultural runoff from forest, barren, and range lands contributes the largest percentage to watershed export of nutrients. Atmospheric nitrogen contributes more than a quarter of the total nitrogen export in a majority of watersheds in the northeastern quadrant of the United States, and is the dominant source in the Great Lakes and Mid-Atlantic water-resource regions.

The importance of animal agriculture as a nutrient source in watersheds is presented in the regional summaries in Tables 1 and 2 and also in map form in Figures 1a and 1b, which show the percentage contribution to TN and TP export from each of the 2,057 hydrologic units. For total nitrogen, the median contributions of animal agriculture in the

Table 1. Point- and nonpoint-source contributions to total nitrogen export from watersheds in major water-resource regions of the conterminous United States. Total export is the median export from hydrologic cataloging units in each region. The median and quartile values for the source contributions within each region are expressed as a percentage of the total export.

Region	Total Export ($kg\ ha^{-1}\ yr^{-1}$)	Percentage of Total Export									
		Point Sources		Fertilizer		Animal Agriculture		Atmosphere		Nonagricultural Runoff	
		Median	Quartiles	Median	Quartiles	Median	Quartiles	Median	Quartiles	Median	Quartiles
Northeast	6.7	4.3	1.2 – 19	6.2	3.6 – 13	5.8	2.8 – 9.3	30	25 – 38	38	27 – 52
Mid Atlantic	9.0	4.1	1.6 – 20	14	10 – 22	16	8.2 – 23	32	20 – 40	22	14 – 28
Southeast Atlantic-Gulf	5.9	2.7	1.1 – 8.2	26	17 – 38	14	8.8 – 21	21	15 – 28	26	19 – 34
Great Lakes	8.0	4.3	1.3 – 12	22	7.8 – 41	10	4.6 – 17	25	16. 34	17	6.5 – 40
Ohio	11	1.8	0.7 – 7.0	30	9.2 – 58	14	9.2 – 20	25	15 – 42	13	6.4 – 23
Tennessee	8.3	2.7	0.6 – 7.2	22	16 – 29	19	15 – 25	26	21 – 33	24	18 – 28
Upper Miss.	13	0.8	0.5 – 1.6	55	40 – 66	21	15 – 27	13	11 – 17	3.6	2.1 – 10
Lower Miss.	7.6	2.3	1.0 – 11	40	14 – 64	6.3	3.2 – 10	22	14 – 28	18	8.0 – 28
Red Rainy	3.5	0.3	0.1 – 0.6	75	57 – 81	5.2	2.8 – 9.0	9.3	7.4 – 14	7.2	3.4 – 20
Missouri	2.1	<0.1	<0.1 – <0.1	30	8.8 – 51	20	15 – 25	16	12 – 20	29	9.5 – 55
Ark-Red	3.9	0.8	0.2 – 1.9	29	20 – 46	23	17 – 29	18	14 – 23	20	12 – 28
Texas-Gulf	3.7	0.9	0.1 – 5.3	30	18 – 41	19	14 – 26	18	14 – 21	23	13 – 37
Rio Grande	1.0	<0.1	<0.1 – <0.1	1.7	0.6 – 5.1	11	6.2 – 14	13	8.9 – 16	71	63 – 80
Upper Colorado.	1.9	0.1	<0.1 – 0.4	2.0	1.1 – 4.5	8.7	5.3 – 12	17	14 – 20	72	64 – 76
Lower Colorado	0.7	<0.1	<0.1 – 0.4	1.6	0.7 – 16	6.6	2.9 – 10	8.6	5.0 – 9.9	78	65 – 84
Great Basin	0.9	<0.1	<0.1 – <0.1	3.6	0.9 – 9.2	9.3	5.6 – 15	6.4	5.4 – 8.1	78	61 – 86
Pacific NW	4.2	<0.1	<0.1 – <0.1	12	5.5 – 30	11	7.3 – 14	13	8.0 – 16	57	34 – 69
California	4.8	1.2	0.3 – 6.7	21	8.9 – 52	12	7.6 – 17	8.7	5.5 – 13	35	16 – 62
United States	4.7	0.8	0.5 – 3.4	22	7.5 – 45	14	8.2 – 21	16	11 – 23	28	13 – 56

Table 2. Point- and nonpoint-source contributions to total phosphorus export from watersheds in major water-resource regions of the conterminous United States. Total export is the median export from hydrologic cataloging units in each region. The median and quartile values for the source contributions within each region are expressed as a percentage of the total export.

Region	Total Export ($kg\ ha^{-1}\ yr^{-1}$)	Point Sources		Fertilizer		Animal Agriculture		Nonagricultural Runoff	
		Median	Quartiles	Median	Quartiles	Median	Quartiles	Median	Quartiles
Northeast	0.41	18	7.1 – 55	7.7	4.1 – 13	6.8	3.1 – 14	54	28 – 68
Mid Atlantic	0.68	14	5.6 – 45	19	14 – 26	25	13 – 38	22	12 – 37
Southeast Atlantic-Gulf	0.54	7.9	3.4 – 22	23	11 – 32	30	20 – 40	29	18 – 40
Great Lakes	0.49	13	4.7 – 24	26	12 – 41	18	8.0 – 29	22	6.6 – 59
Ohio	0.93	7.3	2.6 – 21	30	15 – 45	30	20 – 41	14	6.2 – 33
Tennessee	0.67	7.2	2.3 – 16	24	18 – 30	33	26 – 43	23	16 – 34
Upper Miss.	1.1	3.6	1.7 – 6.8	37	30 – 44	47	35 – 55	3.6	2.1 – 10
Lower Miss.	0.53	9.6	4.7 – 27	29	6.7 – 58	15	8.9 – 25	27	13 – 39
Red Rainy	0.22	1.7	0.9 – 3.6	64	45 – 77	12	7.8 – 22	11	5.1 – 23
Missouri	0.19	1.0	0.3 – 2.4	20	6.8 – 30	42	28 – 55	29	14 – 60
Ark-Red	0.36	2.7	1.0 – 5.4	18	10 – 29	48	38 – 56	24	13 – 33
Texas-Gulf	0.38	2.7	0.5 – 14	18	8.6 – 25	39	29 – 49	29	16 – 46
Rio Grande	0.12	<0.1	<0.1 – <0.1	0.9	0.3 – 2.7	16	11 – 20	81	73 – 87
Upper Colorado	0.14	0.4	<0.1 – 1.8	1.1	0.6 – 2.6	16	10 – 21	81	75 – 88
Lower Colorado	0.10	0.3	<0.1 – 1.3	1.0	0.4 – 7.7	12	4.9 – 18	83	69 – 91
Great Basin	0.09	0.2	<0.1 – 2.1	3.5	1.7 – 8.1	14	8.4 – 20	79	65 – 88
Pacific NW	0.30	1.5	0.2 – 8.9	7.1	2.9 – 18	19	12 – 25	65	43 – 78
California	0.41	4.0	1.1 – 19	9.5	3.7 – 30	19	12 – 29	40	20 – 71
United States	0.37	3.0	0.7 – 11	17	5.5 – 31	26	15 – 42	33	15 – 65

a Total Nitrogen

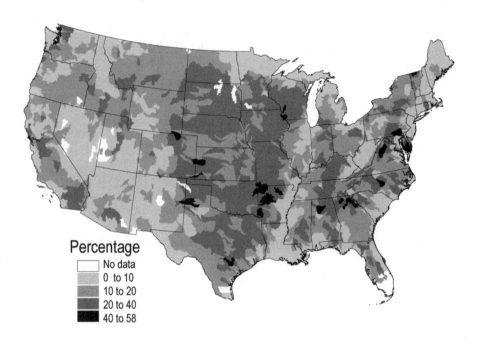

Percentage

	No data
	0 to 10
	10 to 20
	20 to 40
	40 to 58

b Total Phosphorus

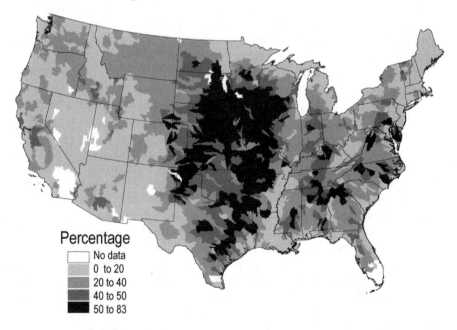

Percentage

	No data
	0 to 20
	20 to 40
	40 to 50
	50 to 83

Figure 1. Contributions (percentage) of animal agriculture to nutrient export from hydrologic cataloging units in the conterminous United States.

water resource regions range from about 5 to 23 percent (Table 1). The highest contributions (median=18 to 23 percent) are found in the Tennessee, Upper Mississippi, Missouri, Arkansas-Red, and Texas-Gulf regions. Figure 1a indicates that in many watersheds in the states of Wisconsin, Iowa, Missouri, Oklahoma, and Texas, animal agriculture contributes from 20 to 58 percent of TN export. Although animal agriculture in the Mid-Atlantic and Southeast regions contributes a median of only about 8 percent to export (Table 1), farm animals contribute more than 20 percent of the exported nitrogen from many watersheds within these regions (Figure 1a). The lowest contributions of animal agriculture (i.e., less than 10 percent) are found in the Northeast, Great Lakes, and many western water-resources regions.

For total phosphorus, the median percentage contributions of animal agriculture in the major water-resource regions range from about 7 to 48 percent (Table 1) or approximately twice the contributions estimated for total nitrogen. The highest contributions occur in the Upper Mississippi, Arkansas-Red, Missouri, and Texas-Gulf regions, including watersheds in the states of Wisconsin, Iowa, Nebraska, Missouri, Kansas, Arkansas, Oklahoma, and Texas (Figure 1b). The water-resource regions with the lowest contributions of animal agriculture to stream phosphorus (i.e., less than 20 percent) are similar to those found for total nitrogen, and include the Northeast, Great Lakes, and many western regions.

Summary and Conclusions

Estimating the importance of animal agriculture as a source of nutrients in watersheds is made difficult by the diverse and ubiquitous nature of nitrogen and phosphorus forms in the environment. The relative importance of nutrient sources is most meaningfully expressed in "in-stream" terms rather than as raw inputs. However, the contribution of individual nutrient sources to in-stream water quality is not directly measurable in large watersheds, and must therefore be estimated using a watershed model. The results of recent research indicate that spatial referencing of variables improves the accuracy of watershed nutrient models. SPARROW models of TN and TP have been calibrated with stream monitoring records from more than 370 locations across the conterminous United States. These models were used here to estimate nutrient contributions from five source categories for the 2,057 cataloging unit watersheds comprising the major water-resources regions.

The relative importance of the different categories of nutrient sources in watersheds vary greatly from one region to another reflecting differences in human activities. Point sources generally contribute little to nutrient export from most of the nation's watersheds, but contribute a majority of the total phosphorus export from many watersheds in the densely populated northeastern United States. Atmospheric deposition is the largest contributor to stream export of nitrogen in more than half of the watersheds in the northeastern United States. In the southwestern United States, nonagricultural runoff is the predominant source of both nitrogen and phosphorus in watershed export. Fertilizer is an important contributor to nutrient export in many watersheds throughout the central United States, and is the largest contributor in most watersheds in the Ohio Valley and Midwestern states. Animal agriculture is also an important contributor of both nitrogen and phosphorus in watersheds in the same regions, but animal wastes constitute a much larger fraction of phosphorus export than nitrogen export in these areas.

References

Alexander, R.B., Smith, R.A., and Schwarz, G.E., Effect of stream channel size on the delivery of nitrogen to the Gulf of Mexico, *Nature* **403**, 758-761 (2000).

Alexander, R.B., Smith, R.A. Schwarz, G.E., Preston, S.D., Brakebill, J.W., Srinivasan, R., and Pacheco, P.A., Atmospheric nitrogen flux from the watersheds of major estuaries of the United States: An application of the SPARROW watershed model, *in Assessing the Relative Nitrogen Inputs to Coastal Waters from the Atmosphere*, American Geophysical Union Water Monograph, Valigura, R., Paerl, H.W., Meyers, T., Castro, M.S., Turner, R.E., Alexander, R.B., Brock, D., and Stacey, P.E. (editors) *in press*.

Caraco, N.F., and Cole, J.J., Human impact on nitrate export: An analysis using major world rivers, *Ambio.* **28**, 167-170 (1999).

Carpenter, S.R., Caraco, N.F., Correll, D.L., Howarth, R.W., Sharpley, A.N., and Smith, V.H., Nonpoint pollution of surface waters with phosphorus and nitrogen, *Ecological Applications.* **8**, 559-568 (1998).

Howarth, R.W., G. Billen, D. Swaney, A. Townsend, N. Jaworski, K. Lajtha, J.A. Downing, R. Elmgren, N. Caraco, T. Jordan, F. Berendse, J. Freney, V. Kudeyarov, P. Murdoch & Zhu Zhao-liang, Regional nitrogen budgets and riverine N & P fluxes for the drainages to the North Atlantic Ocean: natural and human influences, *Biogeochem.* **35**, 75-139 (1996).

Jaworski, N.A., Groffman, P.M., Keller, A.A. & Prager, J.C., A watershed nitrogen and phosphorus balance: the Upper Potomac River basin, *Estuaries* **15**, 83-95 (1992).

Johnes, P.J. Evaluation and management of the impact of land use change on the nitrogen and phosphorus load delivered to surface waters: the export coefficient modelling approach, *J. of Hydrology* **183**, 323-349 (1996).

Johnes, P.J. & Heathwaite, A.L., Modelling the impact of land use change on water quality in agricultural catchments, *Hydrological Processes* **11**, 269-286 (1997).

Preston, S.D. and Brakebill, J.W., *Application of Spatially Referenced Regression Modeling for the Evaluation of Total Nitrogen Loading in the Chesapeake Bay Watershed* (U.S. Geological Survey Water Resources Investigations Report 99-4054, Baltimore, Maryland, 1999), http://md.water.usgs.gov/publications/wrir-99-4054/ .

Puckett, L. J., Identifying the major sources of nutrient water pollution, *Environ. Sci. Techn.* **29**, 408-414 (1995).

Seaber, P.R., Kapinos, F.P. & G.L. Knapp, *Hydrologic Units Maps* (U.S. Geological Survey Water Supply Paper, 2294, Reston, Virginia, 1987).

Smith, R.A., G.E. Schwarz & R.B. Alexander, Regional interpretation of water-quality monitoring data, *Wat. Resour. Res.* **33**, 2781-2798 (1997), http://water.usgs.gov/nawqa/sparrow/wrr97/results.html .

Session 2

Animal Agriculture and Nutrients

●●●●●●●●●●●●●●●●●●●●●●●●●●●●●●●●●●●

A Changing Animal Agriculture: Implications for Economies and Communities in the Northeast and Middle Atlantic States

Charles W. Abdalla, Ph.D.
Associate Professor of Agricultural Economics
Department of Agricultural Economics and Rural Sociology
The Pennsylvania State University, University Park, Pennsylvania

Biographies for most speakers are in alphabetical order after the last paper.

••

Introduction

The structure of animal agriculture in the US is rapidly changing. These structural changes include the number and size of farms, marketing arrangements for coordinating the vertically linked food system, and the location of animal agriculture (Martin and Norris, 1998). The changes are occurring as part of an industrial transformation involving technological change, specialization and changing organizational arrangements that has been underway in agriculture for more than four decades, but that has recently accelerated (Abdalla, Lanyon, and Hallberg, 1995). A number of authors have written on the reasons behind these changes and their consequences for agriculture (Barkema and Cook, 1993; Pagano and Abdalla, 1994; Lanyon and Thompson, 1996; Drabenstott, 1998). To set the stage for discussions about nutrient loadings from animal agriculture, this paper describes the changes occurring in the structure of animal agriculture in the Northeast and Middle Atlantic States. Emphasis is placed upon the economic role of animal agriculture within states in this region and highlighting trends and regional patterns of change. This description is followed by a discussion of several important issues for communities and economies.

A Changing Animal Agriculture: The Northeast and Middle Atlantic Regions

For the purpose of helping to set the stage for discussions at the conference, a profile of the animal agriculture sectors in the Northeast and selected states in the Middle Atlantic region was assembled. These 14 states included: Connecticut, Delaware, Maine, Maryland, Massachusetts, New Hampshire, New Jersey, New York, North Carolina, Pennsylvania, Rhode Island, Vermont, Virginia and West Virginia. Information was obtained primarily from the US Agricultural Censuses and supplemented with other sources (see references).

Human Population

Since some environmental issues and many nuisance and related issues have to do with the proximity of people and animals, information on human population patterns and trends is important. Table 1 presents information on states' population in the region and its growth from the 1970s to the present. The most populated states are New York, Pennsylvania, New Jersey, North Carolina, Massachusetts, Virginia, and Maryland. The fastest growing states over the 1970 - 1990 period (North Carolina, Virginia, Delaware) were in the southern part of the region with the exception of New Hampshire. The population in the two most populated states, New York and Pennsylvania, remained almost unchanged over this period.

Table 1 - Human Population by state				
	Year			
State	1970	1980	1990	1999
Maine	993,722	1,125,043	1,227,928	1,253,040
New Hampshire	737,681	920,610	1,109,252	1,201,134
Massachusetts	5,689,170	5,737,093	6,016,425	6,175,169
Rhode Island	949,723	947,154	1,003,464	990,819
Connecticut	3,032,217	3,107,564	3,287,116	3,282,031
New York	18,241,391	17,558,165	17,990,455	18,196,601
Vermont	444,732	511,456	562,758	593,740
New Jersey	7,171,112	7,365,011	7,730,188	8,143,412
Pennsylvania	11,800,766	11,864,720	11,881,643	11,994,016
Delaware	548,104	594,338	666,168	753,538
Maryland	3,923,897	4,216,933	4,781,468	5,171,634
West Virginia	1,744,237	1,950,186	1,793,477	1,806,928
Virginia	4,651,448	5,346,797	6,187,358	6,872,912
North Carolina	5,084,411	5,880,095	6,628,637	7,650,789

General Agricultural Economy

To illustrate the size of each state's agricultural economy, states' total annual agricultural receipts are presented in Table 2. In 1974, the top ranking states were North Carolina, New York, Pennsylvania, Virginia and Maryland. In 1997, North Carolina remained number one - with over $8.2 billion in total receipts - but Pennsylvania - with about $4.1 billion annually overtook New York for the number two spot.

Table 2 - State Total Cash Farm Receipts ($1000)

State	1974	1987	1997
Maine	418,397	408,549	489,376
New Hampshire	70,384	136,023	152,531
Massachusetts	187,463	385,333	531,136
Rhode Island	22,994	75,631	63,104
Connecticut	202,585	364,424	501,087
New York	1,505,428	2,620,935	2,835,548
Vermont	221,403	434,316	499,632
New Jersey	343,140	628,114	794,263
Pennsylvania	1,504,894	3,236,697	4,131,936
Delaware	271,991	486,912	754,239
Maryland	632,553	1,128,957	1,534,792
West Virginia	142,044	267,699	396,840
Virginia	958,086	1,738,242	2,406,046
North Carolina	2,582,471	3,767,443	8,230,000

Farm-related Employment. One measure of the economic importance of the agricultural economy is the number of farm and farm-related jobs. This measure was obtained for 1996 only (US Bureau of Labor) and ranged from 11 percent to 19 percent. The state with the most farm-related employment were North Carolina (19.1), Vermont (16.8) and Maine (16.7). Those states with the least included Connecticut(11.0), New Jersey (11.9), New York (12.1) and Massachusetts (12.3).

Animal Agriculture's Role in States' Agricultural Economies

Several states in the region have significant animal agriculture sectors. One way to obtain a better picture of animal agriculture's role is examine revenues from crop and livestock receipts separately. Since the 1970s, North Carolina has been the leading state in terms of crop cash receipts. In 1974, Virginia was second followed by New York, Pennsylvania and Maryland. By 1987, Pennsylvania overtook New York and Virginia fell to fourth place, followed by Maryland. The relative rankings were the same in 1997 with the exception of New Jersey replacing Maryland at fifth place.

The relative size of state animal agriculture sectors has also changed significantly over the past 25 years. In 1974, Pennsylvania ranked first followed by New York, North Carolina, Virginia, and Maryland. By 1987, North Carolina had overtaken New York and by 1997, this state surpassed Pennsylvania to become the leading livestock state in the region.

The size and emphasis of a state's agricultural economy toward animal agriculture can be represented by the ratio of a state's livestock receipts to its total farm receipts. Table 3 gives this ratio for the 14 states in the region for three periods: 1974, 1987 and 1997. Alternately, this measure represents the degree of diversification of a state's agricultural economy.

Table 3 - Percent of Total Cash Farm Receipts from Livestock

State	Year 1974	1987	1997
Maine	54.6%	54.5%	56.5%
New Hampshire	72.5%	51.0%	44.8%
Massachusetts	52.9%	31.2%	21.5%
Rhode Island	50.6%	16.8%	13.8%
Connecticut	58.4%	52.2%	44.5%
New York	68.5%	68.4%	64.5%
Vermont	92.1%	87.1%	82.9%
New Jersey	33.1%	31.0%	21.1%
Pennsylvania	69.3%	72.0%	67.9%
Delaware	58.6%	76.1%	76.7%
Maryland	58.1%	65.0%	60.5%
West Virginia	71.0%	77.7%	82.5%
Virginia	47.8%	72.0%	64.1%
North Carolina	35.6%	56.1%	57.4%

In both 1974 and 1997, Vermont, New Hampshire, West Virginia, Pennsylvania, and New York were among that states with agricultural economies that were heavily reliant (i.e. a ration of greater than .75) on animal agricultural receipts. In 1997, Delaware, Virginia and North Carolina also appeared in this group.

Several trends are evident from the data. In several New England states (New Hampshire, Massachusetts, and Rhode Island) the ratio declined dramatically over this period. It also fell in Vermont and New Jersey by 10 percent. The ratio remained fairly steady in three states in the central part of the region, New York, Pennsylvania, and Maryland. Growth in the relative proportion of animal receipts in total revenues was most pronounced in the southern part of the region, particularly in Delaware, Virginia and North Carolina. Most increases in livestock dependence for these three states occurred from 1974 to 1987. The ratio increased steadily over the period only for West Virginia, a state which started the reference period as a very livestock dependent state.

Farmland Loss and Farm Exits

Farms. In 1974, total number of farms in the Northeast and Middle Atlantic Region was about 307,000. By 1997, the number of farms had decreased significantly (by almost one-quarter) to about 234,000 farms. At the same time the amount of land acreage in farms declined from about 51 million acres in 1974 to 42.5 million acres in 1997, a loss of about 16 percent. A large proportion of the total quantity of land that went out of agriculture in this period was located in four principal states: New York, Pennsylvania, Virginia and North Carolina. Other states with significant amounts of land transitioning out of agricultural use included Maine, New Hampshire, Massachusetts, Connecticut, Vermont, New Jersey, and Maryland. West Virginia was the only state in the region that reported relatively little loss of their agricultural land base over this time period.

Farm Exits. The change in farm numbers was most dramatic in North Carolina, with a loss of more than 40,000 farms (Table 4). In addition, significant proportions of farms went out of business in Delaware, New York, Virginia, Maryland, Pennsylvania and Maine. Several states with small or medium sized agricultural economies, including Connecticut, New Hampshire, New Jersey, Massachusetts, Rhode Island and West Virginia, actually increased their farm numbers over time. (For many of these states, the increase in number of farms may not reflect the number of "working" farms since in many cases, hobby farms or farmettes are included in these figures.) Moreover, states with the largest agricultural economies tended to be the same states with largest number of farm losses, suggesting a industrial transformation in these economies to a more consolidated agriculture based on smaller number of larger farming units was underway.

Table 4 - Number of Farms by State

State	Year		
	1974	1987	1997
Maine	6,436	6,269	5,810
New Hampshire	2,412	2,515	2,937
Massachusetts	4,497	6,216	5,574
Rhode Island	597	701	735
Connecticut	3,421	3,580	3,687
New York	43,682	37,743	31,757
Vermont	5,906	5,877	5,828
New Jersey	7,409	9,032	9,101
Pennsylvania	53,171	51,549	45,457
Delaware	3,400	2,966	2,460
Maryland	15,163	14,776	12,084
West Virginia	16,909	17,237	17,772
Virginia	52,699	44,799	41,095
North Carolina	91,280	59,284	49,406

Livestock Farms

The number of farms producing animals or animal products, such as eggs, in the Northeast and Middle Atlantic Region has declined significantly. This trend can be seen by looking at number of farms reporting livestock inventories in each state (calculated by adding together the number of farms with livestock inventories of major livestock categories in three years). Using this as indicator of animal production operations, the following six states lost more than 50 percent of livestock farms: North Carolina, Delaware, Maryland, New York, and Virginia (Table 5). Other states with significant losses of such farms included Pennsylvania, Maine, Vermont, and West Virginia.

Animal Production

From the standpoint of environmental and nuisance issues the number of animals is important. Information was collected on the annual inventories of milk cows, swine (hogs and pigs), and poultry (layers/pullets and broilers/meat chickens) for several time periods. An overview of trends for these animal species emphasizing the dominant production states and any relative changes over the 1974-1997 period is presented on the next page.

Table 5 - Number of Livestock Enterprises by State 1/

State	1974	Year 1987	1997
Maine	4,813	3,654	2,615
New Hampshire	2,276	1,773	1,523
Massachusetts	3,021	3,198	2,175
Rhode Island	394	381	327
Connecticut	2,495	2,214	1,688
New York	39,339	26,518	18,309
Vermont	6,591	5,045	3,748
New Jersey	3,574	3,390	2,593
Pennsylvania	55,965	41,471	28,872
Delaware	1,509	861	575
Maryland	11,388	7,538	5,038
West Virginia	22,423	15,419	12,810
Virginia	56,047	32,505	26,182
North Carolina	69,374	33,217	25,420

1/ Calculated by adding the number of farms reporting inventories of beef, milk cows, hog & pig inventory, and layer & pullets. Due to multiple enterprise, the figures do not reflect the actual number of livestock farms in each state.

Dairy. Milk production is significant for several states in the Northeast and Middle Atlantic Region. The major production states (in order of their milk cow numbers) are New York and Pennsylvania, followed by Vermont, Virginia and Maryland. The relative rank was unchanged from 1974 to 1997 (Table 6). The number of dairy farms declined very significantly in all states. While the number of cows also declined over time, milk production generally expanded due to adoption of productivity enhancing technologies on most remaining farms.

Table 6 - State Ranking for Number of Milk Cows

Ranking	1974	Year 1987	1997
#1	New York	New York	New York
#2	Pennsylvania	Pennsylvania	Pennsylvania
#3	Vermont	Vermont	Vermont
#4	Virginia	Virginia	Virginia
#5	Maryland	Maryland	Maryland

Poultry. Poultry production is very significant in the Northeast and Middle Atlantic region. The dominant states for egg production are Pennsylvania and North Carolina with Maine, Virginia, Maryland and New York also having significant production. Table 7 shows the relative shares of production increasing for Pennsylvania and Maine and decreasing for New York and Virginia. With the exception of Maine, production generally declined in the New England states. Several states, including West Virginia, Maryland, Pennsylvania and Delaware, rapidly expanded their inventories of layers and pullets over this time period. Over the region as a whole, the number of egg producing farms declined significantly.

Table 7 - State Ranking for Layers and Pullets Inventory

Ranking	Year 1974	1987	1997
#1	North Carolina	Pennsylvania	Pennsylvania
#2	Pennsylvania	North Carolina	North Carolina
#3	New York	Maine	Maine
#4	Maine	Virginia	Virginia
#5	Virginia	New York	Maryland

The portion of the poultry industry raising chickens for meat is very significant in the region. A measure of this growth is the number of broiler and meat chickens sold annually. The leading state in both 1974 and 1997 was North Carolina (Table 8). Other current major states include Virginia, Maryland, Delaware, and Pennsylvania. Virginia has increased its relative position since 1974, with Maryland and Delaware declining slightly in their relative shares. The shifts in broiler production within states in the region are noteworthy. For example, Maine was fifth in sales in 1974 and now has very little production. Pennsylvania more than doubled its sales over the past 25 years and now is fifth in the share of sales. Production in several New England states and New York and New Jersey declined over the period. West Virginia is a state that steadily expanded its production and sales. Farm numbers expanded only in the major production state of Virginia and in a few states with much smaller production levels (e.g. New Hampshire, New Jersey and Vermont).

Table 8 - State Ranking for Broilers/Meat Chickens Sold

Ranking	Year 1974	1987	1997
#1	North Carolina	North Carolina	North Carolina
#2	Maryland	Maryland	Virginia
#3	Delaware	Delaware	Maryland
#4	Virginia	Virginia	Delaware
#5	Maine	Pennsylvania	Pennsylvania

Hogs and Pigs. The region contains North Carolina, where many innovations in hog production systems occurred in the last decade and where significant rapid expansion has happened. In 1997, North Carolina's production was almost 10 times the magnitude of the next largest production state, Pennsylvania, in terms of inventories. Pennsylvania's swine production has been increasing and in 1987 it replaced Virginia as the second leading state in the region. Relatively smaller amounts of production occur in New York and Maryland. Both states exhibit a long-term trend of declining hog and pig inventories and sales.

The rapid decline in hog farm numbers illustrates the impact of technological and market change on farm structure. In the leading state of North Carolina, 20,000 of the approximate 23,000 swine farms exited from 1974 to 1997. In Pennsylvania, about 7,700 of the more than roughly 11,000 hog farms went out of business over this time period.

Table 9 - State Ranking for Hog and Pig Inventory

	Year		
Ranking	1974	1987	1997
#1	North Carolina	North Carolina	North Carolina
#2	Virginia	Pennsylvania	Pennsylvania
#3	Pennsylvania	Virginia	Virginia
#4	Maryland	Maryland	Maryland
#5	New York	New York	New York

Discussion

Based on the information provided in the previous section, three important questions for discussion can be identified.

1.)   Given the trends and patterns of the past three decades within the region, what will the future structure of animal agriculture look like? What expansion and/or regional shifts in location of animal agriculture are likely within or outside of the region? Also, what will be the environmental and other costs of such future growth?

2)   What are the economic benefits of the emerging animal agriculture to communities and economies? Given the potential negative environmental from loadings of nutrients from animal agriculture, communities are increasingly asking what do they get out of animal facilities.   The nature of this tradeoff will affect communities' and perhaps states' willingness to host additional animal facilities and/or allow expansion.

3)   What public policy innovations could be developed to allow the beneficiaries of animal agriculture to compensate those potentially harmed so that orderly and appropriate development of animal agriculture might be possible?

Emerging Patterns of Animal Production

As described earlier, there are clearly important changes occurring in patterns of animal production. In general, animal agriculture has become relatively less economically important in the northern parts of the region and become relatively more important in the southern part of the region. Several states, most notably Pennsylvania and New York, are holding their own or expanding slowly.

The process of structural change is a complex one. USDA researchers (Reimund et al 1981) identified three sets of external forces-new mechanical, biological or organizational technology; shifting market forces and demand; and new government policies and programs-that initiate the structural change process. Technological factors that were changing in the broiler industry in the 1950s and 1960s included mechanical and engineering advances in bird housing, materials handling, and processing, and adaptable organizational technology such as contracting and vertical integration. Important market-related factors were the existence of alternative production areas eager to accept new enterprises, potential for expanded consumption, high product market risks with respect to both price and access, high input risk in the form of difficulty in accessing capital, and ease of entry into production. Policy factors conditioning these market shifts included reduced feed grain costs due to the USDA commodity programs, federal tax provisions favorable to agriculture, and antitrust rules that were not prohibitive of industry activities. The researchers concluded that structural change is catalyzed by one or more external factors prompting an adjustment process that occurs in four stages: (1) *technological change*-innovators adopt new technology; (2) *locational shifts*-production of the commodity moves to areas more amenable to changed methods than to traditional ones; (3) *growth and development*-output rises as a result of new efficiencies; and (4) *adjustment to risks*-new institutions for coordination emerge and relationships within the sector evolve to manage new risks. The shift of the poultry industry out of New England and to the Delmarva Peninsula and other areas of the country can be explained by this progression.

Another concept that is helpful to explaining animal industry development and location is agglomeration economies or clustering (Pagano and Abdalla, 1994). The need to achieve economies of scale in processing appears to have been the factor that has driven vertical integration in the swine and poultry industries. For example, in Delaware, Maryland, and Virginia, eight firms with annual production of approximately contract over 500 million chickens in about 6,000 production units operate within a 16,000 square kilometer region Normally, poultry contract producers are located within 25 miles of the vertical integrator's processing facilities in the Delmarva peninsula (Narrod et al., 1994). Following a pattern similar to the poultry cluster to the north, North Carolina increased its hog processing capacity and used its new coordination arrangements to rapidly expand the size of production units and overall production volume in the first half of the 1990s.

More recently, a new factor - ecological concern - appears to be acting as a constraint to future expansion of existing animal agriculture clusters or possibly to development of new ones. Due to a major swine manure spill in North Carolina in June 1995, the state legislature enacted tougher water quality protection rules and a multi-year moratorium on certain-sized hog facilities. These public policy developments, as well as evidence about risks of locating large animal facilities in the flood plain revealed by Hurricane Floyd in September 1999, has caused some observers to predict that hog production may shift to areas with fewer environmental rules such as the Great Plains or southwest US (Drabbenstott, 1998; Bernick, 2000). However, the little systematic research conducted on this issue to date shows that economic, business climate and natural endowment factors are more important than environmental rules in determining growth and expansion in swine production among the leading states. The fish kills and related concerns about pfiesteria in the Pocomoke river in Maryland in summer 1997 and the subsequent enactment of a phosphorus-based state nutrient management law may also be a bellwether of stronger state or federal environmental rules for the Middle Atlantic Region. Such change will inevitably cause the environmental costs of animal agriculture to be factored into firms' decision-making, possibly affecting future expansion or location decisions.

Economic Benefits of Animal Agriculture to Communities

As noted earlier, animal agriculture provides significant benefits to economies in states in the Northeast and Middle Atlantic region in terms of income and jobs. An important question is: as animal agriculture becomes more industrialized, what is the distribution of those benefits to individuals and communities? And are those benefits large enough to offset potential or actual environmental costs from excessive nutrient loadings or other costs that are incurred by communities that host large-scale facilities?

Due to its need for feeds and other inputs, traditional animal agriculture was closely linked local on-farm enterprises or to other farmers and economic agents. As such it benefitted from and provided benefits to the local or regional agricultural economy. With industrialization, animal agriculture has been transformed into a specialized and generally larger capital-intensive activity that is linked to economic agents far from the local community or region. Since larger-scale vertically integrated farms are more closely linked to companies further from the local area, a logical hypothesis is that industrialized agricultural provides less local economic benefits that traditional agricultural production systems.

Little empirical work has been completed to shed light upon this question. One important study is by Chism (1993) who examined the spending patterns among a sample of smaller and larger crop and livestock farmers in southwest Minnesota. When farms were divided by operation type, crop-intensive farms showed a very weak relationship between farm size and local expenditures. This suggested larger crop farms replicate the expenditure patterns of the smaller farms they replace. However, livestock-intensive farms showed a different expenditure pattern with local spending on a percentage basis rapidly declining with size. However, local spending per-acre changed relatively little with the livestock-intensive farm size. For livestock operations, a certain base amount of spending occurs locally. Increases in spending after a certain point due to size seem to occur outside the local area. The differences between crop and livestock operations was also clear in input purchase patterns. The major crop inputs (crop chemicals, fertilizer, fuel and seed) were from local sources at least 85 percent of the time. Major livestock inputs, livestock purchases and feed, were from local sources 11 and 59 percent of the time. Part of livestock-intensive farming spending patterns may be due to the specialized technology used in larger operations. Local suppliers rarely can supply all the specialized equipment and construction techniques used in modern large scale livestock operations. Local suppliers may also be unable to provide the consistent quality and high volume of inputs needed in a large scale livestock operation. Specialists in breeding, along with discounted medicines and feeds are more likely to be found in distant markets. While a limited sample, these findings suggest that as livestock farms increase in size more economic benefits accrue beyond the local community's boundaries. More research is needed to validate these findings and to determine if they are applicable to other areas.

Public Policy Innovations

In some communities and states, expansion of animal agriculture has been blocked as a result of local community concern about the potential adverse environmental, health or nuisance effects. One approach to such opposition has been that of education and community relations. However, if the risks or economic damages (e.g., property value declines) to the community are real, it is unlikely that education and or promises of good behavior will be sufficient to resolve a stalemate. There appears to be a need for public policy innovations that permit the beneficiaries of changes in agriculture to compensate those potentially harmed. Thus far, there does not appear to be either a formal or informal institutional arrangement that can facilitate this negotiated transaction or exchange. One potential difficulty is that the environmental, and especially the nuisance, effects of animal operations are relatively focused upon neighbors or local community members while the economic benefits are more diffuse and broadly enjoyed by a region or state, or more generally, by consumers. Another challenge is that many of these costs are not reflected in markets and thus we have few available indicators of their magnitude. New and creative public policies that reconcile these disparate patterns of costs and benefits resulting from industrialized animal production will be needed for a more orderly transition of the agricultural economy and appropriate development of rural areas.

References

Abdalla, C. W., L. E. Lanyon and M. C. Hallberg. "What We Know about Historical Trends in Firm Location Decisions and Regional Shifts?: Policy issues for an Industrializing Animal Agriculture Sector." *American Journal of Agricultural Economics* 77(5) 1229-1236, 1995.

Barkema, A., and M. L. Cook. "The Changing U.S. Pork Industry: A Dilemma for Public Policy." *Economic Review, Federal Reserve Bank of Kansas City* (2nd Quarter, 1993):49-66.

Bernick, J. "A Farewell to Farms: Geographic Shift in Livestock Production is in the Wind" Farm Journal, January 2000. http//:farmjournal.com

Chism, J.W. "Local Spending Patterns of Farm Businesses in Southwest Minnesota." M.S. thesis. Department of Agricultural And Applied Economics, Univ. of MN, St. Paul. 1993.

Drabenstott, M. "This Little Piggy Went to Market: Will the New Pork Industry Call the Heartland Home?" *Econ Rev, Fed Reserve Bank of Kansas City* (3rd Quarter), 1998:79-97.

Lanyon, L. E. and P. B. Thompson, "Changing Emphasis of Farm Production." pp. 15-24 in *Animal Agriculture and the Environment: Nutrients, Pathogens and Community Relations.* Ithaca, NY: NRAES Publication 96, 1996.

Martin, L. L. and P. E. Norris. "Environmental Quality, Environmental Regulation and the Structure of Animal Agriculture." Paper presented at the USDA Agricultural Outlook Forum, February 24, 1998. Also published in *Feedstuffs*, Vol. 70, no. 18. May 4, 1998.

Narrod, C., Reynnells, R., and H. Wells. "Potential Options for Poultry Waste Utilization: A Focus on the Delmarva Peninsula." Unpublished paper, jointly sponsored by the Univ. of PA, USDA, and US EPA, Office of Pollution Prevention and Toxics. 1994.

Pagano, A. P. and C. W. Abdalla. "Clustering in Animal Agriculture: Economic Trends and Policy." In *Balancing Animal Production and the Environment.* Proceedings of the Great Plains Animal Agriculture Task Force. Denver, CO, October 1994.

Reimund, D. A,. J. R. Martin, and C. V. Moore. *Structural Change in Agriculture: The Experience for Broilers, Fed Cattle and Processing Vegetables.* USDA-ERS Technical Bulletin No. 1648. Washington, D.C. (1981).

U.S. Agriculture Census, http://govinfo.library.orst.edu/cgi-bin/ag-list?01-state.

U.S. Bureau of Labor Statistics

U.S. Census Bureau, www.census.gov/

U.S. Department of Agriculture, Economic Research Service, http://econ.ag.gov

U.S. Dept. of Agriculture Nat'l Ag Statistics Service, http://www.nass.usda.gov/census/census97

Farm Management and Nutrient Concentration in Animal Agriculture

Les E. Lanyon
Professor of Soil Fertility
Department of Agronomy
The Pennsylvania State University

Biographies for most speakers are in alphabetical order after the last paper.

Introduction

Animal agriculture has been implicated as a source of nutrients that are degrading groundwater and surface water and increasing atmospheric nutrient levels. Classification of animal agriculture operations as CAFOs and AFOs and national strategies for change are attracting a lot of attention. Many suggestions have been made and a wide range of regulations have been adopted to change farm management in response to concerns about nutrient losses to water and air. These suggestions and regulations usually rely heavily on the voluntary adoption of farm practices or development of nutrient management plans by farmers that include explicit criteria for environmental protection. Locally-led conservation efforts by agency staff and educators are viewed as the delivery mechanism for the information about the improved practices or the enhanced plans. Can these actions based on the assumptions of voluntary action meet the challenges faced by an animal agriculture that attempts to balance production with environmental protection? Do the approaches to nutrient management that are being promoted match the scope of the farm management change required? How will these issues influence the success of nutrient management consultants, extension educators, and producer advisors?

Current Status of Nutrient Management

Definitions of nutrient management vary widely in their scope and detail. Nutrient balance in field applications of nutrients from all sources with the nutrient requirements of the planned crops is generally part of most definitions. This criterion of nutrient balance is assumed to prevent the application of nutrients at rates that will exceed the capacity of the soil and planned crops to assimilate nutrients and prevent pollution (USDA/USEPA, 1999). Nutrient management is often described as sustaining an increase in agricultural production while protecting the environment (Hrubovcak et al., 1999). Under these circumstances, promoting changes in farm management is considered to be a transfer of technology problem — what is needed is known and it is in the best interests of potential adopters to accept it.

Farm structure, natural resource variability, and the economic risk of new practices, along with potential profitability, are often described as barriers to the adoption of "green" technologies such as nutrient management (Hrubovcak et al., 1999). Farm structural barriers include farm size and the availability of capable management and adequate labor. Diverse soil resources can increase the effort required to implement new practices by increasing the information required and the factors to be considered as well as the success of implementing selected practices that depend on soil characteristics. Because nutrients are required for crop growth, farmers who are not comfortable with risk may be slow to reduce the nutrient supply they provide and to depend on other sources. Typical incentives for farmers to overcome these barriers are economic incentives such as cost-sharing and regulations requiring their use. Has a comprehensive perspective of the situation been developed that will be the basis for effectively overcoming the barriers?

Why Do Nutrients Concentrate in Animal Agriculture?

Biologically, animals rely on the energy captured by plants through photosynthesis. Every plant contains minerals and other nutrients required by animals in addition to fixed, high-energy carbon. Animals burn the energy of plants to grow and maintain themselves. They excrete unavailable, unused, or "used" minerals that were part of their bodies, but replaced in the process of maintaining themselves. Plants must reuse these excreted minerals as they fix more energy that will be used by animals. Mineral excretion in nature is critical to the process of energy capture by plants and animal growth. Domesticated animals also depend primarily on the high-energy carbon of plants and they excrete minerals in the same ways. Wherever animals are concentrated and supported by concentrated feed supplies, nutrients will tend to accumulate also.

Socially, a set of technologies were developed and policies implemented, especially since the end of WW II, to deliver feed that was not produced on the farm where the animals are located to animals. Obviously, transportation technologies are essential to the movement of bulky animal feeds from where they are produced to where they will be consumed. Perhaps more importantly, is the technology to manufacture fertilizers that can replace the minerals harvested in the crops and exported from the crop production areas. Without fertilizers, the farms producing the crops would be mined of their reserves and ultimately not be able to supply the feed chain. This depletion of farm nutrient supplies was a serious concern in many countries during the 19th century and the early part of the 20th century. Fertilizers are currently substituting for the cycling of nutrients that normally occurs in nature and offsetting the exporting of nutrients from crop producing farms. Animals can now be far removed from the cropland that produced their feed and concentrated in numbers that exceed the crop production capacity of the available land where they are located. (Lanyon, 1995).

Additional sets of technologies and policies contributed to the reorganization of agriculture and the development of specialized production locations (Lanyon, 1994). Breaking the bonds of nutrient deficiency was essential and the ability to transport feeds was critical, but policies that encouraged the growth of specific crops for animal feed through farm programs or investment policies that made gave preferential treatment to capital investments in animal facilities also contributed to the concentration of animals. With the concentration of animals in specialized production locations came the concentration of nutrients in those same places (Lanyon, 2000).

In addition to technologies and policies promoting specialization in agriculture, several complementary farm management pathways to concentrated animal production exist. For instance, some agricultural areas have been faced with declining prosperity as competition for their traditional commodities developed from new production regions. Opportunities for businesses to create integrated animal production arrangements meant these regions could develop replacement or complementary agricultural enterprises to stabilize their agricultural economy (Hart, 1991). Some regions settled by farmers with a cultural heritage that favors establishing the next generation on a farm have intensified animal production based on the wide range of external factors that made investments in animal agriculture feasible and attractive. This intensification was often facilitated by the local agribusiness infrastructure that was prepared to provide the necessary inputs and the product marketing. Finally, another set of "producers" with financial capital to invest has been taking advantage of the opportunities to consolidate animal production in new production locations. This has been the case in the concentration of the historically dispersed hog production on many fewer units with capital intensive facilities from breeding to finishing and processing.

The Importance of Different Levels of Management

The concentration of nutrients in animal agriculture seems to have less to do with the "need" of local crops for nutrients than with the consequences of animal biology and the technical and policy factors that made a new pattern of organization for agriculture possible. The current focus of nutrient management in animal agriculture on the use of nutrients in crop production is not consistent with the reasons behind the concentration of animal agriculture (and nutrients) that resulted from the variety of external factors.

One way to better understand the current situation is to compare proposals for nutrient management with the different levels of management that are at work. Operational management deals with the day-to-day activities of how things are done. Tactical management emphasizes short-term planning and allocation of internal resources to the activities of the organization. Strategic management is externally oriented. It attempts to determine what the organization should be doing and what external resources should be acquired in order for the organization to survive over the long-term. (Lanyon and Beegle, 1993).

Many nutrient management efforts have focused on operational management (soil testing, manure testing, spreader calibration, keeping records, etc.). New efforts aimed at reducing the levels of nutrients in animal feeds are also in this category. Nutrient management planning has typically been a form of tactical management with its emphasis on the distribution of manure according to potential crop requirements and the supplementation of manure and other biologically based nutrients with fertilizer. The Phosphorus Index is a tactical tool to incorporate landscape and hydrologic features into the development of these plans. Few approaches to strategic management that determines farm characteristics, how farms are organized, and how farms function have been developed or implemented.

Instead of applying nutrients "better" (operational) or planning the allocation of internal farm resources (manure and land) according to new criteria (tactical), strategic management — reconciling the external requirements of successful competition in the business of agriculture with the protection of environmental resources — seems to be a critical management level. Approaching the challenges of dealing with nutrients in excess of crop requirements within a farm relying heavily on animal production at the strategic level would be consistent with reasons behind the specialization of the farm. Attempting to impose constraints at the operational or tactical level of management that are inconsistent with the strategies of farmers will limit the voluntary adoption of the practices and will perhaps increase the need for enforcement action for compliance with regulations. Furthermore, these constraints will become part of the set of external factors influencing the future direction of each farm. Operational and tactical constraints imposed that add to the costs of farming may actually accelerate the concentration of animals as farmers attempt to cover the sunk costs associated with compliance. Nutrient management consultants, extension educators, and producer advisors are likely to find themselves amidst the dilemmas created by the inconsistencies of programs that address only part of the range in management that has contributed to concentrated animal agriculture.

Contrasting Worldviews

The apparent contradiction between the strategic reasons for increasing concentration of animal production and the adoption of practices to limit the impacts of animal manure on local water resources reflects basic differences in the ways different people, groups, or organizations view the world (Jacobs, 1994). Some, the guardians, are inclined, or explicitly charged, to consider the consequences of all actions for the conditions of a specific place or territory. Impacts of nutrients from animal agriculture on groundwater or surface water in a particular location are examples of guardian concerns. Others with a commercial view focus on the transactions of business and the functioning of interconnected nodes of production activity. How activity at the nodes is conducted or the consequences of their integration into a coordinated network often is influenced little by the consequences for specific locations or territory (Friedman and Weaver, 1979). Farming tends to be at the intersection of these worldviews with a connection to the land, but a need to succeed in the marketplace. As agriculture specializes, the tendency is for the commercial view to dominate over the guardian perspective. As the scope of farming widens to include so many off-farm activities associated with specialized agriculture, resolution of the tension between the guardian and commercial views will affect agribusinesses in addition to farmers. Taking actions to reduce the impacts of the nutrients in an area that are in excess of potential crop utilization will affect the viability of agribusinesses in the targeted areas.

Dealing with Change

Current expectations for nutrient management in animal agriculture will actually require farmers, and the associated agribusinesses, to manage for two bottomlines - one commercial and the other guardian. This expanded perspective on farm performance that includes the social concerns of territory, such as nutrient management for environmental protection, will be compatible with management criteria for preserving the ecological integrity of a specific location. It may be less effective in maintaining a vibrant, vigorous farm-based economy if the commercial bottomline is compromised in order to achieve the guardian demands. The key to successful change will be the ability of farmers and agribusinesses to deal effectively, according to both sets of criteria, with the nutrients in animal manure that are in excess of the potential crop utilization on their farms or in the region where they do business.

Excess nutrients on a farm can be reduced by decreasing the imports of nutrients in excess of the potential crop utilization in the area. This can be done by limiting the nutrients in animal diets or by limiting the number of animals that are supported by external inputs so that nutrient balance goals for land application of manure are achieved. This approach can affect the competitiveness of individual farms by limiting their potential income (Westphal et al., 1989; Pease et al., 1998) and may make it difficult for agribusnesses to survive. Excess nutrients can be reduced by relocating some of the animal production away from an area. This approach can be good for the producers who move with the production to other locations, but it can have unfavorable consequences for the businesses, community, and other producers left behind. (Clouser et al., 1994). Internalizing the costs of dealing with excess nutrients on individual farms can increase the costs of production for those operations compared to other producers who are not required to have those costs. This may mean that some producers will actually intensify their operations even more in order to reduce the costs of compliance for each unit of production and/or some producers will not be able to cover the additional expenses and will cease production. Another alternative that has not been extensively developed is the "marketing" of environmental performance.

Environmental management systems (EMS) are becoming a part of industry management programs in fields outside of production agriculture in response to the interests of customers and suppliers, regulators, communities, special interest groups, and investors. An EMS involves setting targets for environmental performance and then developing specific actions that will contribute to meeting those targets. It will typically involve an environmental policy, planning and implementation, monitoring and corrective action, and management review (Nadler, 1997). The idea is to create a mechanism for pollution prevention as an integral part of management. The challenge is to cover the costs of the system. In some cases this can be done with internal cost savings. However, there are other situations in which economies of efficiency are not sufficient. In these situations the role of those parties interested in the environmental impacts of businesses may develop reward systems for those that meet the additional expectations of pollution prevention. Because the EMS movement has been generated in response to these same outside groups, this may be an opportunity for external, strategic forces to exert a positive influence on the production process through novel, perhaps even market-based, approaches. These incentives will recognize performance and outcomes. They will not be tied to transfer and adoption of specific technology. They will search out and reward the businesses doing the best job in both production and protection. They will not approach each and every situation as a problem to be fixed with a known answer. They will meet the interests of both commerce and guardians.

Conclusion

Nutrients accumulate in locations where animals are concentrated because of the biology of animals, the technological accomplishments that created abundant nutrient supplies, and the favorable policies promoting specialization in agriculture. The environmental consequences of this pattern of agricultural organization are now generating conflicts between those who view the world as guardians and those in farming and agriculture who must survive in the world of commerce. The challenge for nutrient management consultants, extension educators, and producer advisors is to work with clients to develop fully integrated business management systems that will support environmental protection goals. As new perspectives on the concentration of nutrients in animal agriculture emerge, these systems can contribute to value-added agricultural activity by providing pollution prevention. Under these new circumstances the systems will be consistent with and can be included in the strategic goals of all who participate in agriculture. Nutrient management consultants, extension educators, and producer advisors will participate in management and provide essential assurance to farmers, agribusinesses, and other stakeholders about management outcomes.

References

Clouser, R. L., D. Milkey, B. Boggess, and J. Holt. 1994. The economic impact of regulatory decisions in the dairy industry: A case study in Okeechobee County, Florida. J. Dairy Sci. 77:325-332.

Friedmann, J. and C. Weaver. 1979. Territory and function: The evolution of regional planning. Univ. Calif. Press, Berkeley, CA.

Hart, J.F. 1991. The land that feeds us. W.W. Norton & Co., NY.

Hrubovcak, J., U. Vasavada, and J. E. Aldy. 1999. Green technologies for a more sustainable agriculture. Resource Economics Div., Economic Research Service, USDA, Agri. Info. Bull. No. 752.

Jacobs, J. 1994. Systems of survival: A dialogue on the moral foundations of commerce and politics. Vintage Books, Random House, Inc., NY.

Lanyon, L. E. 1994. Dairy manure and plant nutrient management issues affecting water quality and the dairy industry. J. Dairy Sci. 77:1999-2007.

Lanyon, L.E. 1995. Does nitrogen cycle?: Changes in the spatial dynamics of nitrogen following the industrialization of nitrogen fixation. J. Prod. Agric. 8:70-78.

Lanyon, L.E. 2000. Nutrient management: Regional issues affecting the Bay. p.145-168. In Sharpley, A.N. (ed.) Agriculture and phos[horus management - The Chesapeake Bay. Lewis Publishers, Baco Raton, FL.

Lanyon, L.E. and D. Beegle. 1993. A nutrient management approach for Pennsylvania: Nutrient management decision-making. Agronomy Facts 38-C. Department of Agronomy, The Pennsylvania State University, University Park, PA.

Nadler, S. 1997. Environmental management systems: A practical overview. p. 1-14. Environmental Management Systems: Strategies for improving environmental performance. Pennsylvania Bar Institute, Harrisburg, PA.

Pease, J., R. Parsons, and D. Kenyon. 1998. Economic and environmental impacts of nutrient loss reductions on dairy and dairy/poultry farms. Pub. 448-231, Virginia Cooperative Extension, Virginia Polytechnic Institute and State University, Blacksburg, VA.

USDA/USEPA. 1999. Unified national strategy for animal feeding operations. U.S. Department of Agriculture and U.S. Environmental Protection Agency. (http://www.epa.gov/ovm/finafost.htm).

Westphal, P. J., L. E. Lanyon, and E. J. Partenheimer. 1989. Plant nutrient management strategy implications for optimal herd size and performance of a simulated dairy farm. Agricultural Systems 31:381-394.

●●●●●●●●●●●●●●●●●●●●●●●●●●●●●●●

Advances in Animal Waste Management for Water Quality Protection: Case Study of the Delmarva Peninsula

J. Thomas Sims
Professor of Soil and Environmental Chemistry
Department of Plant and Soil Sciences
University of Delaware

Biographies for most speakers are in alphabetical order after the last paper.

●●

The Delmarva Peninsula[1]

The Delmarva Peninsula is a lowland coastal area in the Mid-Atlantic region of the eastern coast of the United States (Figure 1). Delmarva includes all of Delaware and parts of Maryland and Virginia and is bordered on the west by the Chesapeake Bay and on the east by the Delaware Bay and the Atlantic Ocean. The peninsula is mainly located within the Atlantic Coastal Plain physiographic region and is dominated by flat to gently rolling topography. Altitudes on the peninsula range from sea level to a maximum of ~30 m. The northernmost portion of Delmarva is within the Piedmont physiographic province. Upland areas of the peninsula are bordered by low plains that slope toward coastal areas which contain tidal rivers, wetlands, and marshes. There are also many freshwater wetlands in upland forests and riparian wetlands in stream valleys. Rainfall is plentiful (115 cm/yr) and rather constant, averaging 7.5 to10 cm/month in the winter and spring and 10 to 12.5 cm/month in the late summer. Soils on Delmarva range from well-drained, highly productive silt loams in the Piedmont to well- and excessively well-drained sandy loams and loamy sands in the coastal plain. Significant areas of poorly drained soils are also present, particularly in southeastern Delaware.

Delmarva is fairly rural in character, with no large metropolitan areas and a total population of about 750,000. The two largest cities are in Delaware (Wilmington and Dover, populations of 72,000 and 30,000). However, urbanization of the peninsula is proceeding rapidly, particularly in northern Delaware and the coastal areas of Delmarva. In addition to permanent population growth, coastal areas annually experience large influxes of tourists; about five million tourists per year visit the Delaware beaches which have a resident population of < 25,000.

---

[1]Information on the Delmarva peninsula obtained from Shedlock et al., 1999.

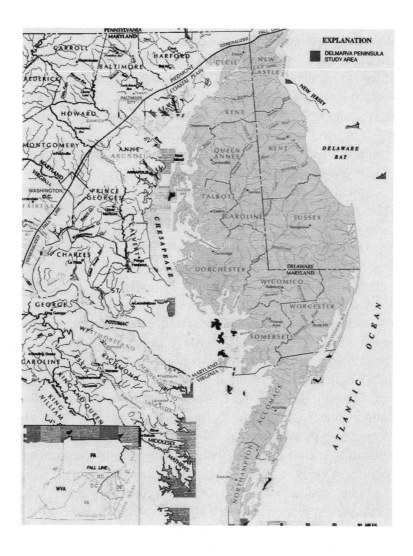

Figure 1. The Delmarva Peninsula (from Shedlock et al., 1999).

The hydrology of the Delmarva peninsula has been studied extensively and is well characterized. Most of the peninsula (~70%) drains westward to the Chesapeake Bay; the remaining lands either drain into the Delaware Bay or the Atlantic Ocean. Drainage from the central upland of the peninsula begins with many small streams that eventually enter tidal rivers and marshes. A notable characteristic of Delmarva is the absence of large watersheds; most drainage basins are < 25 km^2 in area. A rather extensive network of drainage ditches has been constructed throughout the peninsula to lower the water table and discharge surface runoff, especially in the poorly drained areas, such as southeastern Delaware. Groundwater on Delmarva is found in nine major confined sand aquifers. In much of Delmarva the confined aquifers are overlain by a surficial, water-table aquifer. The surficial aquifer serves as a source of water supply to most land uses and as a recharge source to underlying aquifers. Most of the rural population obtains its drinking water from domestic wells drilled into the surficial aquifer. Annual recharge to the surficial aquifer is about 40 cm/yr. Water-table depths in this aquifer range from near land surface to about 7 m in most areas but can be as deep as 10 m in well-drained areas.

Overview of Agriculture on the Delmarva Peninsula

Agriculture is the predominant land use on Delmarva. Of the 15,500 km^2 total land area, on the peninsula about 48% is in agriculture, 31% is in woodlands, 13% is in wetlands (fresh and tidal), 7% is in urban and residential use, and 1% is in barrier beaches and islands. Major crops grown on Delmarva are soybeans, corn, small grains (wheat/barley), grain sorghum, hay/alfalfa, commercial vegetables, and fruits (Table 1). The "green industry" (greenhouses, and nurseries, landscaping) contributes significantly to the agricultural income in parts of Delmarva (e.g., for Delaware, cash receipts of ~$30 million, compared to $40 million for soybeans, the highest valued agronomic crop) and is growing rapidly especially in urbanizing areas

Delmarva's agriculture is dominated by a large and geographically concentrated poultry industry that is vital to the economy of the region. Approximately 600 million broiler chickens are produced each year on Delmarva, mainly in eight counties located in the southern portion of the peninsula (Table 1; DPI, 1999). Note that estimates by the poultry industry of the number of broilers produced on Delmarva each year (602 million; DPI, 1999) are substantially greater than those by the National Agricultural Statistics Service (502 million; NASS, 1997). The cause of this discrepancy is unknown. The poultry industry reported that the total value of broilers "processed and delivered" from Delmarva was $1.63 billion and that the industry as a whole had an annual payroll of > $350 million (DPI, 1999). In Delaware alone, $560 million in agricultural cash receipts (72% of the state total for agriculture) was associated with the production and marketing of broiler chickens (DDA, 1998). Only limited production of other animals (beef, dairy, swine) occurs on Delmarva, hence the discussion in this paper focuses primarily on the poultry industry.

Poultry production increased markedly on Delmarva from the 1960's to early 1990's but has stabilized in recent years (see Figure 2 for Delaware). Today on Delmarva, there are about 2,700 contract poultry growers working with five integrated poultry companies. On average, each grower has 2.1 poultry houses, each with a production capacity of ~23,000 broilers. As mentioned above, poultry production is not uniformly distributed around Delmarva, but is localized in eight counties in close proximity to each other (Sussex, DE; Caroline, Talbot, Dorchester, Somerset, Wicomico, and Worcester, MD; and Accomack, VA). Seven of the eight counties have < 40,000 ha of cropland; Sussex County, DE has the largest agricultural land base (~100,000 ha) and the largest annual production of broilers by far (190 million per year based on NASS, 1997; local estimates are higher, in the range of 220 million per year). Animal densities, in livestock units/ha (LU/ha; all species, but predominantly poultry) on the peninsula range from < 0.2 to 1.65 LU/ha (Table 1; Sims et al., 2000). By way of comparison animal densities in the 12 countries of the European Union, where concerns about environmental pollution from confined animal production are long-standing, range from 0.5 LU/ha in Spain to 4.0 LU/ha in the Netherlands and average 0.9 LU/ha. The Delmarva poultry industry imports large quantities of corn and soybeans from other regions for use in feed. Approximately 69 million bushels of corn and 35 million bushels of soybeans are used by the industry each year. Based on recent (1993-1998) average yields for corn and soybeans grown in Delaware (115 and 30 bu/acre, respectively; DDA, 1998) and cropland data for Delmarva as a whole (Table 1), about 20 million bushels of corn and 15 million bushels of soybeans must be imported to meet the nutritional requirements of Delmarva's poultry industry. These feed imports, which also represent nutrient imports, have major implications for regional nutrient management and water quality issues, as discussed below.

Table 1 Overview of agriculture on the Delmarva Peninsula (NASS, 1997).

County	Cropland[†]	Corn	Soybeans	Small Grains	Hay	Vegetables	Broilers	Animal Density[‡]
	----------			ha		----------	--# sold x 10^6--	---LU/ha---
Delaware								
New Castle	27,000	9,900	11,100	5,700	1,600	400	<1¶	0.08¶
Kent	68,000	17,300	32,400	17,400	2,800	6.700	34¶	0.34¶
Sussex	100,000	36,400	46,800	20,600	2,000	11,300	190	1.22
Maryland								
Caroline	38,300	8,300	20,100	12,100	900	2,700	35	0.50
Cecil	25,600	7,900	5,800	3,400	3,400	40	n/a	0.34
Kent	39,600	15,900	14,300	7,100	1,700	10	4	0.16
Queen Anne's	58,900	20,600	28,200	14,900	1,200	1,000	10	0.15
Talbot	37,600	13,600	19,500	9,600	500	500	12	0.16
Dorchester	40,200	7,700	23,400	9,600	150	3,300	20	0.30
Somerset	16,100	4,900	9,000	2,900	60	300	42	1.65
Wicomico	28,700	8,900	14,000	4,900	900	900	76	1.38
Worcester	35,600	14,700	16,300	4,400	400	100	57	0.99
Virginia								
Accomack	30,300	5,500	17,500	10,000	150	<10	22	n/a
Northampton	20,200	1,100	10,000	7,900	30	<10	0	n/a
Delmarva	566,100	172,700	268,400	130,500	15,790	27,250	502	n/a

[†]Total cropland area. Values for crops are hectares harvested in 1997. [‡]LU/ha=livestock units/ha, based on all animal types produced in each county (poultry, dairy, beef, swine). From Sims et al. (2000); ¶Estimated from local data as information was not available in NASS 1997 statistics due to confidentiality requirements. n/a=not available.

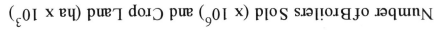

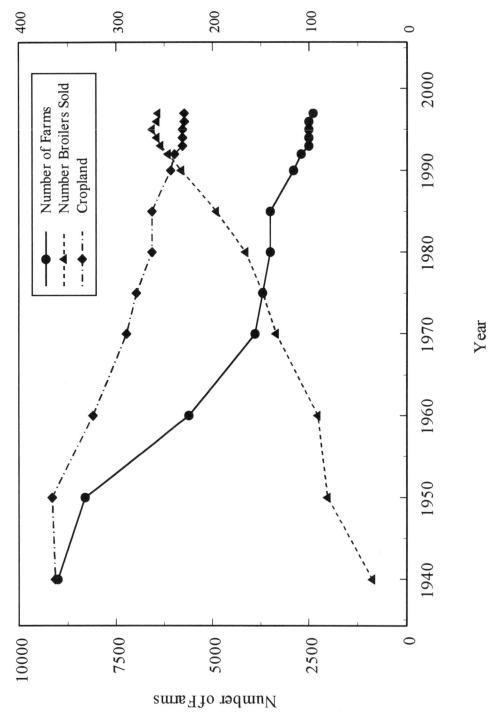

Figure 2. Poultry production trends for Delaware (1940 - 1998).

Environmental Issues Facing Agriculture on the Delmarva Peninsula

There are a number of significant environmental problems related to water quality on Delmarva. For example, a recent report by the Delaware Water Resources Center identified the following priority areas for water quality research and education programs:

- point and non-point source pollution of ground and surface waters and coastal estuaries by nutrients, sediments, pathogens, organics and trace metals, etc. -- of increasing importance as total maximum daily loads (TMDLs) are now being established in Delaware for many surface waters as a result of court-ordered state efforts to protect and improve these waters as required by the Clean Water Act.

- urban stormwater runoff and sediment delivery to surface waters, increasing in importance because of the rapid urbanization of the state

- impacts of septic systems and wastewater irrigation systems on ground water quality

- degradation of habitats, especially stream habitats

- an improved understanding of the responses of aquatic ecosystems to pollutant inputs

- identification and protection of recharge areas for ground water aquifers

- restoration and protection of wetlands

- leaking underground storage tanks, landfills, and chemical spills (past and present)

While agriculture may have some degree of involvement in many of these issues (e.g., wetlands, stream habitats, wastewater irrigation), the most pressing water quality problem faced by agriculture in the past, and still today, is its role in the nonpoint source pollution of ground and surface waters by nutrients, particularly nitrogen (N) and phosphorus (P). Federal and state environmental agencies and environmental advocacy groups have long been concerned about nitrate-N contamination of drinking water supplies and the potential role of N and P in agricultural runoff and groundwater discharge in surface water eutrophication. These concerns have been driven by documented instances of ground water pollution and degradation of surface water quality (Hamilton and Shedlock, 1992) and years of research that have shown the potential for N and P from agriculture to contribute to water quality degradation.

In particular, there has been serious concern that the geographic intensification of the poultry industry, which has resulted in farm, county, state, and regional nutrient surpluses, such as those calculated for Sussex County, Delaware, one of the nation's most concentrated poultry producing areas (Table 2; Cabrera and Sims, 2000), has created an agricultural setting that is prone to non-point source pollution. Simply put, the agricultural land base on Delmarva is not adequate to support the environmentally efficient use of the by-products of the poultry industry (manures, litters, composts). There are several reasons for this. Nutrient surpluses on farms (or at larger scales) primarily result from the fact that nutrient inputs in feed and fertilizers exceed outputs in

Table 2. Simplified annual mass balance for N and P in Sussex County, Delaware (Adapted from Cabrera and Sims, 2000).

Nutrient Input or Output	Nitrogen	Phosphorus
	---------------------------Mg/County------------------	
Nutrient Inputs:		
Animals	330	45
Feed	37240	7510
Fertilizer	8050	1230
Nitrogen fixation	3700	n/a
Total nutrient inputs	49320	8740
Nutrient Outputs		
Animals	16680	2090
Harvested crops	12870	1730
Total nutrient outputs	29550	3820
Nutrient surplus (Mg/county/yr)	19770	4920
Nutrient surplus (kg/ha/yr)	198	49
% of nutrients that are "surplus"	40	56

Assumptions:

1. There are 220,000,000 broiler chickens produced per year in Sussex County.

2. The county has 100,000 ha of cropland used to produce full-season soybeans (34,000 ha), corn (36,000 ha), wheat (20,000 ha), and double-cropped soybeans (12,000 ha). Average yields for corn are 7.8 Mg/ha, for full-season soybeans are 2.7 Mg/ha, for wheat are 4.0 Mg/ha, and for double-cropped soybeans are 1.7 Mg/ha.

3. Fertilizer inputs are based on statewide averages for N and P use (DDA.1998).

4. Nutrient inputs and outputs for animals were calculated using standard poultry feeding programs and animal composition data (Sims and Vadas, 1997).

animal products and crops. Note that these surpluses do not result from animal agriculture alone; commercial fertilizer use is a significant cause of the nutrient surpluses shown in Table 2. The surplus nutrients from feed are concentrated in animal manures which are heterogeneous in composition and have somewhat unpredictable rates of release once incorporated into soils. Poultry manures also have an unfavorable N:P ratio relative to most grain crops, resulting in over-application of P when manures are applied to meet crop N requirements, the long-standing agronomic practice in this region. Many soils in the poultry producing region of Delmarva are now considered "excessive" in P (Figure 3; Sims et al., 2000) relative to crop P needs and are sufficiently saturated with P to be of concern for soluble P losses in leaching and runoff (Pautler and Sims, 2000). Manures and other animal wastes (e.g. composts) are also difficult to store properly and apply uniformly in a timely manner that is well-synchronized with plant uptake patterns. This combination of nutrient surpluses and logistical constraints to efficient use of manure nutrients has created a situation where nonpoint source pollution is prone to occur. The likelihood of ground and surface water pollution by agricultural nutrients is further enhanced by the nature of the topography, soils, hydrology, and climate on Delmarva. Abundant rainfall, easily leached or ditch-drained soils, and shallow aquifers that are interconnected with surface waters (streams, rivers, and estuaries) form a setting that facilitates nutrient transport from land to water.

In summary, agriculture on Delmarva, and especially animal agriculture faces serious environmental challenges today. Public concerns about nonpoint source pollution of ground and surface waters has resulted in recent nutrient management legislation in all three states (discussed below) that will significantly impact agriculture by regulating nutrient use. At the same time efforts are underway in all three states, and regionally, to develop new and more efficient approaches to nutrient management and confined animal production. The remainder of this paper focuses primarily on the research and education programs now being considered, or implemented, to sustain agriculture and protect Delmarva's environment.

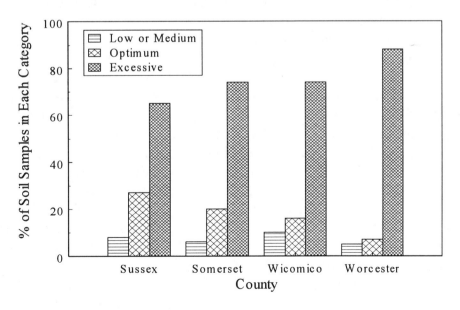

Figure 3. Soil test P status in Delmarva counties where poultry production is concentrated.

Recent Environmental Legislation and Policy on the Delmarva Peninsula

In the past three years a series of events have led to the passage of nutrient management laws in Delaware, Maryland, and Virginia and to increased federal involvement in the regulation of confined animal agriculture in the region. The approximate chronology of these actions is as follows; some specific details on each state's action are provided in Table 3.

The first significant step of relevance to Delmarva was a lawsuit filed in1996 by a consortium of environmental groups that sued the U.S. Environmental Protection Agency (USEPA) for "failure to perform its mandatory duties under the Clean Water Act to identify and then improve water quality" in Delaware. In 1997 the state of Delaware, through the Department of Natural Resources and Environmental Control (DNREC), negotiated a Total Maximum Daily Load (TMDL) agreement with USEPA. This agreement established a 10-year schedule to develop TMDLs for affected waterbodies and to then promulgate "pollution control strategies" to ensure that pollutant loadings are below TMDL values. Virginia entered into a TMDL agreement with USEPA in 1998, adopting a 12-year schedule to set TMDLs and, subsequent to this, to implement plans to reduce pollutant inputs to levels needed to meet the desired water quality. Maryland does not have a TMDL agreement but a lawsuit has been filed to compel USEPA to establish TMDLs for impaired water bodies in that state.

The first state water quality legislation that impacted Delmarva was Maryland's Water Quality Improvement Act of 1998. Passage of this act was stimulated by public concerns over fish kills in the summer of 1997 that were reportedly caused by *Pfiesteria spp.* a toxic dinoflagellate that had been implicated in earlier, massive fish kills in North Carolina and also in human health problems. Detailed information on the events that led to Maryland's law are provided by Simpson (1998). However, it is fair to say that the Maryland law, which passed in a politically-charged atmosphere, stimulated similar efforts in Virginia and Delaware, under pressure from the USEPA, to move away from the voluntary nutrient management practices advocated in the past and in the direction of regulated programs, especially for large confined animal feeding operations (CAFOs, usually those operations with >1000 animal units). The states of Delaware and Virginia worked throughout 1998 and into 1999 to draft legislation addressing nutrient management and water quality. Virginia was the next state to pass legislation, in the form of a poultry waste management bill approved in January of 1999 (see Table 3 for details). In Delaware the Governor appointed an Agricultural Industry Advisory Committee on Nutrient Management, consisting of ten farmers, to develop recommendations for state actions. The efforts of this committee led to the passage in June of 1999 of Delaware's state nutrient management act. Subsequent to the passage of these state laws, committees or commission were appointed to draft the regulations required by each state's legislation. For example, in Delaware a Nutrient Management Commission (DNMC) has been established to develop and implement a state nutrient management program that will protect and improve water quality. Specific information on Delaware's act and the responsibilities of the DNMC is provided by Sims (1999).

National policy initiatives are also underway that impact animal agriculture on Delmarva. By far the most significant is the USEPA-USDA Unified National Strategy for Animal Feeding Operations (AFOs), adopted in March of 1999 after lengthy discussion and public review. This document contains nine "guiding principles" for the joint effort between the nation's lead regulatory agency (USEPA) and its lead technical agency for agriculture (USDA) to "..address the water quality and public health impacts associated with AFOs" (USDA&USEPA, 1999):

Guiding Principles in the USDA-USEPA Unified National Strategy for AFO's

1) Minimize water quality and public health impacts from AFOs.

2) Focus on AFOs that represent the greatest risk to the environment and public health.

3) Ensure that measures to protect the environment and public health complement the long-term sustainability of livestock production in the U.S.

4) Establish a national goal and performance expectations for AFOs.

5) Promote, support, and provide incentives for the use of sustainable practices and systems.

6) Build on strengths of federal agencies and their state/local partners and make appropriate use of diverse tools including voluntary, regulatory, and incentive-based approaches.

7) Foster public confidence that AFOs meet performance expectations and that federal and state/local governments are ensuring the protection of water quality and public health.

8) Coordinate activities among USDA, USEPA, and state agencies and other organizations that affect or influence the management and operation of AFOs.

9) Focus technical and financial assistance to support AFOs in meeting national goal and performance expectations established in the Unified National Strategy.

Following up on this strategy, in the summer of 1999 the USDA Natural Resources Conservation Service released a "national nutrient policy" which will require more comprehensive nutrient management planning and implementation of plans for farmers receiving technical assistance and cost-sharing funds. In September of 1999 the USEPA released for public comment the "Guidance Manual and Example NPDES Permit for Concentrated Animal Feeding Operations" and in December of 1999 the USDA-NRCS released for public comment the "Technical Guidance for Developing Comprehensive Nutrient Management Plans (CNMPs)"[2].
As described in the USDA-NRCS guidance, the objective of a CNMP is "..to combine management activities and conservation practices into a system that, when implemented, will minimize the adverse impacts of animal feeding operations on water quality". Six elements are proposed to be considered in a CNMP: (1) Animal outputs - manure and waastewater collection, handling, storage, treatment, and transfer; (2) Evaluation and treatment of sites proposed for land application; (3) Land application; (4) Records of CNMP implementation; (5) Inputs to animals; and (6) Other utilization activities (e.g., power generation, composting, pelletization).
State agencies, farm organizations, universities, and environmental advocacy groups are now working to understand and then to integrate federal guidance and regulations with that mandated by new state laws. Most anticipate that this process will take from five to ten years.

---

[2] Information on the USDA and USEPA documents described in this paper is available on the web sites of the two agencies (USDA: http://www.nhq/nrcs/usda/gov; USEPA: http://www.epa.gov/owm).

Table 3. Summary of nutrient management legislation in Delaware, Maryland, and Virginia (from Sims, 1999 and Simpson, 1998).

State	Overview of Key Components of the Legislation
Delaware (Nutrient Management Act of 1999)	• Establishes a 15-member, politically appointed Delaware Nutrient Management Commission (DNMC) to develop and implement a state nutrient management program. Anyone with > 8 animal units or who applies nutrients to > 4 ha of land must have and implement a nutrient management plan. Plans will be reviewed beginning in 2003 and the state nutrient management program must be implemented by 2007. For agronomic plans N applications will be limited to that required for a "realistic yield" (defined as best 4 yields from the past 7 years); application of P to "high P" soils (remains to be defined by the DNMC) shall not exceed a 3 year crop removal rate. The act also proposes that the state's NPDES program for CAFOs be delegated to the Department of Agriculture who will rely on the act to the greatest extent practical in the development of CNMPs for CAFOs. Municipalities applying biosolids to cropland are exempt from the act (except for reporting) because they are already permitted by the state through other programs. Financial penalties are included for those who do not comply with the act. • The DNMC will also: (i) consider the establishment of critical areas for targeting voluntary and regulatory programs; (ii) establish BMPs to reduce nutrient losses to the environment; (iii) develop educational and awareness programs; (iv) consider the need for a transportation and alternative use incentive program to move nutrients from areas of overabundance to areas where they are needed; (v) establish a state certification program for four classes of nutrient users and a method to evaluate an applicant's suitability for certification; (vi) cooperate with state agencies to provide cost-share funds for NMP development and BMP implementation; (vii) work with "commercial processors" (e.g. poultry integrating companies) to provided technical assistance and educational programs for contract growers to improve the storage and management of animal wastes; and (viii) keep records and provide annual reports to the legislature on the progress of the DNMC in implementing the requirements of the act. The DNMC is now developing the regulations required by the act.

Table 3 (cont). Summary of nutrient management legislation in Delaware, Maryland, and Virginia (from Sims, 1999, Simpson, 1998).

Maryland (Water Quality Improvement Act of 1998)	• Requires that any agricultural operation with > $2500 in gross annual income or > 8 animal units must develop and implement a NMP. The law clearly includes municipal biosolids application to cropland and impacts non-agricultural nutrient users (e.g. commercial lawn care, nurseries, turf grass producers) who apply nutrients to more than 1.25 ha of land. The WQIA requires that anyone who only uses commercial fertilizer must "develop and implement a N and P based plan by 12/31/02; those using biosolids or animal manures must comply by 12/31/05. The act includes financial penalties for those who do not comply. The addition of restrictions on the amount of P that can be applied is a major change from nutrient management planning efforts in Maryland. • All NMPs must be developed and approved through the Maryland Department of Agriculture; cost-sharing is provided for plan development. A state Nutrient Management Advisory Committee with wide-ranging representation developed and recently released (December, 1999) the proposed regulations that will implement this act. These proposed regulations are now under review and may be subject to additional public hearings.
Virginia (Poultry Waste Management Bill of 1999)	• Requires the development and implementation of nutrient management plans for "any person owning or operating a confined poultry feeding operation". These NMPs will govern the storage, treatment, and management of poultry waste. Also provides for poultry waste "tracking and accounting". • States that N application rates in mandated NMPs shall not exceed crop nutrient needs as determined by the Virginia Department of Conservation and Recreation (DCR). Poultry wastes shall also "be managed to minimize runoff, leaching, and volatilization losses and reduce adverse water quality impacts from nitrogen". • After 10/1/01 P application rates in mandated NMPs shall not exceed the greater of crop nutrient needs or crop nutrient removal as determined by DCR. Poultry wastes shall "be managed to minimize runoff and leaching and reduce adverse water quality impacts from P". Prior to 10/1/01 comprehensive soil conservation plans consistent with USDA-NRCS guidelines will be required for soils with a soil test P (Mehlich 1) value > 55 mg/kg (ppm)

55

Advances in Animal Waste Management to Protect and Improve Water Quality on the Delmarva Peninsula

Clearly in the past three years dramatic changes have occurred in the approach that will be used in the future for agricultural nutrient management on the Delmarva peninsula. These changes will impact many nutrient generators and users, but will have the greatest effect on animal agriculture. Most changes are still in progress, or are still being hotly debated, and will likely not be finalized for several years. Given this, it is somewhat difficult to predict, or describe, what the future holds for nutrient management and water quality on Delmarva. Some things seem certain to occur, while others that seem necessary or promising today may be abandoned after further research and economic, or political, evaluation. Two major changes, however, are likely to occur that will permanently alter the way agriculture is practiced on Delmarva:

Nutrient Management Planning

Nutrient management planning is not a new agricultural activity; farmers and their advisors have developed and implemented plans for the profitable use of nutrients for decades. Environmental considerations, such as preventing soil erosion and reducing nitrate leaching have always played a role in these voluntary plans. It seems apparent, however, that NMPs developed in the future on Delmarva will be more formal, even regulatory, in nature. The passage of the three state laws described in Table 3 and the development of regulations to implement these laws will require most nutrient users (agricultural and non-agricultural) to develop and implement NMPs for their operations. Record-keeping will be required and penalties can be assessed for failure to comply with state and federal requirements. What can be expected as a result of these state and federal actions? Several changes seem likely:

- Expansions in funding and activities of government, university, and private sector advisors involved in educational efforts on NMPs, the design and implementation of NMPs and associated BMPs, and cost-sharing to facilitate NMP implementation can be expected in the future. This should result in the more efficient use of nutrients by agriculture which should in turn contribute to the long-term improvements in water quality hoped for by those that passed the legislation in these three states.

- Perhaps the most significant specific change that will occur will be the re-design of NMPs to include P management practices that are protective of the environment. A regional effort is now underway to develop a holistic approach to P management that integrates site properties related to P transport, water-body sensitivity to P, soil P status, and P management of fertilizers and manures into a field-scale risk assessment tool (the *Phosphorus Site Index;* Leytem et al., 2000). Once a reliable P Site Index has been developed areas requiring more intensive management to prevent P losses in runoff can be prioritized and BMPs implemented most effectively.

- Wide-scale development of NMPs will likely confirm the exact magnitude and specific geographic locations of nutrient surpluses on Delmarva. This is critical because farm-scale NMPs are designed to use the optimum amount of nutrients needed to achieve realistic yields; not to economically dispose of surplus nutrients that are not needed on the farm. Individual farmers do not have the economic wherewithal to design and construct the infrastructure needed to economically use nutrient surpluses of the magnitude estimated to be present on Delmarva. State or regional efforts, involving private industry, as described below, will be required for long-term resolution of the nutrient imbalance problems facing Delmarva agriculture today.

Alternative Uses for Animal Wastes

For the past two years a concerted effort has been underway in this region to critically analyze alternative uses for animal wastes based on the premise that application to agricultural cropland cannot remain the sole end use for the by-products of animal production. Most recently a task force was established by the Chesapeake Bay Executive Council with the charge to "..recommend appropriate procedures pertaining to the interstate distribution of animal wastes". This task force conducted a thorough review of all proposed alternatives to land application of animal wastes and made the following recommendations (Chesapeake Bay Program, 1999):

1) A memorandum of understanding should be signed by the six states in the Chesapeake Bay watershed (DE, MD, PA, NY, VA,WV) that will formalize a long-term commitment to ensure the proper land application of animal wastes regardless of their final destination.

2) The six states should adopt technical guidelines addressing animal waste transport procedures, NMPs, storage of wastes, biosecurity, and monitoring/tracking of animal waste transfers.

3) Potentially feasible alternatives to land application of animal wastes include expanded land application, composting, pelletization or granulation for fertilizers, use in animal feeds, and energy generation. The task force provided detailed analyses of each option and stated:

   - The most appropriate mix of alternative uses and incentive solutions will vary by area. Areas with slight oversupply should expand the use of more distant land application of animal wastes and secondary uses such as composting for the landscape industry. Areas with significant oversupply should pursue waste reuse options such as pelletization and shipment to nutrient deficit areas, energy generation, and minor use options.

   - Nutrient reduction through more effective feeding strategies should be encouraged for liquid wastes which are more difficult to transport for alternative uses.

   - Incentive programs should be developed to promote alternative uses, with preferences given to recurring incentives; incentives with a phase-out period; start-up incentives such as grants and low-interest loans; and insurance mechanisms to reduce the risk of adopting a new practice. Market-based solutions and additional research on alternative uses should be encouraged.

   - Long-term success will require good faith cooperation, funding, risk assumption, and personnel resources from integrating animal production companies, livestock growers, manure brokers, state and federal agencies, utilities involved in energy generation from wastes, and groups such as the Chesapeake Bay Program.

Future Challenges and Directions for Delmarva's Animal Agriculture

Agriculture on the Delmarva peninsula is at a crossroads of sorts. Fundamental questions are being asked about the sustainability of the animal agriculture that dominates the economy of this region. The voluntary approaches to nutrient management used in the past are being replaced by regulatory programs that will require farmers, and other nutrient users, to follow more intensive, and thus costly, approaches to nutrient management planning. Many in agriculture question the need and scientific validity of these changes. Others ask for delays in their implementation to allow for more research to justify the actions required in state laws and more time for an economic evaluation of their impact on agricultural profitability. At the same time, federal agencies, environmental advocacy groups, and the media are pressing states to move forward more rapidly to protect and improve water quality.

Given the rather contentious nature of this issue, what steps should be taken today to address the need to sustain agriculture and protect environmental quality on Delmarva? First, there is a need for all parties involved to recognize that the issue of nutrient management on water quality requires serious attention, in the short and long-term. Sufficient information is available from research on Delmarva and in other states of countries to justify the immediate need for improved nutrient management practices, particularly for animal agriculture. Consequently, there is a need for reasoned public debate on the issue to identify the most appropriate changes that should be required to reduce agriculture's impact on water quality. Second, some changes should occur now, specifically concerted efforts to implement nutrient management practices that are accepted by agriculture, but not widely used due to economic constraints. Examples include the use of diagnostic nitrogen tests to increase fertilizer use efficiency, more extensive efforts to identify soils that are "high" enough in P to be of environmental concern, use of modified animal diets and manure treatment technologies to reduce the potential for nutrient losses, wider implementation of soil conservation practices (e.g. buffer strips), and avoiding poorly timed or badly located applications of animal wastes. Finally, most scientists agree that more information is needed on nutrient cycling and management, particularly for phosphorus. A significant research effort is underway today and all indications are that support for nutrient management research (basic and applied) will continue to be strong for the next decade. This argues for patience and flexibility in the implementation of regulations until the research is completed and clearly understood by all interested individuals.

References

Cabrera, M.L. and J. T. Sims. 2000. Beneficial uses of poultry by-products: Challenges and opportunities. *In* J. Power (ed.) Beneficial uses of agricultural, municipal, and industrial by-products. SSSA, Madison, WI. (in press).

Chesapeake Bay Program. 1999. Task force report on interstate animal waste distribution and use technology. Chesapeake Bay Program, Annapolis, MD.

Delaware Department of Agriculture (DDA). 1998. Delaware agricultural statistics summary for 1998. Delaware Agric. Statistics Service, Dover.

Delmarva Poultry Industry (DPI). 1999. Look what the poultry industry is doing for Delmarva: 1998 Facts about Delmarva's broiler industry. DPI, Georgetown, DE.

Hamilton, P. A. and R. J. Shedlock. 1992. Are fertilizers and pesticides in the ground water? A case study of the Delmarva peninsula. U. S. Geol. Surv. Circ. 1080, Denver, CO.

Leytem, A. B., J. T. Sims, and F. J. Coale. 2000. Developing a reliable site index for the Delmarva peninsula. Proc. Conf. Managing Nutrients and Pathogens from Animal Agriculture, NRAES, Ithaca, NY.

National Agricultural Statistics Service (NASS). 1997. Census of agriculture, 1997. U.S. Department of Agriculture, Washington, D. C. (http://www.usda.gov/nass).

Pautler, M. C. and J. T. Sims. 2000. Relationships between soil test phosphorus, soluble phosphorus, and phosphorus saturation in soils of the Mid-Atlantic region of the U. S. Soil Sci. Soc. Am. J. (in press).

Shedlock R. J., J. M. Denver, M. A. Hayes, P. A. Hamilton, M. T. Koterba, L. J.Bachman, P. J. Phillips, and W. S. L. Banks. 1999. Water-quality assessment of the Delmarva peninsula, Delaware, Maryland, and Virginia: Results of investigations, 1987-1991. U. S. Geol. Surv. Water Supply Paper 2355-A. USGS Information Service, Denver, CO.

Simpson, T. W. 1998. A Citizen's Guide to Maryland's Water Quality Improvement Act. Univ. of MD Coop. Extension, College Park, MD.

Sims, J. T. 1999. Overview of Delaware's 1999 nutrient management act. Fact Sheet NM-02. College of Agric. Nat. Res., Univ. of DE, Newark, DE.

Sims, J. T. and P. A. Vadas. 1997. Nutrient management planning for poultry-grain agriculture. Fact Sheet ST-11. College of Agricultural Sciences and Cooperative Extension. University of Delaware, Newark, DE.

Sims, J. T., A. C. Edwards, O. F. Schoumans, and R. R. Simard. 2000. Integrating soil phosphorus testing into environmentally-based agricultural management practices. J. Environ. Qual. (in press).

Session 3

EPA and NRCS Goals in Nutrient Management

EPA Programs for Attaining Water Quality Goals from Nonpoint Sources

Roberta Parry
Senior Agriculture Policy Analyst
Office of Policy and Reinvention
US EPA

Biographies for most speakers are in alphabetical order after the last paper.

Introduction

Agricultural sources of pollution–cropland, animal feedlots, livestock grazing–are the primary sources of pollution to lakes and rivers nationwide. (U.S. EPA, 1998a) Under the Clean Water Act, one agricultural activity, confined animal feeding operations, is defined as a point source of pollution. Discharges to waters of the United States from any point source are regulated under the U.S. Environmental Protection Agency's (EPA) water permit program. All other sources of agricultural pollution are considered nonpoint sources. EPA has several plans and on-going programs to help control pollution from agriculture and other nonpoint sources of pollution.

Clean Water Action Plan

The Clean Water Action Plan is a blueprint for restoring and protecting the nation's waters. It contains many actions for water quality programs in EPA and other federal agencies which strengthen on-going efforts. However, the focus of the plan is on four key tools to achieve clean water goals: 1) a watershed approach; 2) water quality standards; 3) natural resource stewardship; and 4) improved public information.

1)   The watershed approach encompasses a collaborative effort by federal, state, tribal, and local governments; the public; and the private sector to restore polluted watersheds and sustain healthy conditions in other watersheds. This approach helps to ensure that strategies to control pollution are more site-specific and cost-effective. It is based on an assessment of

the condition of the watersheds, restoration strategies, and pollution prevention efforts.

2) Strong federal and state water quality standards are needed to protect public health, prevent polluted runoff, and ensure accountability. The key actions include ensuring safe shellfish and beaches, better control of storm water runoff, enforceable authorities for polluted runoff, reduced pollution from animal feeding operations, and the establishment of quantitative nutrient water quality criteria.

3) Federal agencies will enhance natural resources for the protection of water quality and the health of aquatic systems on federal lands and for federal resource management. To meet this goal, federal land managers will improve water quality protection for over 2,000 miles of roads and trails each year through 2005 and decommission 5,000 miles each year by 2002. Federal land managers will also accelerate the cleanup rate of watersheds affected by abandoned mines and will implement an accelerated riparian stewardship program to improve or restore 25,000 miles of stream corridors by 2005.   In addition, federal agencies will work with private land owners to protect and restore wetlands and coastal waters and create two million miles of conservation buffers by 2002. New incentives for private land conservation will be explored.

4) Effective management of water resources requires reliable water quality data. The data must be communicated with the public in a meaningful way. The U.S. Geologic Survey will lead an effort improve monitoring and assessment, focusing on nutrients and related pollutants and communicate the information to the public.

In June 1999, the Wyoming Association of Conservation Districts (WACD) filed a lawsuit against EPA and the other federal agencies involved in the development of the Clean Water Action Plan. WACD alleges that the Clean Water Action Plan should have been subjected to a formal public review and comment process since it is a major federal action. The U.S. Department of Justice has filed a motion to dismiss the lawsuit arguing that the plan is not a final federal action and is therefore not subject to review by the courts. Individual regulations called for in the plan will be subject to the rulemaking process.

Nutrient Water Quality Criteria

Water quality standards are a basic building block of the Clean Water Act. States are required to set water quality standards as a tool to evaluate the condition of their water bodies. Water quality standards identify the uses for each waterbody, for example, drinking water supply, swimming, and fishing, and the scientific criteria to support that use.

Nutrients are the major pollutants in lakes and estuaries and the second leading source of pollution in rivers (U.S. EPA, 1998a). However, states often use subjective water quality criteria to assess the seriousness and extent of the nutrient problem. This lack of quantitative criteria can result in widely varying assessments. 17 states have no water quality criteria for nitrogen and 21 states have no water quality criteria for phosphorus (U.S. EPA, 1998b). Many of the other states only use a narrative criteria for nutrients. It is difficult to use these criteria to determine when a water body is clean or polluted. Research to improve the basis for understanding and assessing nutrient over-enrichment problems is critical to better control of nutrient levels in waters and to meet the nation's clean water goals.

In order to provide a scientific basis for quantitative water quality criteria, EPA has developed a national nutrient strategy (U.S. EPA, 1998b). This strategy requires states and tribes to adopt numerical nutrient criteria into their water quality standards by the end of 2003. The strategy identifies 14 eco-regions and 4 waterbody types (lakes, rivers, estuaries, and wetlands) that will require different criteria. The criteria will be based on the technical guidance that EPA will finalize in early 2000 (U.S. EPA, 1999a, 1999b). The draft technical guidance focuses on assessing total phosphorus, total nitrogen, algal biomass, and turbidity to establish nutrient criteria. States and tribes will also be responsible for monitoring and evaluating the effectiveness of programs to control nutrients as they are implemented.

Nonpoint Source Program–Section 319

Under Section 319 of the Clean Water Act, EPA provides grants to the states to carry out regulatory and nonregulatory programs to control nonpoint sources of pollution. Nonpoint sources include any activity that is not defined as a point source under the Clean Water Act. Nonpoint sources include most agriculture operations (except concentrated animal feeding operations), forestry, and urban runoff. The EPA grant money may be used for a variety of projects: enforcement, technical assistance, financial assistance, education, training, technology transfer, demonstration projects, and monitoring. The government has allocated over one billion dollars for the nonpoint source program since its inception. 200 million dollars have been allocated for the program in both fiscal years 1999 and 2000. President Clinton's fiscal year 2001 budget calls for a $50 million increase to $250 million.

In fiscal year 2000, $100 million of the allocation was given to states to support implementation of their nonpoint source programs. The other half of the allocation will be distributed to states based on the extent to which they have incorporated nine key elements into their programs. The elements include:
1. explicit short- and long-term goals, objectives and strategies;
2. strengthened partnerships with appropriate government entities (including conservation districts), private sector groups, and citizens groups;
3. emphasis on both state-wide programs and management of individual watersheds;
4. abatement of known water quality impairments and prevention significant threats to water quality from present and future activities;
5. identification of impaired waters and important threatened waters and a process to implement watershed plans;
6. review, upgrade, and implementation of all components required by Section 319 and establishment of flexible, targeted, and iterative approaches to expeditiously achieve and maintain beneficial uses of water;
7. identification of federal lands and activities which are not managed consistently with state nonpoint source program objectives;
8. efficient and effective management and implementation of the program; and

9. periodic review, evaluation, and revision of the program

The success of the nonpoint source grants in protecting water quality has been difficult to evaluate from a national perspective because of the wide variety in individual state programs. These key elements will help to make the state programs more consistent. In addition, EPA established the Section 319 National Monitoring Program to provide credible documentation of the feasibility of controlling nonpoint sources, to improve the technical understanding of nonpoint source pollution, and to evaluate the effectiveness of nonpoint source control technology and approaches. 20 of the 23 monitoring projects focus on agricultural sources.

Coastal Zone Act Reauthorization Amendments of 1990

The Coastal Zone Act Reauthorization Amendments of 1990 (CZARA) are an additional federal tool to address nonpoint sources of pollution in the coastal zone. CZARA requires coastal states to develop coastal nonpoint source pollution control programs that include technology-based management measures to control nonpoint pollution in accordance with EPA's guidance (U.S. EPA, 1993a and 1993b). The state CZARA program must implement the management measures in their coastal zones in six categories: agriculture; forestry; urban; marinas and recreational boating; hydromodification; and wetlands, riparian areas, and vegetated treatment systems. It must contain enforceable policies and mechanisms to ensure implementation of the management measures. CZARA required EPA and the National Oceanic and Atmospheric Administration (NOAA) to approve state programs. If the state programs did not meet the standards, both agencies were required to withhold a percentage of grant money. The NOAA and EPA guidance required state programs to be implemented by January 1999.

No state submitted a fully approvable program by the deadline. NOAA and EPA conditionally approved state programs with no penalties for up to five years. The three most common reasons that NOAA and EPA delayed the full approval of state programs were that: individual management measures were not in place, enforceable policies and mechanisms were not in place, and coastal zone boundaries excluded areas with significant impact on coastal waters. Currently, of the 34 states and territories in the coastal zone program, only Maryland has a fully approved CZARA program. 28 states have conditional approval and two states submissions are currently under review

Total Maximum Daily Loads

Where water quality standards are not being met, even after the implementation of programs to control point and nonpoint sources, the Clean Water Act requires states to develop a total maximum daily load (TMDL) to determine the assimilative capacity of a water body and develop a plan to bring the water body into compliance with the standards. The importance of this program is emphasized by the fact that 218 million Americans live within 10 miles of a polluted waterbody. 300,000 river and shore miles and 5 million lake acres are not meeting water quality standards. Waste water treatment plants and industry are the sole cause of pollution in only about ten percent of impaired waters.

The focus on TMDLs has increased dramatically within the past few years as a result of about 45 lawsuits filed against states and EPA for not appropriately implementing this portion of the Clean Water Act. (EPA is responsible for developing TMDLs where states fail to act.) Almost half of these lawsuits have been settled, setting deadlines to implement TMDLs for impaired water bodies. Over 2000 TMDLs are currently under development.

In response to these lawsuits and the findings of an advisory committee, EPA has proposed revisions to the TMDL regulations. The proposed regulations set a more definitive structure for state TMDL programs. The proposal would give states flexibility in priority setting, although high priority waters would include public drinking water supplies and water that supports endangered or threatened species. The states would have 15 years to establish their TMDLs. The TMDL must include an implementation plan and provide reasonable assurance that the pollution sources will attain the specified reductions. The reasonable assurance requirement can be filled by voluntary programs for nonpoint sources as long as adequate funding, staffing, and technical assistance are available for implementation. The final TMDL regulations are scheduled to be released in the summer of 2000.

Safe Drinking Water Act

Historically the Safe Drinking Water Act (SDWA) has focused on treatment requirements for public drinking water supplies. Pollution prevention efforts in the watershed were not the domain of this Act. Two new provisions in the 1996 amendments to the Act have shifted the focus somewhat--the Ground Water Rule and Source Water Assessments.

The Ground Water Rule, when it is proposed in the spring of 2000, will require the control of contamination of groundwater public water supplies from microbial sources. Studies of the occurrence of bacterial and viral pathogens or fecal contamination indicators in ground water indicate that the number of ground water sources with fecal contamination is significant. The proposed regulation will specify the appropriate use of disinfection and encourage the use of alternative approaches, including best management practices and control of contamination at the source

SDWA requires states to develop and implement Source Water Assessment Programs (SWAP) to analyze existing and potential threats to the quality of the public drinking water throughout the state. Every state is required to submit a program to the EPA by February 1999 and to complete all the assessments in the state by May 2003. A state SWAP includes:
- delineating the source water protection area,
- conducting a contaminant source inventory,
- determining the susceptibility of the public water supply to contamination from the inventoried sources, and
- releasing the results of the assessments to the public

While the new amendments do not give states any new regulatory or enforcement authorities for drinking water source protection, many of the provisions are intended to encourage states and localities to go beyond source water assessments and implement efforts to manage identified sources of contamination in a manner that will protect drinking water supplies.

Websites

Clean Water Action Plan	http://www.cleanwater.gov/
Coastal Zone Program	http://www.epa.gov/OWOW/NPS/czmact.html
Drinking Water	http://www.epa.gov/safewater/protect.html
Nonpoint Source Program	http://www.epa.gov/owow/nps/Section319/fy2000.html
Nutrient Water Quality Criteria	http://www.epa.gov/ost/standards/nutrient.html
Section 319 Monitoring Program	http://h2osparc.wq.ncsu.edu/99rept319/
Total Maximum Daily Loads	http://www.epa.gov/owow/tmdl/index.html
Animal Feeding Operations	http://www.epa.gov/owm/afo.htm

References

North Carolina State University. 1999. Section 319 National Monitoring Program: 1999 Summary Report.

U. S. Environmental Protection Agency. 1999a. Draft Nutrient Criteria Technical Guidance Manual: Rivers and Streams. EPA 822-D-99-003.

U. S. Environmental Protection Agency. 1999b. Draft Nutrient Criteria Technical Guidance Manual: Lakes and Reservoirs. EPA 822-D-99-001

U. S. Environmental Protection Agency. 1998a. National Water Quality Inventory: 1996 Report to Congress. EPA 841-R-97-008.

U. S. Environmental Protection Agency. 1998b. National Strategy for the Development of Regional Nutrient Criteria. EPA 822-R-98-002.

U. S. Environmental Protection Agency. 1993a. Guidance Specifying Management Measure for Sources of Nonpoint Pollution in Coastal Waters. EPA 840-B-92-002.

U. S. Environmental Protection Agency and U.S. Department of Commerce. 1993b. Coastal Nonpoint Pollution Control Program: Program Development and Approval Guidance.

Water Quality Goals
for Agriculture

Janet K. Goodwin
Environmental Scientist
Environmental Protection Agency
Office of Water
Office of Science and Technology

Biographies for most speakers are in alphabetical order after the last paper.

The Federal Government has ranked agriculture as the principal contributor to Water Quality Impairment in Lakes and Rivers. To address this problem, particularly the contribution made from confined livestock production, USDA and EPA jointly issued the Unified National Strategy for Animal Feeding Operations (AFO Strategy) in March 1999. The AFO Strategy describes the activities that the two Federal Departments plan to take to address water quality and public health impacts associated with Animal Feeding Operations or AFOs. The AFO Strategy also sets forth the goal of minimizing water pollution from the confinement area and land application of manure.

The AFO Strategy envisions about five percent of the AFOs nationwide falling into the Federal regulatory program and the strategy intends that the remainder of the AFOs will voluntarily improve their animal waste disposal practices, assisted by a variety of financial and technical programs. The AFO Strategy's national performance expectation is that all AFOs should develop and implement Comprehensive Nutrient Management Plans (CNMPs) through either the voluntary assistance offered by USDA or through the regulatory program and the issuance of an National Pollutant Discharge Elimination System (NPDES) permit.

EPA is responsible for revising the existing regulations to which as many as five percent of AFOs will be subject. This group of AFOs are defined as Concentrated Animal Feeding Operations (CAFOs). CAFOs are currently defined as those operations with more than 1000 Animal Units (AUs) or smaller operations that meet certain conditions and are subject to NPDES permit requirements to control wastewater discharges.

The regulations that define the term CAFO are found at Title 40 Code of Federal Regulations Part 122.23. These regulations were issued 20 years ago. EPA is currently revising these regulations. The revisions being considered include expanding the definition of CAFO to include land application areas owned or operated by the CAFO to which manure is applied. EPA is also evaluating options for the definition of CAFO for medium sized operations. Currently an AFO can be designated as a CAFO contributing pollution to surface waters on a case-by-case basis if: 1) it houses more than 1000 AUs, or 2) if it houses between 300 AUs and 1000 AUs and also has a manmade conveyance, or 3) has water that originates outside of the AFO and run through or pass over the AFO, or 4) if animals come in direct contact with these waters. In addition to these criteria EPA is considering other criteria that could reflect the potential of medium-sized AFOs to discharge waste and wastewater to Waters of the U.S. EPA is also considering clarifying the definition of CAFO to specifically include poultry operations with dry manure systems.

EPA issued technology based performance standards or effluent limitations guidelines regulations for the feedlots point source category in 1974. These regulations are found at Title 40 Code of Federal Regulations Part 412 and apply to large CAFOs or feedlots with greater than 1000 AUs. The regulations established zero discharge except when chronic or catastrophic rainfall events cause an overflow from a facility which has been designed, operated and maintained to contain all waste and wastewater and the runoff volume associated with a 25 year, 24 hour storm.

EPA is revising the Effluent Guidelines Regulations. Some of the revisions being considered include establishing management practices or requirements to control manure or wastewater application to crop or pasture land. In addition, EPA is considering establishing monitoring and recordkeeping requirements, such as collection of soil and manure samples and recording manure applications, crop yields and other factors related to land application of manure. EPA is also considering requiring practices and inspections that help to ensure the manure storage structures are being adequately maintained. Dry poultry requirements are being considered as part of this rulemaking effort, consistent with the inclusion of dry poultry operations under the definition of CAFO. EPA is also considering establishing effluent guidelines requirements for facilities below the 1000 AU threshold.

The revised regulations are scheduled to be proposed late in 2000 and finalized by 2002. In the interim EPA is developing a permit guidance manual intended to provide guidance for permit writers who will issue NPDES permits to CAFOs. This permit guidance will describe the types of requirements that should be included in NPDES permits issued to CAFOs. Included in these requirements will be the development of a CNMP. The permit guidance will describe what should be included in a CNMP. The CNMP will include activities necessary to ensure that manure storage and animal housing structures are adequately maintained to ensure structural integrity and are operated in such a way as to avoid a discharge. The application of manure and wastewater to crop or pasture land should also be performed in an appropriate fashion. The CNMP should account for the nutrient needs of the crop, the nutrient levels in the soil, the timing of application, the hydraulic loading of the soil and proximity to surface waters.

EPA is working with states that have authorized NPDES programs to issue permits to CAFOs. The AFO Strategy and permit guidance describe two types of permits that can be issued to CAFOs. Most CAFOs are expected to be subject to general permits. A general permit is written to cover a category of point sources with similar characteristics for a defined geographic area (e.g., all CAFOs in a State). General permits offer a cost-effective approach for NPDES permitting authorities because of the large number of facilities that can be covered under a single permit. At the same time, the general permit also provides flexibility for the permittee to develop and implement pollution control measures that are tailored to the site-specific situation of the permittee. To receive coverage by a general permit, facilities submit a notice of intent (NOI), which would include descriptive information about the facility including legal name and address of owner or operator, facility name and address and contact person, physical location type, and number of animals, and receiving stream information. EPA has stated that all CAFOs should develop Comprehensive Nutrient Management Plans as a requirement of their NPDES permits which would address the site-specific features of each CAFO.

The other type of permit would be an individual permit which is developed especially for the affected facility and is written specifically for the conditions at the facility. EPA is considering requiring individual permits for facilities when they meet the following criteria:

- exceptionally large operations;

- operations undergoing significant expansion;

- operations that have historical compliance problems;

- operations that have significant environmental concerns; and

- new CAFOs.

Comprehensive Nutrient Management Plan from a USDA Perspective

Obie D. Ashford
National Leader for Animal Husbandry
NRCS, Beltsville, Maryland

Biographies for most speakers are in alphabetical order after the last paper.

Abstract

Livestock manure has emerged over the past few years as a major political, as well as an environmental issue. As the Congressional Research Service described the situation in a May 1998 report: "Social and political pressure to address the environmental impacts of livestock production has grown to the point that many policy-makers today are asking what to do, not whether to do something." It added: "The bulk of current policy debate on animal waste issues, both legislative and regulatory, is occurring in states, and that activity is vigorous and multi-faceted. Federal attention followed more recently."

National policy attention had reached a peak over the last few years, as symbolized by the issuance of the USDA/EPA Unified National Strategy on Animal Feeding Operations. Below the national level, no less than 34 states have passed, voted on, or at least debated policies in the last five years that would directly or indirectly affect control of livestock manure. Numerous counties have passed their own ordinances relative to the matter. Finally, in the non-governmental realm, some of the national livestock producer groups have undertaken their own initiatives during the past two years to curb manure runoff and related environmental problems.

Market forces, technological changes, state and local regulations, and industry adaptations have produced unprecedented increases, concentrations, and geographic shifts in confined livestock production in the United States. Public perception of the impacts that this concentration of livestock may have on the environment is based on the Pfiesteria outbreaks along the mid-atlantic; urban water supplies contaminated by Giardia, Cryptosporidium parvum, and nutrients; and increased in-stream nutrient, pathogen, and organic loading correlated with livestock numbers. Two issues of growing concern are: 1) non-point source pollution of the Nation's water from AFOs, and 2) the inadequacy of traditional land-based manure nutrient management strategies as livestock operations surpass the carrying capacity of the land in some geographic areas.

Introduction

"Technical Guidance for Developing Comprehensive Nutrient Plans" is a document intended for use by the Natural Resources Conservation Service (NRCS) and conservation partner state and local field staffs, private consultants, landowners/operators, and others that will be developing or assisting in the development of the comprehensive nutrient plans (CNMPs). This technical guidance is not intended as a sole source of reference for developing CNMPs. Rather, it is to be used as a tool in the process of providing technical assistance in identifying the conservation practices and management activities that will be included in a CNMP.

A Comprehensive Nutrient Management Plan (CNMP) is a component of a conservation plan that is unique to animal feeding operations. A CNMP is a grouping of conservation practices and management activities which, when combined into a system, will help to ensure that both production and natural resource goals are achieved. It incorporates practices to fully utilize animal manure and organic by-products as a beneficial resource. A CNMP addresses natural resource concerns dealing with nutrient and organic by-products and their adverse impacts on water quality. A CNMP needs to be in compliance with all applicable local, tribal, State, and Federal regulations. For certain unique, impacted watersheds or water bodies, special management activities or conservation practices may be incorporated to meet specific local, tribal, State, or Federal regulations.

The conservation practices and management activities in a CNMP for which NRCS maintains technical standards are to meet these standards. Components of a CNMP for which NRCS does not currently maintain standards are to meet criteria established by local, tribal, State, Federal government or others recognized by NRCS. Ultimately, it is the producer's responsibility as the decision-maker to select the system of conservation practices and management activities that best meet their needs from the alternatives available.

The goal of a CNMP as described in the Unified National Strategy for animal feeding operations (AFOs) is to minimize the adverse impacts of (AFOs) on water quality and public health. To accomplish this goal will require a significant increase in the intensity and comprehensiveness of technical assistance provided to producers.

Potential Workload

The work load analysis (WLA) conducted by Natural Resources Conservation Service and the conservation partnership shows that there are about 1.4 million animal feeding operations with 0.1 of an animal unit or greater base on the 1997 Census of Agriculture. Of this universe of 1.4 million, it is estimated that 298,500 would need technical assistance to develop CNMPs. There are slightly more than 500,000 operations with goats, horses, sheep, mules and rabbits included in the 1.4 million universe.

A CNMP is to be developed by the client for the client's use to record decisions for production, natural resource protection, conservation, and enhancement.

Decision and resource information needed during implementation and maintenance of the plan are recorded. The narrative and supporting documents provide guidance for implementation and may serve as a basis for compliance with state, tribal, and federal regulations.

A CNMP is to include all land units, on which manure and organic by-products will be generated, handled, or applied, that the client either owns or has decision-making authority over.

ELEMENTS TO CONSIDER WHEN DEVELOPING A COMPREHENSIVE NUTIRENT MANAGEMENT PLAN -

1. **Animal Outputs - Manure and Wastewater Collection, Handling, Storage, and Treatment, and Transfer**

A manure and wastewater management system for a given animal feeding operation (AFO) should include all the components and management activities necessary to minimize degradation of water quality. A system may consist of a single component or as many components as necessary to meet the objectives of the owner/operator while minimizing the environmental impacts. An on-site visit is required to identify existing and potential resource concerns, problems, and opportunities in the siting of manure and wastewater management system components. It is also important during this site visit to document the existing in-place infrastructure, equipment available, and transfer processes being used.

2. **Evaluation and Treatment of Sites Proposed for Land Application**

An on-site visit is required to identify existing and potential resource concerns, problems, and opportunities for the conservation management unit (CMU). This process will be used to identify and assess operations and activities needed to address existing and potential natural resource problems.

3. Land Application

The potential long and short–term impacts of planned land application of all nutrients and organic by-products (e.g., animal manure, waste water, commercial fertilizers, crop residues, legume credits, irrigation water, etc.) must be evaluated and documented for each Conservation Management Unit (CMU).

4. Record of CNMP Implementation

If the landowner/operator is to adequately apply and assess their CNMP, it is critical that they maintain a record of their activities and the functionality of the system. A record keeping plan should be developed that addresses key elements of the CNMP to aid in the application and assessment documentation.

5. Inputs to Animals – Feed Management

Feed management activities may be used to reduce the nutrient content of manure making it easier to manage in a land application scenario. These activities may include phase feeding, amino acid supplemented low crude protein diets, and the use of low phosphorus grain and enzymes such as phytase or other additives. When used, feed management activities shall be in accordance with recommendations by Land Grant Universities, Industry, and others recognized by NRCS.

6. Other Utilization Activities

Using manure and organic by-products to provide for alternative, environmentally safe, uses and solutions should be an integral part of the overall CNMP. This is especially true where past land application of manure and organic by-products are a problem because of residual soil nutrient content, and future land application will make conditions worse.

Summary

In 1998, NRCS conducted a study titled "Nutrients Available from Livestock Manure Relative to Crop Growth Requirements". Using farm-level data from the 1992 Agriculture Census, the authors presented county estimates of pounds of manure nitrogen and phosphorus potentially generated from confined livestock, and compared these estimates to the potential for nitrogen and phosphorus uptake/removal by crops and application on pasture land. This study indicated that the number of counties where manure nutrients exceed potential plant uptake and removal, including pasture land applications, has increased dramatically since 1949. For instance, the number of counties with excess manure production that have the potential to cause water quality problems (nitrogen and phosphorus) more than doubled between 1982 and 1997.

Also, the aggregate effect of odors and gaseous emissions from land application of manure, manure handling, decomposition of dead animals, and, to some extent, from wet feed pose nuisance and public health problems. Current science and technology offer approaches to minimize the problems, but not to entirely eliminate them.

Most of the efforts by USDA and EPA regarding CNMP development have been on land application of manure and organic by-products. To effectively address the manure management concerns, especially in highly concentrated animal areas, other utilization options, such as converting to high-value products need to be a part of the overall management and utilization process.

References

Unified National Strategy for Animal Feeding Operations – March 9, 1999

Natural Resources Conservation Service and Conservation Partnership Workload Analysis, 1999

Charles H. Lander, David Moffitt, and Klaus Alt, U.S. Department of Agriculture, Natural Resources Conservation Service, February 1999, Resource Assessment and Strategic Planning Working Paper 98-1

Census of Agriculture, 1997

General Manual, Title 450, Technology, Part 401, Technical Guides

General Manual, Title 190, Ecological Sciences, Part 4402, Nutrient Management

●●●●●●●●●●●●●●●●●●●●●●●●●●●●●●●

Environmental Water Quality Regulations Applied to Animal Production Activities

John C. Becker
Director of Research
The Agricultural Law Research
and Education Center
The Dickinson School of Law
The Pennsylvania State University

Gene C. Brucker
Graduate Research Assistant
The Agricultural Law Research
and Education Center
The Dickinson School of Law
The Pennsylvania State University

Biographies for most speakers are in alphabetical order after the last paper.

●●●

Introduction

When designing the Clean Water Act's regulatory structure in the 1970's, Congress chose the law's objective to be the restoration and maintenance of the chemical, physical, and biological integrity of the Nation's waters[1]. In selecting regulatory tools to achieve this goal, Congress introduced several key concepts that have evolved over time to become central issues in today's debate about the future of federal and state water quality regulation affecting agriculture, particularly animal production. Two of these key concepts which are discussed in this paper are the distinction between point and nonpoint sources of pollution[2] and water quality standards for regulation of pollution sources.[3]

Discharge of pollutants into waters of the United States from a point source must meet the fundamental regulatory requirement of complying with Clean Water Act requirements before the pollution occurs.[4] Within the definition of a point source are other key concepts such as that of a discernible, confined and discrete conveyance from which pollutants are or may be discharged. At the time it was adopted, the Act put in place these statutory provisions that were the subject of agency interpretation or later explained in regulations. In most cases

---

[1] 33 U.S.C. section 1251 (West, 1998).

[2] Id. Section 1362 (14) defines a "point" source. Whatever does not meet the definition of a "point" source is considered a "nonpoint" source of pollution.

[3] Id. Section 1313 (d).

[4] Id. Section 1311 (a) establishes the central regulatory requirement of the Act, "Except as in compliance with this section and [other] sections of this title, the discharge of any pollutant by any person shall be unlawful. In this context, "discharge" is defined to include the addition of any pollutant from a point source into waters of the United States, section 1362 (12).

these regulations have existed unchanged for a considerable period. The contextual basis upon which these statutory and regulatory provisions were built has changed considerably over that period. How should change that has taken place be reflected in applying existing provisions to changed situations?

In regard to water quality standards, the Clean Water Act mandates that states identify those waters within the state's boundaries for which effluent based limitations are not stringent enough to implement any water quality standard for such water.[5] States are further mandated to establish a priority ranking for restoration of identified impaired waters based on the severity of the pollution identified and the uses to which the waters are put. States quantify the Total Maximum Daily Allowable Loads for pollutants in each listed water body and to allocate the maximum loads among the contributing sources, whether they be classified as point or nonpoint in origin. Determining Total Maximum Daily Loads establishes the maximum amount of pollutants that a water body can receive. However, despite such mandates, states have been slow to move forward with programs to accomplish these goals. Litigation to force states to move forward has been the result of the inaction.

In the nearly thirty years since major federal water quality legislation was passed, considerable progress has been made toward achieving the law's stated purpose. However, the job is far from finished. As noted in the 1995 General Accounting Office Report to the Senate Committee on Agriculture, Nutrition, and Forestry, significant portions of America's surface waters are impaired as a result of agricultural activities.[6] This report, which was based on data taken from state water quality assessments, led GAO to conclude that agriculture is the main source of groundwater pollution in the United States. As to surface water pollution the report noted that soil and nutrient run off, as well as animal waste run off from animal feeding operations, contributes greatly to the impairment of streams and lakes through introduction of excess nutrients, organic matter and pathogens. Excess nutrient loadings stimulate algae development which when decomposed robs water of oxygen needed by fish and aquatic animals for their own survival.[7] Pathogenic contamination can impact on water quality for personal use, recreation and wholesomeness of fish and shellfish obtained from the water source. In watershed areas studied by the U.S. Geological Survey's National Water Quality Assessment, increases of stream loadings of nitrogen and phosphorous have a strong statistical correlation, in part, to increases in the concentration of livestock in the watershed area.[8] The study noted this statistical correlation exists despite the fact that most nutrients applied to land do not end up in lakes, rivers and streams as they are taken up by plants in the watershed or returned to the atmosphere as gas.

While all of this was taking place on the regulatory front, substantial changes were taking place in the agricultural sector of the economy. Several prominent commentators describe this change as the application of a manufacturing mentality to traditional agricultural

---

[5] Id. Section 1313 (d).
[6] See GAO Report, "Animal Agriculture, Information on Waste Management and Water Quality Issues" (GAO/RCED-95-200BR), at p.9. (Hereafter, GAO Report).
[7] Id., p. 11.
[8] Id., p. 13.

production practices.[9] No longer do agricultural producers ask the question whether consumers will buy what they produce. Producers now actively seek to determine what consumers want to buy and then design production practices and structures that deliver to consumers precisely what consumers want. Traditional production practices are also modified through increased use of production contracting arrangements for livestock production, greater integration of production, processing and distribution channels, increased investment in assets devoted to production and movement away from the so called "traditional' farm. By the end of the 1990's, an increasing share of total agricultural production was being generated by a relatively small, but growing, number of large-scale producers. A description of the modern agricultural sector would include a fairly small number of very large producers; a significant number of small-scale producers; and a like number of mid-sized producers. It is the mid-sized producers who wrestle with the question of whether the producer can afford to continue to operate at present levels, or must a choice be made between opposite ends of the spectrum just described[10].

Despite the fact that the small number of large producers generate a large part of the total output of the sector, small producers who face the competitive pressures posed by large-scale producers have not accepted their lot in life without objection. In several parts of the country, active opposition to proposals for placement of large animal production facilities is being mounted on grounds that allowing such facilities in rural areas adversely affects "family farmers", the group which Jefferson considered to be the very basis of a democratic society. Of important significance from a political and economic policy perspective is that the agricultural community now appears fragmented into groups that favor or oppose expansion of production facilities.

While these changes in the agricultural economy continue to evolve, the Courts also have a role in shaping the debate. The Second Circuit's 1994 decision in *Concerned Area Residents for the Environment v. Southview Farms, Inc.*[11] examined the question of when a given set of production practices or management approaches could be viewed as subject to mainstream environmental regulation. In many ways the Southview Farms case presents a number of key issues to address. First, as Southview Farms can be described as a model of what large animal production facilities ought to be (in that case a dairy farm), is the decision an indictment of the manufacturing mentality applied to new types of production facilities? How much weight did the Court's decision put on the factual record of poor manure management practices by Southview employees? Are those factual elements crucial to evaluating the impact that the decision could have? Why didn't the court simply rely on the numbers animals alone to determine that Southview met the definition of a CAFO and, therefore, needed a permit that it failed to have at the time the discharge occurred?

---

[9] See Michael Beohlje, *Industrialization of Agriculture: What are the Policy Implications?* Contained in *Increasing Understanding of Public Problems and Policies,* edited by Steve A. Holbrook and Carrol E. Myers, The Farm Foundation, Oak Brook, Illinois, 1996.

[10] Milton C. Hallberg and Dennis R. Henderson, Structure *of Food and Agricultural Sector,* contained in *Food, Agriculture, and Rural Policy into the Twenty-First Century, Issues and Trade-Offs,* edited by Milton C. Hallberg, Robert G.F. Spitze, and Darryl E. Ray, Westview Press, Boulder Colorado, 1994, at p. 58,59.

[11] 34 F.3d 114 (2nd Cir., 1994).

In this article we explain the principal water quality oriented provisions of federal environmental law that directly apply to animal agricultural production practices and facilities. Included in this discussion is the recently developed joint USDA/EPA National Strategy for dealing with pollution threats posed by large-scale animal production facilities. Concerns about pollution from such sources are not merely idle speculation. Over the past five years events across the country, and particularly in North Carolina[12], Maryland and Virginia have made these environmental consequences the subject of headline news at the national level.

1.   The Federal Water Pollution Control Act

The *Federal Water Pollution Control Act* of 1972 was meant to restore the integrity of the nation's navigable water to fishable and swim-able quality by 1983, with a total elimination of pollutant discharges by 1985. Although this goal was never realized, the Federal Water Pollution Control Act, commonly referred to as the Clean Water Act, has done much in terms of cleaning up the nations rivers and streams.[13] The Clean Water Act's primary tool in regulating the discharge of pollutants into the nation's navigable water is the National Pollutant Discharge Elimination System (NPDES) found in Title IV of the Clean Water Act. Under the Act water pollution sources are classified as either *point sources* or *nonpoint* sources of pollution. The primary regulatory focus of the NPDES is controlling the discharge of pollutants from point sources.[14] Under the Act a point source is defined as:

> Any discernable confined and discrete conveyance, including but not limited to any pipe, ditch, channel, tunnel, conduit, well, discrete fissure, container, rolling stock, concentrated animal feeding operation, or vessel or other floating craft, from which pollutants are or may be discharged.[15]

In order to discharge any pollutant from a point source, a permit is required under the NPDES. A discharge of a point source pollutant without a permit is illegal under the Clean Water Act. To be granted a permit under this section of the Clean Water Act, a point source polluter must conform to a certain set of technology based standards that are tied to the pollutant source and the nature of the pollutant being discharged. In addition to meeting technology-based effluent limits, point sources of pollution may also be subject to regulation based on water quality standards. Under the Act, states are directed to identify those waters within their boundaries for which the technology based effluent limitations required under the NPDES are not stringent enough to implement a water quality standard that is applicable

---

[12] Reference here is made to a release of raw swine manure from a treatment lagoon located next to the Neuse River in North Carolina and the recent fish kills along the Pocomoke River in Maryland and Virginia that is attributed to the microbe, *Pfiesteria piscicida*.

[13] U.S. Department of Agriculture and Environmental Protection Agency Draft Unified National Strategy for Animal Feeding Operations, 63 Fed. Reg. 50192, 50195 (1998). (Hereinafter, Unified Strategy).

[14] Under 33 U.S.C. section 1311, except as in compliance with the Clean Water Act, the discharge of a pollutant is unlawful. "Discharge of a pollutant" is defined in section 1362 (12) as the addition of any pollutant to navigable waters from a point source.

[15] Id., section 1362(14).

to that water.[16] It is through this section of the act that the quality of the receiving waters is taken into consideration. For waters that are identified by the states, a total maximum daily load (TMDL) for specific pollutants is identified for each body of navigable water, or a designated segment of the water course. The TMDL is set at the level necessary to implement the applicable water quality standard. In this way, the TMDL acts as a restriction on all sources of pollution that discharge into the body of water.

In contrast to the detailed definition of a point source, nonpoint sources are largely undefined. A good universal definition of a nonpoint source is anything that is not covered under the point source category. Despite the advances made to regulate point source pollution, nonpoint source pollution has been blamed for a significant portion of the nation's current water pollution problems.[17] Nonpoint sources come from a variety of activities, including storm water runoff, forestry, mining and farming.

Under the Act's original approach, states were directed to develop area-wide waste treatment management plans to identify those areas which have substantial water quality control problems and design and implement plans to address them. In addition, these plans were directed to include a process by which the state could identify agricultural and silvicultural related nonpoint sources of pollution, including runoff from manure disposal areas and from land used for livestock and crop production. In addition such plans were to set forth procedures and methods to control these sources.[18] To carry out these control measures, the Secretary of Agriculture was authorized to enter into contracts with land owners and farm operators who controlled rural land for the purpose of installing and maintaining best management practices to control nonpoint sources of pollution as part of these approved plans.[19]

2. Point Source Discharge Prior to Legislation

The EPA originally held a very broad view of the definition of a point source. Many agricultural activities that generated pollutants could be included in that broad definition. For example, irrigation return flows and rainfall runoff were considered to be point sources if the water was channeled or collected by any man-made activity.[20]

Under this broad view of a point source, many agricultural activities were required to obtain permits under the NPDES program. The EPA estimated the number of silviculture point sources to be over 300,000 and approximately 100,000 separate storm water discharge point sources.[21] Each individual point source would need a permit under the NPDES requirement. This proved to be a heavy administrative burden on the EPA. For this reason, the EPA

---

[16] Id., section 1313(d).
[17] GAO Report, p. 6.
[18] 33 U.S.C. section 1288(b)(1)(A); (2)(F).
[19] Id., section 1288 (j)(1).
[20] Drew L. Kershen, "Agricultural Water Pollution: From Point to Nonpoint and Beyond", Natural Resources and the Environment, Winter, 1995, 3.
[21] NRDC v. Costle, 568 F.2d 1369 (D.C. Cir. 1977), Costle II.

exempted agricultural activities from the NPDES requirement.[22] The EPA justified their exemption of all agricultural activities under section 402 of the Clean Water Act.

In the early 1970's, the Natural Resources Defense Council filed suit against the EPA and asked the Court to order EPA to enforce the NPDES requirement for all point sources, including those originating from agricultural sources. In *NRDC v. Costle*[23], the court ruled against EPA's decision to exempt all agricultural activities. The court held that EPA could not exempt a wide class of point source polluters merely because of administrative burdens. NPDES requirements applied to all discharges of pollutants without exception. Section 402 of the Act directed EPA to issue permits for point source polluters.

As a result of the *Costle* decision, all agricultural and silvicultural activities that discharged pollutants from point sources were expected to meet NPDES requirements. This meant that irrigation return flows and storm water run discharges collected and distributed by man made activities were required to have a permit. Activities of this nature were considered to be nonpoint sources at the time of this decision. This is an example of where the point source regulatory requirements were seen to be moving into the field of regulating nonpoint sources.

3. Later Amendments and Definition Following Litigation

After the *Costle* decision, many agricultural related activities were required to comply with the Act and obtain a permit under the NPDES. In response to it, Congress amended the Clean Water Act in 1977 and redefined the term "point source" to exclude "return flows from irrigated agriculture."[24] As a result of this amended definition, irrigation return flows were now included in the nonpoint source category.[25]

In 1987, Congress amended the Act with the passage of the Water Quality Control Act of 1987. Included in this Act's provisions was a renewed effort to deal with pollution from nonpoint sources through adoption of nonpoint source management programs. Under the amended provisions, the Governors of the states were directed to prepare and submit to the Administrator for approval reports that identified navigable waters within the state's borders which, without additional action to control nonpoint sources of pollution, could not reasonably be expected to attain or maintain applicable water quality standards or meet the goals of the Act.[26] In this report the states were also to identify those state and local programs which would control pollution from nonpoint sources. Principal reliance under such plans would be placed on best management practices and measures designed to reduce pollutant loadings from particular nonpoint sources. To assist in designing these plans, the Administrator was directed to make grants to the States to assist in carrying out the needed groundwater quality protection activities that are part of the State's comprehensive nonpoint source pollution control program.[27]

---

[22] Kershen, infra note 20, p. 4.
[23] NRDC v. Costle, 568 F.2d 1369 (D.C. Cir. 1977), Costle II.
[24] 33 U.S.C. section 1362 (14).
[25] Drew L. Kershen, infra note 20, p.4.
[26] 33 U.S.C. section 1329 (a).
[27] Id., section 1329(I)(1).

The 1987 amendments also adopted a similar exemption from NPDES permit requirements for storm water discharges as those created for irrigation return flows in the 1977 Clean Water amendments. As a result of both the 1977 and 1987 acts, the EPA's position in *NRDC v. Costle* was that such activities, although arguably meeting the definition of a point source, should not be regulated as such was incorporated into the Act. This lifted a serious obligation from the shoulders of agricultural producers.

Not all agricultural activities, however, were intended to be exempted from Clean Water Act regulations. The point source definition, originally enacted and as it stands today, specifically includes *concentrated animal feeding operations* (CAFOs) within the meaning of the term point source.

4. Animal Feeding Operations and Concentrated Animal Feeding Operations

An example of an agricultural activity that has significant potential for detrimental effect upon the nation's navigable waters is an *animal feeding operation* (AFO). For Clean Water Act purposes, an AFO is a facility in which animals are confined and fed for a total of no less than 45 days in a 12-month period;[28] and no crops or vegetation grown or sustained over any part of the facility during the normal growing season.[29] AFOs pose a number of risks to water quality and public health because of the amount of animal manure and wastewater they generate.[30] Manure and wastewater from AFOs have the potential to contribute pollutants such as nutrients (i.e. nitrogen or phosphorous), sediment, pathogens, heavy metals, hormones, antibiotics, and ammonia to the environment. Excess nutrients in water can result in or contribute to eutrophication, anoxia, and, in combination with other circumstances, have been associated with outbreaks of microbes such as *Pfiesteria piscicida*.

There are an estimated 450,000 animal feeding operations within the United States.[31] Approximately 85% of the AFOs are run by small farm operations with fewer than 250 animals. As will be seen in subsequent discussion, a large majority of AFOs is not subject to the NPDES permit requirement. In the classic distinction between point sources and nonpoint sources, those which did not meet the regulatory definition of a point source were regulated under Clean Water Act programs such as the section 208 and 319 programs described above. However, the Clean Water Act definition of point source has always included *concentrated animal feeding operations* (CAFOs) as point sources of water pollution.

A concentrated animal feeding operation (CAFO) first and foremost is an AFO that meets further regulatory requirements that focus on the size of the operation or the manner in which the facility has operated. Under the NPDES program, an AFO is a CAFO if more than 1000

---

[28] 40 CFR 122,23 (b)(1).
[29] Id.
[30] Unified Strategy, 50195.
[31] Office of Wastewater Management, Animal Feeding Operations, January 18, 2000, <wysiwyg://35/http://www.epa.gov/owm/afo.htm>.

animal units are confined at the facility. For Clean Water Act purposes, an "animal unit" is a means of comparing the diverse types of animals raised under confined conditions. The animal unit concept was developed by EPA to make comparisons between species possible. Rather than relying solely on the aggregate number of animals as the determining factor, each livestock type, with the exception of poultry, is assigned a multiplication factor to determine the total number of animal units , or AUs, at a given facility.[32] The number of actual animals located at the facility is multiplied by the appropriate factor for that species to determine the number of animal units at the facility.

Large facilities having more than 1000 animal units are considered to pose a threat to water quality and public health because they produce manure in such volume that the risks are considered to exist whether or not the facilities are well managed.[33] The amount of manure stored at such facilities, if spilled while handling or during a breach of a storage system, can release large amounts of manure and wastewater to the environment with potentially catastrophic results. In addition, land application of large volumes of waste requires careful planning to avoid adverse impacts to the environment.

If a facility has between 301 and 1000 animal units located on the facility, the facility can be considered a CAFO if it meets one of the following specific criteria: (1) Pollutants are discharged into waters of the United States by means of a ditch, flushing system, pipe or other man made device; or (2) pollutants are discharged directly into waters of the United States which originate outside of and pass over, across, or through the facility or some other direct contact with animals confined on the feedlot[34].

Under either of these two CAFO definitions is an exemption from the classification of CAFO if the facility can establish that it would discharge pollution only in the event of a 25 year, 24 hour, or larger, storm event.[35] It is EPA's position that to be eligible for this exemption, the facility owner or operator must demonstrate to the permitting authority that the facility has not had a discharge.[36]

Generally, AFOs that have up to 300 animal units are not considered to be CAFOs. However, EPA can also designate an AFO as a CAFO on a case by case basis if the AFO is determined to be a significant source of pollution.[37] These determinations are made on the basis of a visual observation and results of water quality monitoring.

5. Litigating the Definitions

Many AFOs operated under the notion that they did not meet the CAFO definition and therefore were not subject to NDPES requirements. In *Concerned Area Residents for the*

---

[32] 40 CFR Part 122 Appendix B.

[33] Unified Strategy, page 50200.

[34] 40 C.F.R. Section 122 app. B.

[35] Id., app. B(a).

[36] EPA Guidance manual and Example NPDES Permit for Concentrated Animal Feeding Operations, Review Draft, August, 1999., page 2-9.

[37] 40 C.F.R. part 122.23(c).

Environment v. Southview Farms, the U.S. Court of Appeals for the Second Circuit ruled against a large dairy operation and designated it as a CAFO.[38] This case is one of the most influential court decisions in terms of agricultural regulation under the Clean Water Act.

In *Southview Farms*, the plaintiffs were a group of landowners who lived near the defendant, Southview Farms, a large dairy operation. Southview Farms' methods of raising dairy cattle were unlike traditional means. Dairy cattle were kept in a confined area and not allowed to graze in a pasture.[39] The handling of manure was unlike other dairy operations as well. Through the use of man made devices, the manure was separated into liquids and solids.[40] The liquid manure was pumped into one of five various lagoons located upon the farm. At that point, the manure was dispersed upon the fields by means of a complex irrigation system, which included a piping system and liquid manure spreading vehicles. The whole manure irrigation process at Southview Farms managed millions of gallons of liquid manure.[41] In testimony provided by Southview's neighbors was evidence indicating that on several occasions Southview employees spread liquid manure from vehicles onto the same field. Subsequent to these applications, liquid manure was seen trickling off the field, crossing under a fence by means of a pipe and into a ditch that led to a creek, that ran into a nearby river.[42]

There are important points to be made concerning the Southview decision holding it to be a CAFO that has important ramifications to all livestock producers. First, Southview Farms was found to be a CAFO despite the fact that over 1,100 acres was dedicated to crop production and may have been more than enough land to properly disperse the manure generated by the farm's dairy herd. Under NPDES regulations, a major criteria of AFO designation is that there are no crops or vegetation grown or sustained over any part of the facility where the animals are confined.[43] The court, in determining that Southview was a CAFO, did not look to the farm unit as a whole, but looked specifically to the facility where the animals were confined. Southview Farms' practice of confining its dairy cattle, unlike traditional dairy cattle practices, is a key factor in concluding that the facility is a CAFO that is subject to NPDES regulation.

The second aspect of the Southview Farms decision addressed the fact that the facility was considered a CAFO despite the fact that the discharge occurred during rainfall. As mentioned above, for AFOs with more than 300 animal units, the designation of CAFO will not apply if the facility owner or operator can establish that it has not discharged and has the additional capacity to effectively manage the manure despite a 25 year, 24 hour storm event."[44] In the Southview Farms case, the mere fact that the discharge occurred while it was raining was not enough to trigger application of the storm discharge exemption.

---

[38] Concerned Area Residents for the Environment v. Southview Farm, 34 F3d 114 (1994).
[39] Id., at 116.
[40] Id., at 116.
[41] Id., at 116.
[42] Id., at 118.
[43] 40 CFR 122,23(b)(1).
[44] Id., section 122 app. B(a).

Third, a production facility owner or operator has a continuing responsibility to manage employees to assure that they recognize the importance of performing their job properly. Southview Farms had a considerable sized workforce to manage its operation, but the sloppy performance of some of its employees is a central factor in the dispute and the Court's rationale for imposing liability on the Farm.

With the changing nature of livestock production practices, the designation of CAFO status on Southview farms is very significant. With the declining number of privately owned "family" farms, and a growing number of "industrial" / "corporate" agricultural entities, more agricultural activities employ large-scale methods of animal production that utilize non-traditional production methods to generate desired economic rewards from the intensified management. Those smaller "family" farms find themselves increasingly confronted by the question whether the scale of their production facility will enable them to compete in the market. If the intensified production processes are more widely adopted, more and more farms may meet the CAFO designation and ultimately be subject to regulation as a point source.

6. The Clean Water Action Plan

In February 1998, President Clinton released the Clean Water Action Plan (CWAP), which provided a roadmap to restoring and preserving the Nation's navigable waters. One of the main areas of concern in the plan was the discharge of manure runoff from AFOs. The CWAP called for cooperation between the USDA and EPA in creating the Unified National Strategy to minimize the water quality and public health impacts of AFOs.

The USDA and EPA announced the release of the Unified National Strategy on September 16, 1998. The major goal of the Unified Strategy was for AFOs to minimize water pollution from their confinement facilities and land application of manure.[45] In order to meet its goal, all AFOs are encouraged to develop "technically sound and economically feasible" Comprehensive Nutrient Management Plans (CNMPs).[46]

The general purpose of CNMPs is to outline a plan for disposing of manure from AFOs. The Unified Strategy expects each CNMP to address, at a minimum, feed management, manure handling and storage, and land application of manure. The CNMPs should be tailored to address the individual needs and practices of each individual AFO. CNMPs that are developed to meet the requirements of the NPDES are to be developed by a person who is certified to develop them, by a State agency or by the Natural Resource Conservation Service. In addition, private individuals or nonprofit groups may also be approved to develop such plans.

While CNMPs are voluntary for most AFOs, the implementations of CNMPs are mandatory for facilities that require NPDES permits. All CAFOs, which require a point source discharge permit, must also have a CNMP. As of 1998, approximately 2000 facilities have

---

[45] Unified Strategy, p. 50195 (1998).
[46] Id.

been issued NPDES permits under section 402 of the Clean Water Act. Smaller CAFOs that are required to have NPDES permits, such as those with between 301 and 1000 animal units, can opt to exit the CNMP requirement at the end of the five year permit term by demonstrating the facility has successfully addressed the initial condition which caused it to be designated as a CAFO and it is in full compliance with the plan and other permit requirements.[47]

While the Unified Strategy is aggressive in its goals, the CNMP program is only voluntary for most AFOs. Strongly encouraged, CNMPs are considered the best way to improve the cleaning and preservation of the Nation's navigable waters. Three types of a voluntary programs to help AFOs implement the CNMPs are available: 1) Locally Led Conservation; 2) Environmental Education; and 3) Technical and Financial Assistance Programs. Once approval is granted by the appropriate agency, the AFOs must agree to implement the CNMPs through the voluntary programs before any financial assistance becomes available.

Whether or not the Unified Strategy will be successful in its goals of further cleaning and preserving the Nation's navigable water supply has yet to be seen. While the development of CNMPs are mandatory for a small number of AFOs designated as CAFOs, it is still largely a voluntary program. The Unified Strategy implements no new mandatory regulations, but uses existing regulations in a different way. It is primarily up to each and every individual AFO owner to decide what to do to promote water quality.

7. Regulations of Wetlands

Section 404[48] of the Clean Water Act provides the principal federal authority to regulate wetlands use. Both the Environmental Protection Agency (EPA) and the Army Corps of Engineers use the same definition to describe "waters of the United States." Under their definition traditionally navigable waters are included, as well as all interstate waters, including commerce, impoundments, tributaries, territorial seas and wetlands adjacent to waters other than wetlands. Artificially created wetlands are also generally recognized as being subject to regulation under the Clean Water Act.

The Environmental Protection Agency and the Department of Justice are the principle federal agencies involved with the development of regulations and other enforcement action regarding its provisions. Under section 404, landowners and developers must obtain permits from the Army Corps of Engineers in order to carry out dredge and fill activities in navigable waters, which include adjacent wetlands. The authority of the Army Corp of Engineers to be in such matters is one of historical significance dating back to 1890 and the adoption of the Rivers and Harbors Act.[49]

The Clean Water Act specifically exempts certain activities, including normal agriculture, forestry, and ranching, provided they do not convert areas of U.S. waters to uses which they

---

[47] Id.
[48] 33 U.S.C. section 1344.
[49] Id. section 403.

were not previously subject and do not impair the flow or circulation of such waters or reduce their reach[50]. Of particular interest to agricultural producers are exemptions for normal farming, forestry, or ranching activities, maintenance of dikes, dams, and levees, construction or maintenance of farm ponds or irrigation or drainage ditches. If performance of any of the discharge activities change the use of the waters, impair its flow or circulation or reduces its reach, such an activity may be recaptured and therefore regulated under section 404 of the Act. Therefore, primarily routine activities which have relatively minor impact on waters are considered to be exempt activities. Activities that convert large areas of water into dry land or impede circulation or size of the body of water are outside of the exemptions. In regard to the drain clearing exemption, several issues have arisen in cases where prior drainage ditches, once abandoned, are then re-opened. In such cases, the property owner who desires to re-open the drain must seek a permit to do so.

8. The Food Security Act of 1985

By most estimates, section 404 regulates only about 20 percent of the activities that destroy wetlands. Activities not regulated under section 404 include drainage, ditching, and channelization for agricultural production, which are major causes of past wetland losses. To fill this gap in coverage, the Food Security Act of 1985 – also referred to as the 1995 Farm Bill – included two major wetlands-related provisions, Swampbuster and the Conservation Reserve Program (CRP). Both of these programs involve the U.S. Department of Agriculture in the development of regulations and other enforcement action. The Food, Agriculture, Conservation and Trade Act of 1990 – referred to as the 1990 Farm Bill – amended Swampbuster and CRP. Further amendments were then made again with the Federal Agriculture Improvement and Reform Act of 1996 – also referred to as the 1996 Farm Bill.

9. Swampbuster

Prior to the enactment of the Food Security Act of 1985, federal agricultural policies indirectly encouraged farmers to convert wetlands to farmland by providing credit and commodity price supports. The Swampbuster provision of the Food Security Act of 1985 denied federal farm program benefits to producers who planted an agricultural commodity (defined as an annually tilled crop or sugarcane) on wetlands that were converted after December 23, 1985. Swampbuster violations result in farmers losing eligibility for commodity program benefits, crop insurance, disaster payments, and other federal benefits.

The Food, Agriculture, Conservation and Trade Act of 1990 strengthened Swampbuster by stipulating that a person who drains or otherwise manipulates wetlands for the purpose, or to have the effect, of making the production of an agricultural commodity possible on such converted wetlands (actual planting is not required), is ineligible for farm program benefits for that year and all subsequent years. The act also created a system of graduated sanctions for inadvertent violations and provided that farmers can regain lost federal benefits if they restore converted wetlands.

---

[50] Id. section 1344(f).

Under the Federal Agriculture Improvement and Reform Act of 1996, the reach of Swampbuster was scaled back. Under previous incarnations, farmers were subject to the Swampbuster provisions despite the fact that they had obtained a permit under the Clean Water Act to convert wetlands. Now, a farmer is exempt from the Swampbuster provision provided that a Clean Water Act permit is obtained and measures are taken to mitigate the damages concerning the converted wetlands.[51]

---

[51] Bradley C. Karkkainen, *Biodiversity and Land*, 83 Cornell L. Rev. 1, 67.

The Following Sources were utilized in developing this paper and will be beneficial in further researching this subject.

"Animal Agriculture, Information on Waste Management and Water Quality Issues" (GAO/RECD-95-200BR). General Accounting Office report to the U.S. Senate.

"Animal Feeding Operations," Office of Wastewater Management. World Wide Web Address: "wsyiwig://35/http://www.epa.gov/owm/afo.htm" viewed on January 8, 2000.

Beohlje, Michael. *Industrialization of Agriculture: What are the Policy Implications?* Contained in *Increasing Understanding of Public Problems and Policies,* edited by Steve A. Holbrooke and Carroll E. Myers, The Farm Foundation, Oak Brook, Illinois, 1996.

Clean Water Act, 33. U.S.C. section 1251.

Code of Federal Regulations. 40 CFR part 122.

Concerned Area Residents for the Environment v. Southview Farms, Inc. 34 F.3d page 114 (2nd Cir., 1994). 1994 Second Circuit Court decision.

Halberg, Milton C. and Dennis R. Henderson. *Structure of Food and Agriculture Sector*, contained in *Food, Agriculture, and Rural Policy into the Twenty-First Century, Issues and Trade-Offs,* edited by Milton C. Hallberg, Robert G.F. Spitze, and Darryl E. Ray, Westview Press, Boulder Colorado, 1994.

Karkkainen, Bradley C. *Biodiversity and Land.* 83 Cornell Law Review page 1.

Kershen, Drew L. "Agricultural Water Pollution: From Point to Nonpoint and Beyond," *Natural Resources and the Environment,* Winter, 1995, page 3.

NRDC v. Costle, 568 F.2d page 1369 (D.C. Cir. 1977). 1977 D.C. Circuit Court Decision. Commonly referred to as the Costle II decision.

Rodgers, Walter, H. Jr., Environmental Law, second edition, West Publishing Company, 1994.

"U.S. Department of Agriculture and Environmental Protection Agency Draft Unified National Strategy for Animal Feeding Operations," 63 Federal Register page 50192. Commonly referred to as the "Unified Strategy."

Session 4

Waterborne Pathogens

●●●●●●●●●●●●●●●●●●●●●●●●●●●●●●●●●

Ag-Related
Waterborne Pathogens

Susan M. Stehman, M.S., V.M.D.
Senior Extension Veterinarian
Cornell Diagnostic Laboratory
Department of Population Medicine and Diagnostic Science
College of Veterinary Medicine
Cornell University, Ithaca, New York

Biographies for most speakers are in alphabetical order after the last paper.

●●

Introduction

The focus on livestock farms as potential sources of waterborne pathogens in watershed systems is the result of several converging trends. Media and consumer awareness of foodborne and waterborne illnesses is high. Public health officials have identified livestock manure as a potential source of enteric (intestinal) bacterial and protozoan agents[1, 2]. While manure has the potential to harbor a large number of different animal and human pathogens [3, 4], relatively few microbial agents cause the majority of human waterborne disease outbreaks (see Table 1) [5-7].

This paper will provide an overview of the major pathogens that have been associated with waterborne disease outbreaks in humans and discuss the potential role of livestock or livestock manure as one source of these pathogens in watersheds. Pathogens are defined as any microbial agent such as a virus, bacterium, fungus, or parasite (worms or protozoa) capable of causing infection or disease. Zoonotic pathogens (pathogens transmitted among animals and humans) of concern in watersheds will be emphasized in this paper.

Most pathogens are shed in feces from hosts with active infection. The majority of organisms present in manure and feces are part of the normal microflora of the intestinal tract which includes fecal coliforms including *E. coli*, fecal streptococci and other organisms. The potential for fecal material from a given source to cause infection or disease depends on the types and number of pathogenic organisms present, ability of the pathogen to survive or multiply outside of a host, and the potential for contact with a susceptible host. Strains within

93

a given species of organism can vary in their ability to infect a host. For example, several different strains of *C. parvum* have been identified with different host ranges and different infectious doses for each host [8, 9]. Host susceptibility varies with immune status. The very young, the elderly and individuals with compromised immune system are at greatest risk of infection by most enteric pathogens. *C. parvum* [10] infection in previously infected adults resulted in less severe illness and less shedding of organism. Therefore, the ability of a "pathogen" to cause disease depends on a number of different host, pathogen and environmental variables.

Microbial Agents of Concern in Waterborne Disease Outbreaks

Microbial agents associated with waterborne disease outbreaks (WBDOs) can be divided into three categories: (1) Those associated with intestinal infection and feces from multiple species including humans such as *Cryptosporidium parvum*, *Giardia species (sp.)*, *Escherichia coli* O157:H7, *Campylobacter jejuni*, and *Salmonella sp.*; (2) those associated with human intestinal infection and feces such as *Shigella sp.*, *Salmonella typhi* and human intestinal viruses; and (3) those which live in the environment such as *Pseudomonas* and *Legionella* that are associated with a variety of human illnesses including skin infections (dermatitis), and Legionaire's disease. Intestinal infections are the most common type of waterborne infection and affect the most number of people.

A collaborative surveillance system for reporting waterborne-disease outbreaks in humans has been maintained by The Centers for Disease Control (CDC) and the US Environmental Protection Agency (USEPA) since 1971. The results are used to assess water systems, the type of agents associated with waterborne disease outbreaks (WBDO) and to improve water quality regulations. The pathogens causing gastrointestinal illness and type of water exposure have been summarized from CDC/EPA reports published in the last decade in Table 1[5-7, 11].

Protozoal infections caused by *Giardia sp.* and *Cryptosporidium parvum* were the leading cause of infectious WBDOs for which an agent was identified both in total cases and in number of outbreaks. Most bacterial WBDOs are caused by *Shigella sp.* which are bacteria associated specifically with human feces. *Escherichia coli 0157:H7* (*E. coli* O157) is emerging as the second most important cause of bacterial waterborne disease.

Most drinking water outbreaks have been associated with sewage contamination, drinking water treatment failures, or biofilms in post-treatment distribution systems. Untreated ground water is emerging as an important source of drinking water contamination. Recreational water exposure is also emerging as an important source of waterborne gastrointestinal illness. *C. parvum* outbreaks are increasingly associated with treated (pool) recreational water exposure indicating the need to educate users about the hazards of fecal accidents especially in pools used by diaper-aged children. Untreated recreational water was the most common source of exposure for *Shigella sp.* and *E. coli* O157. The presence of

Shigella sp. indicates human fecal contamination emphasizing the need for education of users.

Information extrapolated from voluntary surveillance systems has limitations [5, 12]. There may be under reporting of the true incidence of water borne disease outbreaks (WBDO) and inaccurate reporting of the relative frequency of disease caused by particular agents. Viral waterborne disease is probably under reported because of difficulties in routine diagnosis of enteric viral infections. Community outbreaks may be investigated more commonly than individual infections which tend to be under reported.

Table 1. Etiology of Waterborne Disease Outbreaks Causing Gastroenteritis 1989-1996

Type of Organism	Etiologic Agent	Number of Outbreaks	Outbreaks associated with Drinking Water		Outbreaks Associated with Recreational Water	
			Surface	Ground	Natural	Pool/Park
Protozoa	*Giardia sp*	27	12	6	4	5
	Cryptosporidium parvum	21	4	4	2	11
Bacteria with Potential for infecting multiple species	*Escherichia coli O157:H7*	11		3	7	1
	Campylobacter jejuni	3	3			
	Salmonella typhimurium	1		1		
	Salmonella java	1				1
	Leptospira grippotyphosa	1		1		
Bacterial infections associated with humans	*Shigella sonnei*	17		7	10	
	Shigella flexneri	2		1	1	
Human Viruses	Hepatitis A	3				3
	Norwalk virus	1		1		
	Norwalk like virus	1				1
	Small Round Structured Virus	1	1			
Acute Gastro-enteritis	Unidentified Etiology – many are consistent with viral epidemiology	60	8	44	7	1
Other	Cyanobacteria like bodies	1	1			

Many of the agents discussed can also be transmitted by direct contact and foodborne exposure [1, 2]. Implication of water as the true source of infection requires timely epidemiological and laboratory investigations. Information presented in the surveillance summaries is used to guide regulatory programs so the key agents (*C. parvum, Giardia sp.*) identified by this surveillance system will be the focus of the rest of this paper. Information about and *E. coli O157* will be provided because of a recent large waterborne and foodborne outbreak implicating cattle as one potential source of infection [13].

Characteristics of Waterborne Pathogens

Protozoa: Cryptosporidium parvum and Giardia sp.

Protozoan parasites such as *Cryptosporidium parvum* and *Giardia sp.* are important causes of waterborne infection that are also potentially present in animal and human manure and feces. *C. parvum* and *Giardia species* are commonly spread by fecal contaminated water or by direct contact [14, 15]. Foodborne transmission is possible but contributes to a smaller percentage of infections [16].

The factors which contribute to their importance as waterborne disease agents include: their relative resistance chlorine water treatment [14], their ability to survive in water for a month or more [14, 17, 18], the wide host range which includes wild mammals, companion animals, livestock and humans [19, 20]; and the relatively low dose required to cause infection [2, 8, 14]. Infection in humans and animals leads to fecal shedding of numbers of cysts (*Giardia*) and oocysts (*Cryptosporidium*) for 1-2 weeks or longer. Infection can be present with or without signs of intestinal illness which include diarrhea. While both infections are generally self limiting, infection can be more severe and life threatening and result in higher rates of fecal shedding in immune compromised individuals.

In a survey of surface water for protozoal contaminants, *Cryptosporidium sp.* and *Giardia* were identified in 87% and 81% of 66 surface water sites tested [21]. Because protozoa do not multiply outside of the host, wildlife, companion animals, humans and livestock serve as the potential sources of contamination to each other and to watersheds. There are additional species of *Cryptosporidium* carried by birds, reptiles and fish, and cattle and mice (*C. muris*). *C. muris* can infect older cattle [22]; however, only *C. parvum* appears to be infective to humans. Differentiation of *Cryptosporidium* sp. requires examination by special tests or by an experienced diagnostician. Determining viability of oocysts requires use of special dyes. Both issues present challenges in assessing environmental contamination with for *C. parvum*. As mentioned above, genetic differences exist within the *C. parvum* species [9, 23]. Current studies suggest that cattle and human strains exist an may be associated with different cattle to human and human to human cycles of waterborne infection [24-26].

Genetic differences also exist within the *Giardia*. Human isolates can infect animals but the potential for *Giardia* from different animals to cause human infection has not been well defined [19].

Bacteria: Escherichia coli O157:H7

Bacteria such as *Escherichia coli* O157:H7 and to a lesser degree, *Salmonella sp.*, *Campylobacter jejuni*, and *Leptospira sp.* are also found in livestock manure and have been associated with waterborne disease. Most human infections caused by enteric bacterial pathogens are spread by foodborne or by direct contact [1, 27]. Hazard analysis critical control point (HACCP) programs have been developed to reduce contamination of meat products and vegetable crops by pathogens potentially present in livestock manure. Although less frequently transmitted by water than by food, *E. coli* O157 is emerging as an important cause of bacterial waterborne infection in untreated and recreational water.

Cattle were originally thought to be the main reservoir of infection but *E. coli* O157 has been isolated from humans, a dog, horse, sheep, goats, deer, and flies [28-33]. Infection can be life threatening especially in the young children and elderly patients. Illness caused by *E. coli* O157 is rare in hosts other than humans. *E. coli* O157 has been identified in a viable but non-culturable state in numerous surface water sources tested in Japan [34, 35]. Unlike protozoa, growth may occur outside of a host in water under certain conditions [28, 36, 37]. Agricultural water sources have been implicated as an important source of infection to cattle [28, 30, 38-40]. It's ability to survive for months in water [36, 37, 41-43] and feces [44, 45] and low infectious dose [46-48] contribute to its infectiousness and potential for waterborne transfer.

Intestinal viruses, which have been strongly implicated in a number of waterborne disease outbreaks (WBDO) are considered to be host specific and farms are not considered to be a source of infection for humans unless human septic sludge is present.

In summary, to cause waterborne disease, microbes suspended in feces and manure need to survive environmental conditions long enough and at high enough doses to infect a host. Dilution and loss of viability occur when pathogen laden feces is mixed with normal feces, transported across landscapes and mixed in water. Waterborne pathogens of greatest concern have the following characteristics:
- the organisms are either shed into the environment in fairly high numbers, or are highly infectious to humans or animals at low doses (*E. coli O157:H7, Cryptosporidium parvum, Shigella spp.*);
- the organisms have the ability to survive and remain infectious in the environment for long periods of time (*C. parvum, Giardia, E. coli O157*) or are highly resistant to water treatment (*C. parvum*); or
- The organisms can potentially multiply in the environment under appropriate conditions (*E. coli O157:H7*).

The Role of Livestock Agriculture as a Source of Pathogens to Watersheds

The role of livestock farms as the source of WBDOs has been documented in a limited number of cases. Two outbreaks of *C. parvum* were traced to contamination of drinking water by cow manure in England [49] . One outbreak of *E. coli O157* in a farm family from well water was reported from Canada [50]. The well was located in area subject to runoff by manure contaminated surface water and *E. coli* O157 was cultured from a large number of the cows in the herd. Most waterborne disease outbreaks that affect more than the individual farmer have not been directly linked with agriculture. However, it is important to identify and control the potential contaminants from agricultural sources to address perceptions of a public health risk to water quality from farms.

Livestock (especially ruminant) manure has been identified as potential source of *C. parvum, Giardia sp.* and *E. coli O157*. Most livestock prevalence surveys have been done on cattle farms. In a national survey in the USA, the presence of *Cryptosporidium* sp. was identified on 59% of farms tested and in 22.4% of calves [51]. Increased risk of calf infection correlated with increased herd size. Close to 50% of calves shed oocysts between 1-3 weeks of age. The infection rate declined to 22% in 3-5 week old calves and to < 15% in 5 week and older calves. However, the diagnostic method used did not differentiate *C. parvum* from *C. muris*. C. muris is found in older calves and adults and is not infective to humans [22, 52, 53].

C. parvum is typically shed by calves between 4 and 30 days of age Most calves stop shedding within 2 weeks of being infected [54]. Oocyst shedding by adult dairy cattle has not been reported in North American studies [53, 55]. Oocyst shedding in beef calves [56, 57] and lambs [58, 59] tends to be seasonal and peaks with presence of youngstock in the herd. Adult sheep shed oocysts around the time of lambing and are believed to be a source of infection to lambs [58, 59]. Preweaned foals have been found to shed oocysts for up to 19 weeks of age [60]. Infection and shedding in foals appears to be less sporadic across farms [61]. Infection in swine has been reported in nursing and weanling groups [62].

While no large prevalence studies have been reported, *Giardia* infections appear to be widespread in livestock [52-54, 63-65]. Most shedding occurs in calves less than 6 months of age but all ages of cattle can shed *Giardia sp.*[52-54]. Compared to *C. parvum* shedding, calves begin to shed Giardia at older ages (average 31 days) and can shed for up to several months [52-54, 66]. Giardia shedding by pigs, sheep and horses appears to occur more frequently in youngstock but adults can present a risk of infection [60, 62, 67].

E. coli O157 farm surveys have been reported in dairy [28, 30, 38, 40, 68-73] and beef cattle [74-76]. Surveys done at one point in time did not accurately assess the prevalence of infection. *E. coli* O157 was found on the majority of farms tested if testing was repeated over time. Fecal shedding was strongly seasonal and peaked during summer months on most premises. Typically 1- 5 % of cattle shed *E. coli* O157 in their feces. The maximum shedding was detected in heifers after weaning. Less than 1% of adults typically shed organisms. Fecal shedding lasted < 1 month in most cattle and usually was not detected beyond 3 months [73, 77]. Numbers shed by heifers were relatively low (10^2 - 10^4/ gram of feces) [40]. Detection and accurate diagnosis of *E. coli* O157 can be diagnostically challenging [78, 79].

In summary, animal manure from young stock represents the highest potential risk to the environment for these selected pathogens. However, *C. parvum* (in sheep only), *Giardia sp.*, and *E. coli* O157 can be shed by adults to a lesser degree. Fecal contamination can be seasonal. This is especially true for *E. coli* O157, and *C. parvum* in management systems where young stock are intermittently or seasonally present. These include horses, beef and sheep farms and dairy herds with seasonal calving.

The risk of animal manure as a source waterborne infection depends not only on pathogen presence but also the ability of these organisms to survive in the environment. Their ability to survive in water has been investigated as mentioned above. Survival in the environment outside of water has been reported in a limited number of studies. *C. parvum* infectivity has been assessed after exposure to environmental parameters [80-83]. Oocysts lost infectivity after 30 minutes of drying under laboratory conditions [81] and after 1-4 days in summer and winter conditions in a barn [83]. Infectivity of moist oocysts was lost after exposure to 37° C for 5 days or longer or 15 - 20° C for 2 weeks or longer [80]. Studies using sentinel chambers containing *C. parvum* oocysts and exposed to different farm environmental conditions indicate that freeze-thaw cycles, higher temperatures, and low moisture contribute to loss of viability of oocysts [82]. *Giardia* survivability has not been as thoroughly studied but is expected to parallel *C. parvum* survivability. *E. coli* O157 has been reported to live for months in manure [44, 45] but spreading manure on pasture or forage crops was not considered a risk factor for infection on farms [28, 84].

While the potential exists, the number of pathogens entering watersheds from farms has not been assessed. Use of indicator organisms such as fecal coliforms or fecal streptococci to assess the risk of fecal contamination in water has been used but has limitations. A lack of correlation between the presence of fecal indicator organisms and the presence of protozoan parasites, and occasionally *E. coli* O157 in contaminated water has been reported [6, 11]. Multiple species shed the indicator organisms in feces and multiple sources may contribute to watershed contamination. The source of fecal contamination cannot be accurately determined without the use of advanced and expensive serological or genetic typing systems. Without source tracking the relative contributions from different activities within watersheds including farms and the effectiveness of specific control measures cannot be accurately assessed.

At the watershed level, a multiple barrier approach has recommended to enhance water quality by decreasing particulates, excess nutrients, chemical, and microbial contaminants. Standard barriers or control points include drinking water treatment and source water protection from community and agricultural runoff. Best Management Practices (BMP) designed to decrease contaminants by enhancing stream-edge and field barriers have been added in agricultural watershed programs [85].

Quality Assurance programs have been promoted on livestock farms to address food safety issues and promote livestock health and production. The New York State Cattle Health Assurance Program (NYSCHAP) is one example of a preventive herd health program.

It is an integrated disease prevention program which utilizes a farm's team of animal health advisors to develop a farm-specific herd health plan to address infections of environmental, animal health, public health, and food safety concern. Risks are evaluated and farm management or disease control interventions are designed by animal health advisors who have training in disease pathogenesis and who are knowledgeable of a farm's health history and disease risks. By integrating the concepts promoted by farm quality assurance programs and the multiple barrier approach used in watershed protection, additional control points can be added to protect water quality.

The first barrier or control point addresses biosecurity issues and the potential for pathogen import to the farm. The second barrier addresses break the cycle of pathogen amplification in the animal operation. The third barrier is the appropriate collection and treatment of animal waste and prevention of contamination of the food and water sources, and the final barrier is control of pathogen export from the farm. BMP's to support the third and fourth barriers are developed in conjunction with additional farm environmental advisors providing expertise in soil and water conservation, crops and nutrient management. The first and second control points will be addressed below.

On- Farm Control Points: Sources and Amplification of Pathogens

Sources to Farms
The wide host range and environmental persistence of *C. parvum, Giardia sp.* and *E. coli O157* put farm animals at risk for infection as well as being potential sources of infection. Because of the difficulties in source tracking pathogens carried by multiple hosts, the risks from different sources cannot be accurately addressed so all potential sources must be taken into account and an evaluation made of their potential presence and importance on a farm. Pathogens can be introduced to a farm by infected animals (livestock or other) or transferred by feed, water, equipment or clothing contaminated by viable cysts or oocysts.

Feces from wild mammals including mice, racoons, white-tailed deer, farm rodents [86, 87], domestic livestock, companion animals, and humans are all possible sources of *C. parvum* to livestock farms [20, 88]. Flies and geese may act as transport hosts [89, 90]. *Giardia sp.* can be spread by an even broader number of hosts. The host range of *E. coli* O157 was listed above. Cattle manure, feces from other wild and domestic ruminants, and potentially fecal contaminated water and feed are potential sources of *E. coli* O157 infection to farms [28].

Amplification on Farms
Amplification of infection occurs as more animals become infected and environmental buildup occurs. Protozoa do not multiply in the environment but can be environmentally stable in water and to a lesser degree, manure; therefore, the animal's environment can become an additional source of infections if not managed properly.

Several studies have attempted to define risks associated with infection of *C. parvum* in calves on dairy farms [71, 91].   Risk of infection on farms increases with increased number of cattle [71, 91], and number of species as well as total number and management of other agricultural animals present [91]. Because *C. parvum* infects most calves at less than 30 days of age, preweaning management factors were more important than post weaning management for decreasing risk of infection [91]. Preweaning farm management factors associated with decreased risk of *C. parvum* infection include [91]:

- use of ventilation in calf rearing areas (drying)
- daily disposal and cleaning of bedding and daily addition of bedding (to prevent oocyst buildup in the calve environment)
- feeding of milk replacer.

Atwill et al. investigated sources of infection to dairy calves in a large persistently infected California dairy [55]. Most calves were infected between 17-21 days of age. Adult periparturient dairy cattle was not detected to be shedding oocysts and the maternity area environment did not appear to be a source of infection. Viable *C. parvum* oocysts were present on porous surfaces of wooden calf hutches but not in maternity pens in a survey of environmental sources. The hutches had been steam cleaned between calves and left to dry for 1- 4 weeks in ambient temperatures ranging from 29-39°C. Spray from cleaning adjacent dirty hutches or flush water which moved manure from under the older calves toward the younger calves could have been a potential source of contamination to the cleaned hutches and to newborn calves entering the facility. The life cycle is direct so oocycts are infective at time of excretion. Contact with other and especially older calves can increase risk of infection. A calf's contact with its own manure could also serve to amplify infection.

Flush water was also found to be a risk for *E. coli* O157 infection on farms [70]. Risks for *Giardia* infection are being studied but have not yet been published [92]. Any potential for fecal-oral contact with manure or an infected environment including feed and water present a risk for this infection also.

Management Suggestions

General management recommendations have been made to prevent and control *C. parvum* infection cycles on farms[88]. These practices focus on decreasing the potential for oral contact with feces or fecal contaminated fomites and generally optimizing health to increase resistance to infection. These general best management practices have the added benefit of decreasing the risk of infection from many different enteric agents - not just *C. parvum*.

Farm biosecurity practices reduce risks of pathogen introduction to farms from outside sources including other farms and livestock, people, wildlife, pests including rodents, birds and insects, and contaminated feed and surface water. Suggested biosecurity practices include:

- Purchase feed from sources which control quality and minimize fecal contamination from rodents other livestock.
- Use a water source for livestock that has a low risk of transfer of pathogens.
- Avoid transport of manure onto the farm by requesting any visitors to use a boot-cleaning bath or wear plastic boots. Avoid contact between calves and clothing. Clothes with obvious fecal contamination should be changed between farms.
- If manure is brought onto the farm, do not allow it to enter the barns or calf housing facilities and protect feed and water sources from fecal contamination by livestock, pets, rodents, or human sewage. Minimize movement of feces between livestock groups of different ages by manure and feed handling equipment, foot traffic and clothing, and surface runoff.
- Purchase replacement livestock from farms with a food health history. Avoid purchase from co-mingled sources such as sales barns.

On farm control management practices prevent amplification of existing or introduced infection. The calf-housing environment, facilities, and feeding and watering equipment, if not cleaned thoroughly to remove manure after use by older calves, can become a source of infection to incoming baby calves. Breaking the cycle of environmental contamination and calf infection depends on minimizing the opportunity for ingestion of feces by basic good hygiene practices as described above.

- For birthing, provide a clean and dry maternity area free of contact with other older calves and their manure.
- An "all-in, all-out" calf rearing strategy may be helpful in breaking a cycle if stringent disinfection is used between groups.
- *C. parvum* oocysts are not killed by common disinfectants and can remain infective in the environment for months under the conditions of moderate temperatures and moisture. Heat and drying limit viability.
- Raise calves in a clean, dry and well ventilated environment. Provide clean bedding and replace bedding frequently to prevent buildup of infectious agents.
- Prevent each calf's contact with its own manure, manure of other calves, and adult manure.
- Segregate calves by age. Prevent direct contact between calves for at least the first 2-3 weeks of life. Feed and handle calves starting with the youngest and working to the oldest calf groups; always segregate and handle sick calves last.
- Provide feed and water free of fecal contamination by livestock, pets, rodents, wildlife, and humans.
- Institute a rodent and fly control program
- Reduce the potential for runoff water to carry contaminated manure into and from areas used to house young animals

General good health and nutrition decrease an animal's susceptibility to infection.

- Provide sufficient colostrum within the first 6 hours of birth and preferably within 2 hours of birth.
- Provide a vaccination program designed by the farm's veterinarian and other animal health advisors. A new vaccine for *C. parvum* is being evaluated and shows promise in decreasing the number of organisms shed in feces but does not prevent infection [93].
- Provide clean, dry, comfortable, well ventilated housing with appropriate animal densities in each area.

The environmental management of E coli O157 on farms has been reviewed [28]. This agent is widespread and has been found in free ranging deer, birds, flies, sheep, dogs, humans, a horse and in cattle herds. In cattle it acts like transient normal flora with short blooms of shedding primarily during the summer months and with intestinal clearance in 1-2 months. A long term reservoir species has not been identified. Environmental proliferation is thought to play a role in livestock infection so the only potential for farm level control is feed and water management. Specific control interventions have not been defined. On farm chlorination of water is being studied as a potential control effort but those studies are not yet completed. Manure spreading on pastures and forages has been assessed as a risk factor but does not appear to be an important part of the on- farm infection cycle.

The third barrier is restriction of movement of contaminated feces into watercourses. To minimize surface and ground water contamination, prevent runoff from animal housing and exercise lots, store manure from animals six months and younger separately from the rest of the herd and apply it to non-hydrologically sensitive areas. If calf manure cannot be safely stored in an area that restricts leaching or runoff from the pile, combine the calf manure with the manure from the rest of the herd, then apply this manure according to best management practices. Combining cow and calf manure under these circumstances may provide sufficient dilution and land treatment of protozoan pathogens while avoiding risks of creating a point source of contamination from the storage of calf manure. Given the heat sensitivity of *C. parvum*, composting processes that heat the entire manure pile above 140°F should kill *Cryptosporidium parvum* and *Giardia* in the manure. Incorporate manure into soil that it is exposed to freeze-thaw cycles. This practice can significantly reduce the viability of *Cryptosporidium parvum* oocysts and *Giardia* cysts in manure.

Summary

While farms may be a potential source of microbial contamination of watersheds, the number of waterborne disease outbreaks directly linked to farms has been limited. Given the number of hosts potentially shedding protozoan and enteric bacterial pathogens into the environment, farms can be recipients and amplifiers of pathogens present in the environment as well as a potential source of infection. Best Management Practices have been reported which can help to decrease levels of infections on farms for *C. parvum* and other enteric pathogens. These used in conjunction with BMP's to enhance field and stream edge protection should decrease the risk on watershed contamination from livestock sources.

References:

1. Tauxe, R.V., *Emerging foodborne diseases: an evolving public health challenge.* Emerg Infect Dis, 1997. **3**(4): p. 425-34.
2. Juranek, D.D., *Cryptosporidiosis: sources of infection and guidelines for prevention.* Clin Infect Dis, 1995. **21 Suppl 1**: p. S57-61.
3. Strauch, D. and G. Ballarini, *Hygienic aspects of the production and agricultural use of animal wastes.* Zentralbl Veterinarmed [B], 1994. **41**(3): p. 176-228.
4. Cole, D.J., *et al., Health, safety, and environmental concerns of farm animal waste.* Occup Med, 1999. **14**(2): p. 423-48.
5. Herwaldt, B.L., *et al., Waterborne-disease outbreaks, 1989-1990.* Mor Mortal Wkly Rep CDC Surveill Summ, 1991. **40**(3): p. 1-21.
6. Kramer, M.H., *et al., Surveillance for waterborne-disease outbreaks--United States, 1993-1994.* Mor Mortal Wkly Rep CDC Surveill Summ, 1996. **45**(1): p. 1-33.
7. Levy, D.A., *et al., Surveillance for waterborne-disease outbreaks--United States, 1995-1996.* Mor Mortal Wkly Rep CDC Surveill Summ, 1998. **47**(5): p. 1-34.
8. Okhuysen, P.C., *et al., Virulence of three distinct Cryptosporidium parvum isolates for healthy adults.* J Infect Dis, 1999. **180**(4): p. 1275-81.
9. Peng, M.M., *et al., Genetic polymorphism among Cryptosporidium parvum isolates: evidence of two distinct human transmission cycles.* Emerg Infect Dis, 1997. **3**(4): p. 567-73.
10. Okhuysen, P.C., *et al., Susceptibility and serologic response of healthy adults to reinfection with Cryptosporidium parvum.* Infect Immun, 1998. **66**(2): p. 441-3.
11. Moore, A.C., *et al., Surveillance for waterborne disease outbreaks--United States, 1991-1992.* Mor Mortal Wkly Rep CDC Surveill Summ, 1993. **42**(5): p. 1-22.
12. Frost, F.J., R.L. Calderon, and G.F. Craun, *Waterborne disease surveillance: findings of a survey of state and territorial epidemiology programs.* Journal of Environmental Health, 1995. **58**(5): p. 6-11.
13. Anon, *Outbreak of Escherichia coli O157:H7 and Campylobacter among attendees of the Washington County Fair-New York, 1999.* MMWR Morb Mortal Wkly Rep, 1999. **48**(36): p. 803-5.
14. Steiner, T.S., N.M. Thielman, and R.L. Guerrant, *Protozoal agents: what are the dangers for the public water supply?* Annu Rev Med, 1997. **48**: p. 329-40.
15. Guerrant, R.L., *Cryptosporidiosis: an emerging, highly infectious threat.* Emerg Infect Dis, 1997. **3**(1): p. 51-7.
16. Rose, J.B. and T.R. Slifko, *Giardia, Cryptosporidium, and Cyclospora and their impact on foods: a review.* J Food Prot, 1999. **62**(9): p. 1059-70.
17. Robertson, L.J., *et al., Removal and destruction of intestinal parasitic protozoans by sewage treatment processes.* International Journal of Environmental Health Research, 1999. **9**(2): p. 85-96.
18. deRegnier, D.P., *et al., Viability of Giardia cysts suspended in lake, river, and tap water.* Appl Environ Microbiol, 1989. **55**(5): p. 1223-9.
19. Thompson, R. and P. Boreham, *Discussion Report: biotic and abiotic transmission,* in *Giardia: From Molecules to Disease,* J.R. RCA Thompson, AJ Lymbery, Editor. 1994, CAB International: Wallingford, UK. p. 131-136.
20. Fayer, R., C. Speer, and J. Dubey, *General Biology of Cryptosporidium,* in *Cryptosporidium and Cryptosporidiosis,* R. Fayer, Editor. 1997, CRC Press: New York. p. 1-41.
21. LeChevallier, M.W., W.D. Norton, and R.G. Lee, *Occurrence of Giardia and Cryptosporidium spp. in surface water supplies.* Appl Environ Microbiol, 1991. **57**(9): p. 2610-6.
22. Anderson, C., *Cryptosporidium muris in cattle.* Veterinary Record, 1991. **129**(1): p. 20.
23. Morgan, U.M., *et al., Variation in Cryptosporidium: towards a taxonomic revision of the genus.* Int J

Parasitol, 1999. **29**(11): p. 1733-51.

24. Ong, C.S.L., *et al.*, *Molecular epidemiology of cryptosporidiosis outbreaks and transmission in British Columbia, Canada.* American Journal of Tropical Medicine and Hygiene, 1999. **61**(1): p. 63-69.

25. Patel, S., *et al.*, *Molecular characterisation of Cryptosporidium parvum from two large suspected waterborne outbreaks. Outbreak Control Team South and West Devon 1995, Incident Management Team and Further Epidemiological and Microbiological Studies Subgroup North Thames 1997.* Commun Dis Public Health, 1998. **1**(4): p. 231-3.

26. Peng, M.P., *et al.*, *Genetic polymorphism among Cryptosporidium parvum isolates: evidence of two distinct human transmission cycles.* Emerging Infectious Diseases, 1997. **3**(4): p. 567-573.

27. Mead, P.S., *et al.*, *Food-related illness and death in the United States.* Emerg Infect Dis, 1999. **5**(5): p. 607-25.

28. Besser, T., *et al. The ecology fo shigatoxic E. coli O157 and propects for on farm control. in 32nd Annual Conference of the AABP.* 1999. Nashville, TN: AABP.

29. Bielaszewska, M., *et al.*, *Human Escherichia coli O157:H7 infection associated with the consumption of unpasteurized goat's milk.* Epidemiol Infect, 1997. **119**(3): p. 299-305.

30. Hancock, D.D., *et al.*, *Multiple sources of Escherichia coli O157 in feedlots and dairy farms in the Northwestern USA.* Preventive Veterinary Medicine, 1998. **35**(1): p. 11-19.

31. Iwasa, M., *et al.*, *Detection of Escherichia coli O157:H7 from Musca domestica (Diptera: Muscidae) at a cattle farm in Japan.* J Med Entomol, 1999. **36**(1): p. 108-12.

32. Chapman, P.A. and et al., *Sheep as a potential source of verocytotoxin-producing Escherichia coli O157.* Vet Rec., 1996. **138**(1): p. 23-4.

33. Chapman, P.A. and et al., *Farmed deer as a potential source of verocytotoxin-producing Escherichia coli O157.* Vet Rec., 1997. **141**(12): p. 314-5.

34. Kogure, K. and E. Ikemoto, *[Wide occurrence of enterohemorragic Eschrichia coli O157 in natural freshwater environment].* Nippon Saikingaku Zasshi, 1997. **52**(3): p. 601-7.

35. Kurokawa, K., *et al.*, *Abundance and distribution of bacteria carrying sltII gene in natural river water.* Lett Appl Microbiol, 1999. **28**(5): p. 405-10.

36. Rajkowski, K.T. and E.W. Rice, *Recovery and survival of Escherichia coli O157:H7 in reconditioned pork- processing wastewater.* J Food Prot, 1999. **62**(7): p. 731-4.

37. Warburton, D.W., *et al.*, *Survival and recovery of Escherichia coli O157:H7 in inoculated bottled water.* J Food Prot, 1998. **61**(8): p. 948-52.

38. Faith, N.G., *et al.*, *Prevalence and clonal nature of Escherichia coli O157:H7 on dairy farms in Wisconsin.* Appl Environ Microbiol, 1996. **62**(5): p. 1519-25.

39. Porter, J., *et al.*, *Detection, distribution and probable fate of Escherichia coli O157 from asymptomatic cattle on a dairy farm.* J Appl Microbiol, 1997. **83**(3): p. 297-306.

40. Shere, J.A., K.J. Bartlett, and C.W. Kaspar, *Longitudinal study of Escherichia coli O157:H7 dissemination on four dairy farms in Wisconsin.* Appl Environ Microbiol, 1998. **64**(4): p. 1390-9.

41. Ackman, D., *et al.*, *Swimming-associated haemorrhagic colitis due to Escherichia coli O157:H7 infection: evidence of prolonged contamination of a fresh water lake.* Epidemiol Infect, 1997. **119**(1): p. 1-8.

42. Wang, G. and M.P. Doyle, *Survival of enterohemorrhagic Escherichia coli O157:H7 in water.* J Food Prot, 1998. **61**(6): p. 662-7.

43. Rice, E.W., *et al.*, *Survival of Escherichia coli O157:H7 in drinking water associated with a waterborne disease outbreak of hemorrhagic colitis.* Letters in Applied Microbiology, 1992. **15**(2): p. 38-40.

44. Kudva, I.T., K. Blanch, and C.J. Hovde, *Analysis of Escherichia coli O157:H7 survival in ovine or bovine manure and manure slurry.* Applied and Environmental Microbiology, 1998: 3166-3174

45. Wang, G., T. Zhao, and M.P. Doyle, *Fate of enterohemorrhagic Escherichia coli O157:H7 in bovine feces.* Appl Environ Microbiol, 1996. **62**(7): p. 2567-70.

46. Keene, W.E., *et al.*, *A swimming-associated outbreak of hemorrhagic colitis caused by Escherichia coli O157:H7 and Shigella sonnei.* N Engl J Med, 1994. **331**(9): p. 579-84.

47. Easton, L., *Escherichia coli O157: occurrence, transmission and laboratory detection.* Br J Biomed Sci, 1997. **54**(1): p. 57-64.

48. Tuttle, J., *et al.*, *Lessons from a large outbreak of Escherichia coli O157:H7 infections: insights into the infectious dose and method of widespread contamination of hamburger patties.* Epidemiol Infect, 1999. **122**(2): p. 185-92.

49. Badenoch, J., *Cryptosporidium in water supplies. Report of the Group of Experts.*, . 1990, HMSO: London. p. 230.

50. Jackson, S.G., *et al.*, *Escherichia coli O157:H7 diarrhoea associated with well water and infected cattle on an Ontario farm.* Epidemiol Infect, 1998. **120**(1): p. 17-20.

51. Garber, L.P., *et al.*, *Potential risk factors for Cryptosporidium infection in dairy calves.* J Am Vet Med Assoc, 1994. **205**(1): p. 86-91.

52. Olson, M.E., *et al.*, *Giardia and Cryptosporidium in dairy calves in British Columbia.* Can Vet J, 1997. **38**(11): p. 703-6.

53. Wade, S., H. Mohammed, and S. Schaaf, *Prevalence of Giardia sp., Cryptosporidium parvum and Cryptosporidium muris in 109 dairy herds in five counties of southeastern New York.* Canadian J. Vet. Res., 1999. **submitted for publication.**

54. O'Handley, R.M., *et al.*, *Duration of naturally acquired giardiosis and cryptosporidiosis in dairy calves and their association with diarrhea.* J Am Vet Med Assoc, 1999. **214**(3): p. 391-6.

55. Atwill, E.R., *et al.*, *Evaluation of periparturient dairy cows and contact surfaces as a reservoir of Cryptosporidium parvum for calfhood infection.* Am J Vet Res, 1998. **59**(9): p. 1116-21.

56. Atwill, E.R., *et al.*, *Age, geographic, and temporal distribution of fecal shedding of Cryptosporidium parvum oocysts in cow-calf herds.* Am J Vet Res, 1999. **60**(4): p. 420-5.

57. Atwill, E.R., E.M. Johnson, and M.G. Pereira, *Association of herd composition, stocking rate, and duration of calving season with fecal shedding of Cryptosporidium parvum oocysts in beef herds.* J Am Vet Med Assoc, 1999. **215**(12): p. 1833-8.

58. Ortega-Mora, L.M., *et al.*, *Role of adult sheep in transmission of infection by Cryptosporidium parvum to lambs: confirmation of periparturient rise.* Int J Parasitol, 1999. **29**(8): p. 1261-8

59. Xiao, L., R.P. Herd, and D.M. Rings, *Diagnosis of Cryptosporidium on a sheep farm with neonatal diarrhea by immunofluorescence assays.* Vet Parasitol, 1993. **47**(1-2): p. 17-23.

60. Xiao, L. and R.P. Herd, *Epidemiology of equine Cryptosporidium and Giardia infections.* Equine Veterinary Journal, 1994. **26**(1): p. 14-17.

61. Cole, D.J., *et al.*, *Prevalence of and risk factors for fecal shedding of Cryptosporidium parvum oocysts in horses.* J Am Vet Med Assoc, 1998. **213**(9): p. 1296-302.

62. Xiao, L.H., R.P. Herd, and G.L. Bowman, *Prevalence of Cryptosporidium and Giardia infections on two Ohio pig farms with different management systems.* Veterinary Parasitology, 1994. **52**(3/4): p. 331-336.

63. Olson, M.E., *et al.*, *Giardia and Cryptosporidium in Canadian farm animals.* Vet Parasitol, 1997. **68**(4): p. 375-81.

64. Ruest, N., G.M. Faubert, and Y. Couture, *Prevalence and geographical distribution of Giardia spp. and Cryptosporidium spp. in dairy farms in Quebec.* Can Vet J, 1998. **39**(11): p. 697-700.

65. Xiao, L.H., R.P. Herd, and D.M. Rings, *Concurrent infections of Giardia and Cryptosporidium on two Ohio farms with calf diarrhea.* Veterinary Parasitology, 1993. **51**(1/2): p. 41-48.

66. Xiao, L. and R.P. Herd, *Infection pattern of Cryptosporidium and Giardia in calves.* Vet Parasitol, 1994. **55**(3): p. 257-62.

67. Xiao, L., R.P. Herd, and K.E. McClure, *Periparturient rise in the excretion of Giardia sp. cysts and Cryptosporidium parvum oocysts as a source of infection for lambs.* Journal of Parasitology, 1994. **80**(1): p. 55-59.

68. Zhao, T., *et al.*, *Prevalence of enterohemorrhagic Escherichia coli O157:H7 in a survey of dairy herds.* Applied and Environmental Microbiology, 1995. **61**(4): p. 1290-1293.

69. Mechie, S.C., P.A. Chapman, and C.A. Siddons, *A fifteen month study of Escherichia coli O157:H7 in a dairy herd.* Epidemiol Infect, 1997. **118**(1): p. 17-25.

70. Garber, L., *et al.*, *Factors associated with fecal shedding of verotoxin-producing Escherichia coli O157 on dairy farms.* J Food Prot, 1999. **62**(4): p. 307-12.

71. Garber, L.P., *et al.*, *Risk factors for fecal shedding of Escherichia coli O157:H7 in dairy calves.* Journal of the American Veterinary Medical Association, 1995. **207**(1): p. 46-49.

72. Hancock, D.D., *et al.*, *A longitudinal study of Escherichia coli O157 in fourteen cattle herds.* Epidemiology and Infection, 1997. **118**(2): p. 193-195.

73. Besser, T.E. and D.D. Hancock. *Surveillance for Escherichia coli O157:H7: seasonal variation and duration of infection in cattle.* in *Proceedings of the 8th International Congress on Animal Hygiene, St. Paul, Minnesota, USA, 12-16 September 1994. 1994 . FS13-FS16.* 1994.

74. Laegreid, W.W., R.O. Elder, and J.E. Keen, *Prevalence of Escherichia coli O157:H7 in range beef calves at weaning.* Epidemiol Infect, 1999. **123**(2): p. 291-8.

75. Hancock, D.D., *et al.*, *Epidemiology of Escherichia coli O157 in feedlot cattle.* Journal of Food Protection, 1997. **60**(5): p. 462-465.

76. Dargatz, D.A., *et al. Escherichia coli O157 in feces of feedlot cattle.* in *Proceedings of the Annual Meeting of the United States Animal Health Association. 1996. 100:16-17.* 1996.

77. Besser, T.E., *et al.*, *Duration of detection of fecal excretion of Escherichia coli O157:H7 in cattle.* J Infect Dis, 1997. **175**(3): p. 726-9.

78. Sanderson, M.W., *et al.*, *Sensitivity of bacteriologic culture for detection of Escherichia coli O157:H7 in bovine feces.* Journal of Clinical Microbiology, 1995. **33**(10): p. 2616-2619.

79. Chapman, P.A. and et al., *An outbreak of infection due to verocytotoxin-producing Escherichia coli O157 in four families: the influence of laboratory methods on the outcome of the investigation.* Epidemiol Infect., 1997. **119**(2): p. 113-9.

80. Sherwood, D., *et al.*, *Experimental cyrptospoidiosis in laboratory mice.* Infection and Immunity, 1982. **38**: p. 471-475.

81. Robertson, L.J., A.T. Campbell, and H.V. Smith, *Survival of Cryptosporidium parvum oocysts under various environmental pressures.* Appl Environ Microbiol, 1992. **58**(11): p. 3494-500.

82. Jenkins, M.B., *et al.*, *Use of a sentinel system for field measurements of Cryptosporidium parvum oocyst inactivation in soil and animal waste.* Appl Environ Microbiol, 1999. **65**(5): p. 1998-2005.

83. Anderson, B.C., *The effect of drying on infectivity of cryptosporidium laden calf feces for 3 to 7 day old mice.* American Journal of Veterinary Research, 1986. **47**: p. 2272-2273.

84. Hancock, D.D., *et al.*, *Effects of farm manure-handling practices on Escherichia coli O157 prevalence in cattle.* Journal of Food Protection, 1997. **60**(4): p. 363-366.

85. Meals, D.W. *Watershed-scale response to agricultural diffuse pollution control programs in Vermont, USA.* in *Water Science and Technology. 1996 . 33 (4/5). 197-204.* 1996.

86. Quy, R.J., *et al.*, *The Norway rat as a reservoir host of Cryptosporidium parvum.* J Wildl Dis, 1999. **35**(4): p. 660-70.

87. Klesius, P., T. Hayes, and L. Malo, *Infectivity of Cryptosporidium sp. isolated from wild mice for calves and mice.* J. Am. Vet. Med. Assoc., 1986. **189**: p. 192-193.

88. Angus, K., *Cryptosporidiosis in Ruminants*, in *Cryptosporidiosis of Man and Animals*, J. Dubey, C. Speer, and R. Fayer, Editors. 1990, CRC Press: Boca Raton. p. 83-103.

89. Graczyk, T.K., *et al.*, *Giardia sp. cysts and infectious Cryptosporidium parvum oocysts in the feces of migratory Canada geese (Branta canadensis).* Appl Environ Microbiol, 1998. **64**(7): p. 2736-8.

90. Graczyk, T.K., *et al.*, *House flies (Musca domestica) as transport hosts of Cryptosporidium parvum.* Am J Trop Med Hyg, 1999. **61**(3): p. 500-4.

91. Mohammed, H.O., S.E. Wade, and S. Schaaf, *Risk factors associated with Cryptosporidium parvum infection in dairy cattle in southeastern New York State.* Vet Parasitol, 1999. **83**(1): p. 1-13.

92. Wade, S., H. Mohammed, and S. Schaaf, *Epidemiologic study of Giardia sp. infection in dairy cattle in southeastern New York State.* **accepted** Vet Parasitology, 1999.

93. Perryman, L.E., *et al.*, *Protection of calves against cryptosporidiosis with immune bovine colostrum induced by a Cryptosporidium parvum recombinant protein..* Vaccine, 1999. **17**(17): p. 2142-9 .

Sources of Pathogens in a Watershed: Humans, Wildlife, Farm Animals?

Daniel R. Shelton, Ph.D.
Research Microbiologist
USDA/Agricultural Research Service
Bldg. 001, BARC-West, 10300 Baltimore Ave.
Beltsville, MD 20705-2350

Biographies for most speakers are in alphabetical order after the last paper.

Introduction

There is increasing concern regarding the impact of water-borne pathogens on human health. In particular, farm animal feces/manures have come under increasing scrutiny as a potential source of pathogens. For the majority of pathogens, there are multiple hosts/sources, including humans, companion animals (*e.g.*, dogs, cats), wildlife, and farm animals, such that few generalization can be made across watersheds regarding the predominant source(s).

<u>Indicator organisms</u>

Fecal coliforms (or generic *Escherichia coli*) are used extensively as indicators of potential pathogen contamination, primarily because specific pathogen testing protocols are too time-consuming, expensive, and/or insensitive to be utilized for monitoring purposes. For example, bacterial pathogens are typically cultured on semi-selective agar media followed by biochemical and serological identification, while protozoan parasites are typically concentrated, stained and enumerated microscopically. Rapid, relatively inexpensive genetic and immunological methods are currently being developed, but are not yet commercially available. A major limitation of fecal coliform testing is that there are no established relationships between fecal coliform and pathogen contamination. In the context of waste water treatment, specific relationships are unnecessary because it is reasonable to assume that, at any given time within a given human population, some percentage of individuals will be shedding pathogens. In the context of watersheds, however, the absence of such relationships seriously limits the reliability of standard microbial testing.

Fecal coliforms (including generic *E. coli*) are excreted by all warm-blooded animals, including birds and mammals. The numbers typically excreted by healthy animals vary from 10^5 organisms/gram feces for cows to 10^7 organisms/gram feces for waterfowl; humans and companion animals typically excrete 10^6 - 10^7 organisms/gram feces (Schueler, 1999). The potential for fecal contamination from any particular source is a function of multiple parameters including: adequacy of septic/sewage treatment, animal population densities and total fecal/manure production, proximity to surface waters, surface hydrology, survival, etc. Consequently, watersheds must be evaluated on a case-by-case basis.

Recent studies suggest that "genetic fingerprinting" techniques may provide a mechanism for identifying the predominant source(s) of fecal contamination in a watershed. One such technique, referred to as ribotyping (see Glossary), has been proposed by several investigators as a means of discriminating between different sources of generic *E. coli*. Ribotyping has recently been used to distinguish human from nonhuman *E. coli* contamination in the Apalachicola Bay, FL (Parveen et al., 1999). Similarly, researchers in Washington State (Checkowitz, 1998) used ribotyping to differentiate generic *E. coli* contamination in a creek from livestock (hobby farms) vs. companion animals vs. humans (septic systems). Note, however, that in this study 1,639 individual isolates were analyzed. The reliability/sensitivity of genetic fingerprinting techniques is dependant on comprehensive sampling protocols which provide adequate representation of all potential sources of fecal contamination. In addition, it is unclear whether there are unique ribotypes associated with specific animals such that data can be extrapolated to other watersheds or geographical areas, or whether data are site specific. This question must await further analysis and development of a ribotype data base.

Escherichia coli O157

The vast majority of *E. coli* strains are harmless. However, a few strains cause diarrhea in humans; these are classified as enterotoxigenic (ETEC), enteropathogenic (EPEC), or enterohemorrhagic (EHEC) *E. coli* (see Glossary). Although there is significant potential for water contamination by ETEC and EPEC strains from farm animals---these are frequently the cause of scours in calves---they are not considered a major health threat for humans because they are self-limiting. EHEC, typically referred to as *E. coli* O157:H7 (see Glossary), is a serious health threat, particularly in children. It causes bloody diarrhea and, if not treated promptly, can result in kidney failure and death.

Although human-to-human contact and contaminated food are the predominant modes of *E. coli* O157 transmission, there does appear to be limited potential for water-borne transmission. It is unclear, however, to what extent dairy/beef herds, or wildlife, are a source of water contamination. Note that in those instances where water-borne outbreaks of hemorrhagic colitis have been documented, they have generally been attributed to direct human contamination of a recreational body of water (Ackman et al., 1997).

E. coli O157:H7 was unknown prior to 1982, when it was associated with a multistate outbreak of hemorrhagic colitis. In 1986, it was recovered from healthy dairy cows, suggesting that dairy/beef herds could serve as a reservoir (Martin et al., 1986). Numerous researchers have since documented the presence of *E. coli* O157 in cows/beef cattle or associated with the farm environment: water, birds, flies, rodents and companion animals (Hancock et al., 1998). Studies indicate, however, that shedding by cows/beef cattle is transient and that concentrations do not exceed approx. 10^5 organisms/gram feces (Shere et al., 1998; Zhao et al., 1995). Limited data suggests that deer also harbor *E. coli* O157 (Rice et al., 1995; Keene et al., 1997), although the author is unaware of any data on prevalence or concentrations. Once in the environment, *E. coli* O157 cells can survive in bovine feces/manure or water for several weeks (particularly at lower temperatures). However, unlike in food, there is no evidence for proliferation; numbers decrease continuously because there are no substrates to support growth. Collectively, these data suggest that, after accounting for dilution and mortality, *E. coli* O157 concentrations in surface waters from agricultural sources are likely to be relatively low; insufficient data exist to draw any conclusions regarding wildlife.

At present, the risk from water-borne transmission, regardless of source, cannot be estimated because no quantitative data are available establishing the minimum number of organisms required to cause infection in humans. Based on studies with infant rabbits, a minimum dose of approx. 10^5 *E. coli* O157 cells are required to cause diarrhea (Center for Disease Control, website). Based on epidemiological data, other researchers have speculated that lower doses may be sufficient to cause disease in children. Consequently, there are currently too many knowledge gaps to assess the risks associated with watershed contamination by *E. coli* O157.

Similar to generic *E coli.*, "genetic fingerprinting" techniques have been developed to track specific *E. coli* O157 strains. The most frequently utilized method is referred to as restriction fragment length polymorphism (RFLP) or pulsed field gel electrophoresis (PFGE; see Glossary). This method produces a series of unique DNA fragment patterns, referred to as restriction endonuclease digestion profiles (REDP). This method has been utilized to identify different *E. coli* O157 strains on one or more dairy farms (Faith et al., 1996; Rice et al., 1999). In at least one instance, it has also been used to document transmission from cattle to humans associated with an on-farm outbreak (Louie et al., 1999). However, the author is unaware of any published studies utilizing PFGE to document sources of water contamination.

Cryptosporidium parvum

Another important water-borne pathogen is *Cryptosporidium parvum*, the causal agent of cryptosporidiosis. *C. parvum* is a widespread protozoan parasite afflicting over 80 mammalian species, including humans. Several outbreaks of cryptosporidiosis have occurred in the past decade, the most severe in Milwaukee, WI where over 400,000 people were infected. Water-borne *C. parvum* oocysts (the infectious agent outside the host) present a serious public health threat because of the low infectious dose (approx. 130 for adults) and their resistance to standard chlorination treatment. Immuno-deficient individuals (*e.g.*, AIDS, cancer patients) are particularly at risk because there are no effective treatments for the disease.

Although infections can result from human-to-human contact, contaminated drinking or recreational waters are believed to be an important mode of transmission, with humans, wildlife, and farm animals all potential contributors. An extensive multi-state survey of dairy farms indicates that virtually all herds with >100 cows are infected with *C. parvum* (Garber et al., 1994). New-borne calves are the most susceptible and can excrete several billion oocysts if they develop scours. Older animals can continue to shed for up to four months, although at lower levels (Atwill et al., 1997; author, unpublished data). Although few studies have been conducted of wildlife, limited data suggests that feral animals are potential contributors to watershed contamination, particularly where animals congregate at the edge of streams, creeks, etc. (Atwill et al., 1997). Watershed monitoring studies appear to confirm the contribution of both farm animals and wildlife, although oocyst concentrations downstream of dairy or beef farms were several fold higher than upstream, with the highest concentrations observed during calving (Hansen and Ongerth, 1991; Ong et al., 1996). Sewage outflows from waste water treatment plants have also been documented to contain substantial numbers of oocysts, particularly in the absence of tertiary treatment (States et al., 1997).

Once in surface waters, oocysts are extremely persistent. Oocysts possess a tough outer wall which is resistant to many common disinfectants. Laboratory studies indicate that oocysts can remain viable for several months in waters, particularly at lower temperatures. Less in known about fate/survival *in-situ*. Limited data suggests that oocysts are susceptible to filtering, predation and/or microbial decomposition. Oocysts appear to be readily filtered from estuarine waters by oysters (Fayer et al., 1999) or, alternatively, ingested/digested by rotifers (Fayer et al., 2000).

Despite the relative abundance of information regarding sources, survival and infectious dose, risk assessment remains problematic for several reasons. One, the majority of published data documenting fecal sources are expressed as percent of animals infected. In the absence of quantitative data, it is difficult to estimate total numbers of oocysts excreted into the environment. Two, several species of *Cryptosporidium*, other than *C. parvum*, have been described which are infectious to birds, reptiles, and/or some mammals. Current standard microscopic detection/enumeration methods, however, cannot differentiate between most of these species. Consequently, oocysts observed in natural water samples must be considered, at best, presumptive *C. parvum*. Three, current microscopic detection/enumeration methods do not distinguish viable from nonviable oocysts. Several viability testing methods have been developed (*e.g.*, viability staining, excystation, mouse bioassay, cell culture), however, all have major limitations or drawbacks which limit their application to routine water monitoring.

Because of the inherent limitations of current detection methods, considerable effort has been devoted to developing genetic protocols to overcome these limitations. The primary genetic method utilized for *C. parvum* detection is referred to as PCR-RFLP (see Glossary). This method utilizes DNA primers to amplify different regions of the genome; there are currently at least sixteen different protocols. Recent comparative studies indicate, however, that not all protocols are equally valid or reliable (Sulamain et al., 1999).

An important result of genetic characterization has been the elucidation of two different *C. parvum* genotypes with distinct transmission cycles. Genotype I ("human") is infectious only to humans, while genotype II ("bovine") is infectious to humans, calves, and laboratory mice (Peng et al., 1997). Utilizing this protocol, analysis of oocysts extracted from oysters (harvested from the Chesapeake Bay) indicates that the predominant genotype in these waters is the "bovine" genotype. By comparison, of the five oocyst samples remaining from the Milwaukee epidemic, all were the "human" genotype. Although these genetic methods allow for differentiation between some species and "human" vs. "bovine" genotypes, the ability to distinguish between oocysts from different mammalian hosts has not yet been demonstrated.

A limitation of PCR-RFLP protocols is the inability to distinguish between different oocyst types in a mixture. An alternative approach, referred to as TaqMan® PCR (see Glossary), allows for detection of a specific DNA sequence unique to a particular strain or genotype. Protocols have been developed with theoretical detection limits of approx. 100 oocysts/mL. Preliminary information, however, suggests that this method may also be unable to distinguish among oocysts from farm animals vs. wildlife (*e.g.*, deer).

Glossary

Enterohemorrhagic *E.coli* (EHEC): This pathogen secretes shiga toxins (verotoxins). Symptoms are bloody diarrhea and hemolytic uremic syndrome.

E. coli O157:H7: The primary serological type of EHEC in the USA and Europe; O111 is the dominant serological type in Australia. O157 refers to the cell membrane lipopolysaccharide antigen while H7 refers to the flagellar antigen. Note that *E. coli* O157 is not detected by the standard water testing protocol for generic *E. coli*.

Enteropathogenic *E.coli* (EPEC): This pathogen invades the microvilli of the small intestine and causes diarrhea.

Enterotoxigenic *E.coli* (ETEC): This pathogen is most commonly responsible for "travelers' diarrhea". It colonizes (but does not invade) the proximal small intestine and secretes enterotoxins.

Polymerase chain reaction-restriction fragment length polymorphism (PCR-RFLP): The polymerase chain reaction is a method for amplifying specific fragments of DNA. In PCR-RFLP specific segments of genomic DNA are amplified, the segments fragmented with one or more restriction enzymes, and the fragments separated by gel electrophoresis. The unique patterns of DNA fragments allows for differentiation of genotypes or strains.

Restriction fragment length polymorphism (RFLP)/pulsed field gel electrophoresis (PFGE): Genomic DNA is fragmented with a specific restriction enzyme (e.g., *Xba* I is commonly used for *E. coli* O157) producing a series of DNA fragments of variable size. These fragments are separated by pulsed field gel electrophoresis to produce a pattern, called a restriction endonuclease digestion profile (REDP). Comparison of REDPs from different locations/animals allows for tracking of specific strains.

Ribotyping: Genomic DNA is fragmented with a one or more specific restriction enzymes producing a series of DNA fragments of variable size. After separation by gel electrophoresis, fragments are hybridized with DNA probes specific to ribosomal RNA genes. Only those DNA fragments which hybridize are detected. Comparison of DNA profiles from different locations/animals allows for tracking of specific strains.

TaqMan®: DNA primers and probes are selected which bind to a very specific segment of the genome. PCR methods are then used to amplify that specific DNA segment in an environmental sample (*e.g.*, water, manure). If the segment is absent, no reaction occurs. In theory, very small differences in DNA sequence can be detected. Note that since this is a direct detection method, gel electrophoresis is not required.

References

Ackman, D.S., S. Marks, P. Mack, M. Caldwell, T. Root and G. Birkhead. 1997. Swimming-associated haemorrhagic colitis due to *Escherichia coli* O157:H7 infection: evidence of prolonged contamination of a fresh water lake. *Epidemiol. Infect.* 119:1-8.

Atwill, E.R., R.A. Sweitzer, M. Perira, I.A. Gardiner, D. van Vuren and W.M. Boyce. 1997. Prevalence of and associated risk factors for shedding *Cryptosporidium* oocysts and *Giardia* cysts within feral pig populations in California. *Appl. Environ. Microbiol.* 63:3946-3949.

Atwill, E.R., E. Johnson, D.J. Klingborg, G.M. Veserat, G. Markegaard, W.A. Jensen, D.W. Pratt, R.E. Delmas, H.A. George, L.C. Forero, R.L. Phillips, S.J. Barry, N.K. McDougald, R.R. Gildersleeve and W.E. Frost. 1999. Age, geographic, and temporal distribution of fecal shedding of *Cryptosporidium parvum* oocysts in cow-calf herds. *Am. J. Vet. Res.* 60:420-425.

Checkowitz, N. 1998. Little Soos Creek microbial source tracking. *Washington Water Resource.* Spring:10-11.

Faith, N.G., J.A. Shere, R. Brosch, K.W. Arnold, S.E. Ansay, M.-S. Lee, J.B. Luchansky and C.W. Kaspar. 1996. Prevalence and clonal nature of *Escherichia coli* O157 on dairy farms on Wisconsin. *Appl. Environ. Microbiol.* 62:519-1525.

Fayer, R., J. Trout, E. Walsh and R. Cole. 2000. Rotifers ingest oocysts of *Cryptosporidium parvum*. *J Eukaryotic Microbiol.* (In press)

Fayer R., E.J. Lewis, J. Trout, T.K. Graczyk, M. Jenkins, J. Higgins, L. Xiao and A. Lal. 1999. *Cryptosporidium parvum* in oysters from commercial harvesting sites in the Chesapeake Bay. *Emerg. Infect. Dis.* 5:706-710.

Garber, L.P., M.D. Salman, H.S. Hurd, T. Keefe and J.L. Schlater. 1994. Potential risk factors for *Cryptosporidium* infection in dairy calves. *J. Am. Vet. Med. Assoc.* 205:86-91.

Hancock, D.D., T.E. Besser, D.H. Rice, E.D. Ebel, D.E. Herriott and L.V. Carpenter. 1998. Multiple sources of *Escherichia coli* O157 in feedlots and dairy farms in the Northwestern USA. *Prev Vet Med.* 35:11-19.

Hansen, J. S. and J. E. Ongerth. 1991. Effects of time and watershed characteristics on the concentration of *Cryptosporidium* oocysts in river water. *Appl. Environ. Microbiol.* 57:2790-2795.

Keene,W.E., E. Sazie, J. Kok, D.H. Rice, D.D. Hancock, V.K. Balan, T. Zhao and M.P. Doyle. 1997. An outbreak of *Escherichia coli* O157:H7 infections traced to jerky made from deer meat. *JAMA.* 277:1229-1231.

Louie, M., S. Read, K. Ziebell, K. Rahn, A. Borczyk and H. Lior. 1999. Molecular typing methods to investigate transmission of *Escherichia coli* O157:H7 from cattle to humans. *Epidemiol. Infect.* 123:17-24.

Martin, M.L., L.D. Shipman, J.G. Wells, M.E. Potter, K. Hedberg, I.K. Wachsmuth, R.V. Tauxe, J.P. Davis, J. Arnold and J. Tilleli. 1986. Isolation of *Escherichia coli* O157:H7 from dairy cattle associated with two cases of haemolytic uraemic syndrome. *Lancet.* 2:1043.

Ong, C., W. Moorehead, A. Ross and J. Isaac-Renton. 1996. Studies of *Giardia* spp. and *Cryptosporidium* spp. in two adjacent watersheds. *Appl. Environ. Microbiol.* 62:2798-2805.

Parveen, S., K.M. Porter, K. Robinson, L. Edmiston and M.L. Tamplin. 1999. Discriminant analysis of ribotype profiles of *Escherichia coli* for differentiating human and nonhuman sources of fecal pollution. *Appl. Environ. Microbiol.* 65:3142-3147.

Peng, M.M., L. Xiao, A,R. Freeman, M.J. Arrowood, A.A. Escalante, A.C. Weltman, C.S.L. Ong, W.R. MacKenzie, A.A. Lal and C.B. Beard. 1997. Genetic polymorphism among *Cryptosporidium parvum* isolates: evidence of two distinct human transmission cycles. *Emerg. Infect. Dis.* 3:567-573.

Rice, D.H., D.H. Hancock and T.E. Besser. 1995. Verotoxigenic *E. coli* O157 colonisation of wild deer and range cattle. *Vet. Rec.* 137:524.

Rice, D.H., K.M. McMenamin, L.C. Pritchett, D.D. Hancock and T.E. Besser. 1999. Genetic subtyping of *Escherichia coli* O157 isolates from 41 pacific northwest USA cattle farms. *Epidemiol. Infect.* 122:479-484.

Schueler, T. 1999. Microbes and urban watersheds:concentrations,sources, and pathways. *Watershed Protection Techniques.* 3:554-565.

Shere, J.A., K.J. Bartlett and C.W. Kaspar. 1998. Longitudinal study of *Escherichia coli* O157:H7 dissemination on four dairy farms in Wisconsin. *Appl. Environ. Microbiol.* 64: 1390-1399.

States, S., K. Stadterman, L.Ammon, P. Vogel, J. Baldizar, D. Wright, L. Conley and J. Sykora. 1997. Protozoa in river water: sources, occurrence, and treatment. *J. Am. Water Works Assoc.* 74:75-83.

Sulaiman, I.M., L. Xiao and A.A. Lal. 1999. Evaluation of *Cryptosporidium parvum* genotyping techniques. *Appl. Environ. Microbiol.* 65:4431-4435.

Zhao, T., M.P. Doyle, J. Shere and L. Garber. 1995. Prevalence of enterhemorrhagic *Escherichia coli* O157:H7 in a survey of dairy herds. *Appl. Environ. Microbiol.* 61:1290-1293.

Session 5

Manure Management Practices

Manure Management on Swine Farms: Practices and Risks

Kenneth B. Kephart
Associate Professor of Animal Science
Department of Dairy and Animal Science
The Pennsylvania State University
University Park, Pennsylvania

Biographies for most speakers are in alphabetical order after the last paper.

Overview of Swine Production Systems

The swine industry began a transition from outdoor, extensive systems to indoor, intensive production in the late '60's and early '70's. Many "farrow-to-finish" operations housed pigs from birth to market weight. Animals were grouped into four production phases: 1) breeding/gestation; 2) farrowing; 3) nursery; 4) grow-finish. Within each production phase, animals were often housed in large rooms or barns with a wide variation in age and body weight. This approach helped to keep building and ventilation costs down, but by the mid '80's producers realized that housing pigs in single age groups greatly improved health and growth performance, even though separate rooms and ventilation systems were required. This production system is referred to as All-In-All-Out (AIAO). Further refinement of the AIAO approach resulted when producers realized that by moving age groups to separate farms, transmission of disease was reduced even further. Throughout the late 90's, this new system, called multiple-site production became a standard practice for the majority of hogs produced in the country.

Over this same time period, swine farms have been increasing in size to help offset diminishing profit margins. In addition, many of the nursery and grower-finisher units are now operated under contract. The contract requires the producer to own the buildings, care for the pigs, pay most of the utilities, and dispose of the manure. But another individual or company owns the pigs and provides the feed.

These trends, which have been driven by economics and the need to maximize health and productivity, have many environmental implications. On the positive side, multiple site production spreads the production of manure over many locations, compared to that of a farrow to finish operation. In addition, the grouping of animals by age and weight means that diets can be formulated precisely, which helps to reduce excretion of nutrients. In the past, only two diets were fed during the grow-finish period (45 lbs. to 250 lbs.); now, diets may be changed as many as seven times (personal communication, Joe Garber, Wenger's Feeds, 2000). Some producers even group pigs by gender and formulate separate diets for each sex. In addition, general improvements in productivity and efficiency mean that less feed is required to produce a pound of pork, which also reduces nutrient wastes.

The recent changes in swine production have also created some environmental challenges. Even though production phases are distributed over large areas, some of the individual units, particularly the sow units, are significantly larger than that of previous farrow to finish operations. Manure output can often surpass several million gallons per year. Odor from the application of manure and from the building can cause conflict in the community. And now, all of the feed and therefore all of the nutrients are imported to the hog production unit, causing a significant nutrient surplus in hog production regions.

Overview of Manure Storage Systems
In the Northeast, swine manure is stored in an anaerobic state. Nursery and grow-finish facilities generally have deep pit storage under totally slatted buildings. Sometimes, the pits are shallow "pull-plug" arrangements that allow producers to drain the manure to outside storage systems on a frequent basis. Sow units almost always have shallow pits with outside storage systems. The storage units are earthen ponds, approximately 12 feet deep, with about two-thirds of the depth below grade. The ponds are lined with high-density polyethylene, typically .060 inches thick. Beneath the liner, a network of drainage tiles carries surface water away from the structure. The outlet from the drainage tile is monitored on a regular basis to watch for leaks.

Because manure is removed completely at least once per year from storage systems in the Northeast, sludge build up is generally not a problem. Solids may accumulate in areas several hundred feet from the pump-out ports, or when gutters are shallow and not recharged after draining, or when feed waste is excessive. Agitation or removing manure from several locations throughout the storage facility will generally prevent solids problems.

In the Southeast, manure is most often stored in anaerobic lagoons. The design and operation of a lagoon is significantly different from that of anaerobic storage system. For a given volume of manure, lagoons are larger than anaerobic storage facilities, and the dry matter content of lagoon liquid is lower because of the frequent addition of flush water. The higher dilution rate and the warmer temperatures in the South support the biological decomposition of organic material. As such, the emission of ammonia and hydrogen sulfide is higher than that of an anaerobic storage system. But the release of volatile organic compounds from well-functioning lagoons is lower, so odors are somewhat lower as well.

Application of Manure from Anaerobic Storage and Lagoons

Lagoons are sometimes built in two stages; solids settle in the primary stage, and liquids drain from the top of the primary stage into the secondary stage. Whether there are one or two stages, only the liquids are land-applied on a frequent basis – usually through an irrigation system onto a warm season grass. If the lagoon is designed and functioning properly, the accumulation of sludge will not require removal for at least 10 to 15 years. Sludge will accumulate more rapidly if manure-loading rates are too high, if insufficient dilution water is added, or if feed wastage is excessive.

The application rate of manure is generally nitrogen-driven for both lagoons and anaerobic storage systems. Because the ratio of N:P in manure is lower than that required by most cropping systems, phosphorus will eventually accumulate in soil when hog manure is applied from either lagoons or anaerobic storage systems. For the anaerobically stored manure, the phosphorus deposition begins in the early years of manure application. For lagoons, since only the liquids are irrigated, phosphorus is less of a problem initially. However, because the phosphorus concentration of sludge is fairly high, producers face a significant challenge when sludge must be land applied.

In the Northeast, most or all of the manure produced on contract nursery or grower-finisher units is applied on the home farm. Large sow units, however, usually produce an excess of manure causing the application of manure to sometimes become a disposal issue. Manure is often surface-applied to increase nitrogen volatilization, and crop yields are projected at the upper limit, in order to maximize manure application rates. For example, many producers will surface apply manure in the fall, to maximize nitrogen volatilization prior to crop planting. Nitrogen losses are often high enough to permit a second application in the spring before planting. This approach may keep nitrogen in balance, but it will hasten the soil deposition of phosphorus.

The environmental concerns associated with phosphorus accumulation in the soil will be addressed in other papers. However, it's important to note that under present swine production systems there are no immediate and completely effective solutions to the problem of excess phosphorus, or for that matter, the excretion of other nutrients.

Variation in Nutrient Content of Swine Manure

Producers are advised to provide their own manure analysis to determine an accurate concentration of nutrients in the manure. Sampling manure after thorough agitation or from several loads throughout the pump-out process will provide a good representation of nutrient values.

Table 1: Estimated Nutrient Concentrations in Swine Manure

	Farrow	Nursery	Grow-Finish	Breeding-Gestation
Dry matter, %	4-6	4-6	4-6	4-6
Total N, lb/1000 gal	15-55	23-56	32-57	25-54
NH_4-N, lb/1000 gal	8	15	19	11
P, lb/1000 gal	12-18	15-18	18-26	18-25
K, lb/1000 gal	11-36	23-35	25-38	25-34

Adapted from Midwest Plan Service (1979) and Sutton et al. (1996).

Table 2: Nutrient Concentrations in Swine Manure from 23 Pennsylvania Farms

	Farrow	Nursery	Grow-Finish	Breeding-Gestation
Dry matter, %	2.4	5.9	6.5	4.2
Total N, lb/1000 gal	26.6	39.2	47.9	31.1
NH_4-N, lb/1000 gal	18.4	27.2	31.8	22.2
P, lb/1000 gal	8.3	17.0	23.9	15.1
K, lb/1000 gal	11.5	20.0	20.7	11.5

Adapted from Kephart et al. (1999).

Note that the data presented in Table 2 from Pennsylvania swine farms are similar to the values in Table 1, but not always within the expected range. This reinforces the observation that nutrient concentration, and in particular, dry matter content, is quite variable. In the 78 samples checked on the Pennsylvania farms, dry matter content averaged 5.2%, but ranged from .59% to 18.2%. In the Pennsylvania study, at least three samples were taken from each storage facility and pooled for the final analysis. Samples were also analyzed for nutrient content according to depth of sample. Concentrations of most nutrients increased with sample depth. This relationship was most evident for phosphorus, as depth of sample was strongly correlated with concentration ($P < .01$, $r = .86$). Correlation coefficients between depth and concentrations of other nutrients were also statistically significant ($P < .01$), but not as high: Total N ($r = .74$), ammonium N ($r = .57$), K_2O ($r = .42$). Note that the concentrations of the most soluble nutrients were not as highly correlated with depth of sample.

Partial Solutions to Nutrient Imbalance and Nutrient Excretion
The use of lysine and other amino acids in swine diets enable the feed manufacturer to decrease the amount of protein, and therefore the nitrogen, in the diet. This practice can reduce nitrogen excretion by more than 20% (Lenis, 1989; Koch, 1990; Hansen and Lewis, 1993). Unfortunately it decreases the N:P ratio in the manure even further. The use of dietary phytase, an enzyme that enhances the digestion of plant-borne phosphorus, can reduce phosphorus excretion by at least 20% (Cromwell, 1990). Some researchers have also shown that phytase may improve the availability of protein, starch, and several trace elements (Kies, 1998).

Injecting manure under the soil or incorporating it at the time of land application will reduce nitrogen losses (volatilization, and possible run-off) resulting in a N:P ratio more closely aligned with what the crops require. This practice also reduces ammonia and odor emissions.

Summary

The use of amino acids can economically reduce nitrogen excretion. Unfortunately, reductions in manure nitrogen further reduce the ratio of N:P, which is already lower than that required by most crops. Phytase can reduce phosphorus excretion, but methods are needed for further reductions, or for extracting P from the manure.

Many producers now surface apply manure to increase nitrogen losses to the air. But in the near future, phosphorus regulations or limits on ammonia emissions may require producers to conserve as much nitrogen as possible. Equipment is needed to provide fast and economical injection or incorporation of manure without destroying conservation practices.

References

Cromwell, G. L. 1990. Application of phosphorus availability data to practical diet formulation. IN Proc Carolina Swine Nutr. Conf. Raleigh, NC Nov. 7-8. pp. 55-75.

Hansen, B. C. and A. J. Lewis. 1993. Effects of dietary protein concentration (corn: soybean meal ratio) and body weight on nitrogen balance of growing boars, barrows, and gilts: mathematical descriptions. J. Anim. Sci. 71:2110-2121.

Kephart, K., B. Berrang, and R. Mikesell. 1999. Nutrient content of swine manure. Progress Report to the Pennsylvania Pork Producers Council.

Kies, Arie K. 1998. The influence of Natuphos Phytase on the Bioavailability of protein in swine. Proceedings, BASF Technical Symposium, June 4, 1998.

Koch, F. 1990. Amino acid formulation to improve carcass quality and limit nitrogen load in waste. IN Proc Carolina Swine Nutr Conf. Raleigh, NC Nov. 7-8 pp. 76-95.

Lenis, N. P. 1989. Lower nitrogen excretion in pig husbandry by feeding: Current and future possibilities. Netherlands. J. Agri. Sci. 37:61-70.

Midwest Plan Service. 1979. Livestock Waste Facilities Handbook. Third Printing. May 1979. pp 4-5.

Sutton et al. 1996. Swine manure as a plant resource. Pork Industry Handbook. Purdue University. PIH 25.

Manure Management on Poultry Farms: Practices and Risks

Eldridge R. Collins, Jr., Ph.D.
Professor and Extension Agricultural Engineer
Department of Biological Systems Engineering
Virginia Polytechnic Institute and State University

Biographies for most speakers are in alphabetical order after the last paper.

Introduction

Good manure management on poultry farms is fundamental in minimizing nutrient losses to the environment. In addition, manure management is important to minimize nuisance litigation related to generation of odors, insects and vermin, runoff, or leachate that offends neighbors or passersby, or otherwise endangers the environment.

Many different systems can be successfully used for storing and managing poultry wastes. The character of waste involved, the level of moisture content, the soil type and other site conditions are important factors in determining suitability of any particular waste management system. The individual grower must weigh the advantages and disadvantages of each method and decide which one best fits into the overall operation, and which best meets local, state, and federal regulations. Regional characteristics such as depth to groundwater, soil type, year-round climate, and other factors also influence methods of manure management chosen. Regardless of the system chosen, success will largely depend upon proper operation and maintenance. It is up to the grower to establish a regular inspection and maintenance schedule to keep the waste management system functioning properly to protect the environment.

Overview of Modern Poultry Waste Management Systems

Relationship of Poultry Housing to Waste Management

Waste management systems are closely related to the type of production house involved. House types are generally designed for layers, pullets, broilers/turkeys, or breeder chickens/turkeys, and are typically either high-rise or single-floor.

The high-rise layer house is normally 40 to 60 feet wide. Sidewalls may be constructed of a drop curtain (natural ventilation), a drop curtain with auxiliary fans, or a windowless wall with light and air control (mechanical ventilation). The watering system typically consists of cups or nipples. Solid or semi-solid manure is collected and stored in a pit under the house. It may be removed any time during the laying or growing cycle but is normally removed when the flock is moved out. Figure 1 shows the high-rise concept, which minimizes the amount of daily labor needed to manage the waste system.

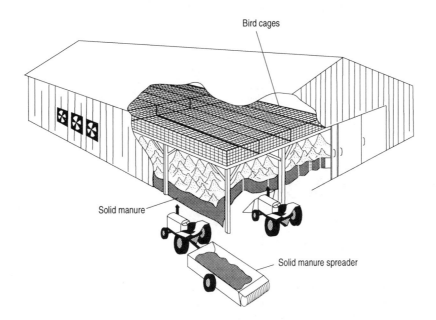

Figure 1. High-rise layer house concept (Source: *Agricultural Waste Management Field Handbook*, Natural Resources Conservation Service)

Waste drops directly to the lower floor of the high-rise layer house. If the storage area (lower floor) is properly ventilated, rows of dry mounds or "cones" of manure will form on paved or unpaved earthen floors for a year or longer before removal. Some states may require paving to provide greater insurance against groundwater contamination. Others may allow compacted clay floors to be used. Control of all excess water is required to help keep manure dry and make clean-out easier. The volume and moisture content of the solid or semi-solid manure can be influenced by leaking waterers, high humidity, or extent of ventilation. Drinker leakage, blowing rains (with open-sided houses), and groundwater intrusion can turn the manure piles into an unmanageable slurry mix.

Design and management of the total building ventilation system can either aid in drying the manure or cause moisture to accumulate to problem levels. The high-rise system works best with lower level exhaust fan ventilation, which pulls fresh air in through controlled inlets on the top floor; over the birds on the top floor; and down past the manure, taking moisture, stale air, and odors out of the building. Interior circulation fans may also be used in the pit area to improve circulation and promote manure drying (Figure 2).

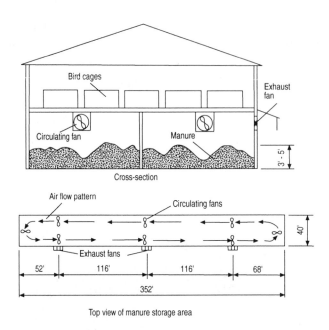

Figure 2. Ventilating the manure pit in a typical high-rise layer building (Source: *Livestock Waste Facilities Handbook*, MWPS–18)

Moisture can also be absorbed into the manure from the earthen floor surface, or wicked from the manure into the soil, depending on surrounding soil moisture conditions. Maintaining the manure cones at less than 45% moisture assists in fly control. Manure from the high-rise building can be directly land applied, composted, ensiled for feed, dried in drying beds, or mixed with additional water and used to generate biogas.

A standard single-story stair-step layer house is 30 to 60 feet wide. Sidewall construction may be a drop curtain, a drop curtain with auxiliary fans, or a windowless wall with light and air control (mechanical ventilation). The watering system may be cups or nipples. Manure is removed with a scraper or by flushing to a lagoon. Flushing is typically done daily for 20 minutes at a rate of 500 to 1,000 gallons of water per minute. Scraping is done two to three times per week. The manure from either mode of removal can be delivered to a storage tank or pond, or to a treatment lagoon. For a flush system, settling tanks or channels can be used to remove the solids, while the effluent is discharged into a holding pond or to a lagoon. Undercage paved collection channels are usually mechanically scraped, with solid or semi-solid manure collected over a two-day period. Liquids may evaporate from these shallow alleys between scrapings. Figure 3 shows a wide-span caged-layer house with collection channels (gutters) for flushing or scraping. Figure 4 shows a mechanical scraper system for removing manure from under the cages.

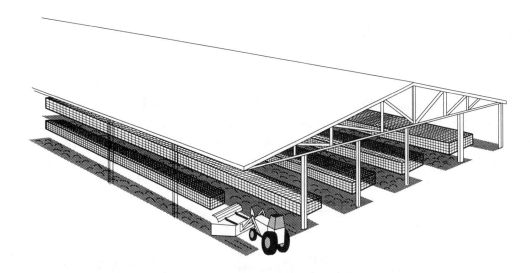

Figure 3. Cage layer building concept with collection alleys (gutters) for scraping or flushing (Source: *Agricultural Waste Management Field Handbook*, Natural Resources Conservation Service)

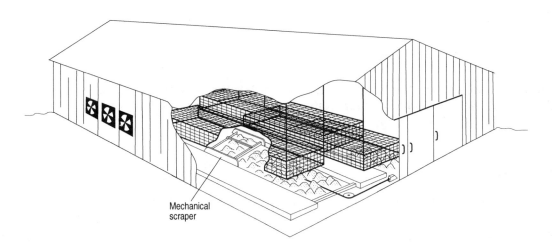

Mechanical
scraper

Figure 4. Mechanical scraper beneath layer cages for manure removal from building (Source: *Agricultural Waste Management Field Handbook,* Natural Resources Conservation Service)

In fully automated cage systems, the manure may be removed by a belt system that runs under each tier of cages. Proper management of the system can greatly reduce manure moisture before the manure is removed from the house. Some systems have special ventilation ducts to assist in drying the manure. This system could be used very effectively in a manure composting operation with minimum labor input. Collection alleys are not necessary with a belt cleaning system.

Wet manure from cage birds over belts, scrapers, or flush manure channels is handled as a sticky semi-solid or a liquid, depending on the amount of water in the manure. Wet manure will become anaerobic very quickly and will contribute to high odor levels, especially when it is disturbed and spread.

Broilers and turkeys are typically grown on a litter floor housing system. The litter base (saw-dust, wood shavings, peanut or rice hulls, or similar materials) may be changed after each flock, or a built-up litter-based system may be utilized. In some parts of the United States, the built-up litter may be used for 2 to 3 years before cleaning. Figure 5 shows a house cross-section that may be used in the production of broilers and turkeys. Clear-span-truss construction facilitates clean-out. Dry litter (manure + litter base) from birds is handled and stored as a solid material.

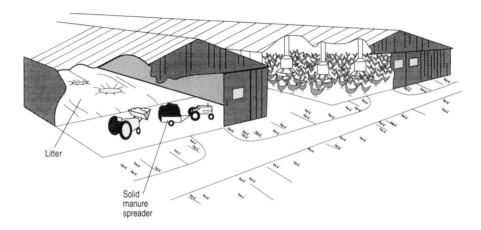

Litter

Solid manure spreader

Figure 5. Building with litter floor system used for broilers and turkeys (Source: *Agricultural Waste Management Field Handbook,* Natural Resources Conservation Service)

Manure Storage [1]

Poultry manure storage allows for optimum use of labor and equipment and provides a means of nutrient retention. Proper manure storage provides environmental protection and may help offset storage costs by allowing more effective use of nutrients as fertilizer. Storage also provides future opportunities for the sale of litter for feed or compost feedstock or other off-farm sales. However, depending on the type of storage method or structure selected, the capital cost can negate any economic gain.

Poultry manure storage systems can range from temporary piles to permanent roofed structures. The type of system most useful to an enterprise is dependent on the quantity of manure to be handled, manure moisture content, the frequency or timing of manure removal, the capital investment required, and outside environmental and social factors. Local Cooperative Exten-

---

[1] A full discussion of both temporary and permanent poultry manure storage facilities is given in NRAES–132, *Poultry Waste Management Handbook.*

sion offices or Soil and Water Conservation Districts usually provide assistance with selecting a manure storage facility. The Natural Resources Conservation Service (NRCS) has developed design standards and specifications for poultry manure storage structures.

Solid Manure Storage: Within the Poultry House

One type of solid manure storage occurs in the poultry house in both floor litter and high-rise cage-type systems. High-rise houses for caged birds allow accumulation of manure beneath the cages in pits that can be entered with cleaning vehicles from the outside of the structure (see Figure 1). With floor litter systems, manure is mixed with the litter by bird foot action, and storage occurs on the floors through a continuous buildup of the dry litter/manure mixture (see Figure 5).

The cleaning frequency of either system is determined by the quality of the manure or manure litter in the house and the amount of remaining storage space available. Wet manures will require more frequent removal than dry manures. Typically, deep-pit and high-rise houses are cleaned once or twice per year. On the other hand, floor systems might be partially cleaned of wet manure "cake" after every flock but not totally cleaned for a year or more. Poultry manure should be maintained in a dry state so that nutrients are conserved, insects and odors are controlled, bird welfare is enhanced, and handling and storage costs are minimized.

Solid Manure Storage: Outside the Poultry House

Storage outside the house is required only when manure must be removed from inside and no land is available for immediate manure application. Cleaning out high-rise pits can usually be scheduled to allow manure field applications when needed without additional storage. Caged bird systems with manure removal by belts or scrapers do not provide in-house storage. Floor litter houses are partially cleaned between flocks, while whole-house clean-out is determined by litter management schedules of poultry integrators.

The storage method chosen must protect manure from prolonged contact with rainwater. This requires a surface that sheds water. A deep, well-rounded stockpile of compacted litter, manure, and associated material will shed water. However, the edges of the pile at the ground surface may become saturated and cause surface water and groundwater pollution. Caged-bird manure will readily soak up moisture and should be stored only under cover with confining walls.

All storage systems should be separated from seasonal high groundwater by a minimum of 4 feet of soil or a water-resistant liner of compacted clay, plastic, or concrete. Locate the storage to avoid wells, normally wet areas, runoff or drainage pathways, and other areas of running or standing water. It is also a good idea to provide a grassed buffer around the entire storage area.

Careful storage site location must consider insects, birds, and rodents that can transmit or transfer avian diseases. Storage receiving manure from many different sites should not be located near a poultry production facility.

Floor manure litter contains both wet and dry organic materials that produce heat when stored in confined piles. Storage structures and compact piles may be subject to spontaneous combustion.

Liquid, Slurry, and Semi-Solid Manure Storage

Wet manure removed from under caged birds by mechanical scrapers and belts, or by flushing with water cannot be stacked and requires containment storage. Manure liquids and slurries that are mostly water (less than 12% solids) require containment in tanks or basins constructed of materials impervious to water transfer. Semi-solid manure (12 to 20% solids) does not readily flow like water but still needs containment walls to keep the manure in a manageable mass, and to prevent pollutant losses to the environment. Semi-solid manures can be handled with bucket loaders and open spreaders. Liquid manures must be pumped and spread with tank trucks, tank wagons, or irrigation equipment.

Manure that is stored in tanks or storage ponds is normally anaerobic and can be expected to generate considerable odor in storage. It is particularly odorous during spreading. If odor is likely to be a problem at the farm site or during spreading, wet storage should not be used, or methods of advanced treatment such as aerobic treatment should be adopted.

Liquid and slurry manure storage can be a constructed concrete or lined steel tank aboveground or below ground or an earthen basin (Figures 6 and 7). Tanks can be open-topped or covered with a roof or top. Earthen basins are usually open-topped but can be covered with a geotextile fabric. Open-topped storage structures collect rainwater, which increases the volume of material to be handled and dilutes nutrients. Open-topped structures are also significant sources of odor, especially when manure enters the structure at the top and when the manure is mixed during unloading. Bottom loading through gravity or pressurized pipe may assist in odor control when manure dries on the storage surface and forms a floating crust.

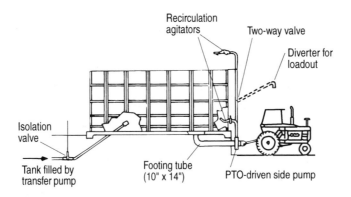

Figure 6. Aboveground manure storage tank

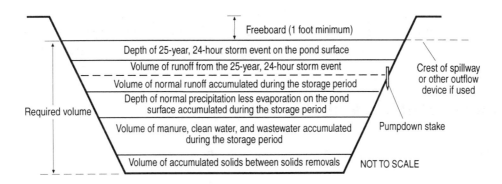

Figure 7. Cross-section of an earthen storage basin showing the volumes that must be accounted for in the design (Source: *Agricultural Waste management Field Handbook,* Natural Resources Conservation Service)

Semi-solid manure storage can be a tank or basin fitted with a sloped driveway to allow tractor or bucket loader entrance for unloading. The storage is usually open at the top. The entrance of rain and runoff may cause some of the manure to become liquid, which cannot be easily handled with the loader bucket. Methods of allowing water to flow away from the manure, through directed floor slopes and perforated dams, are not very effective with poultry manures. A roof, combined with site grading to direct runoff away from the structure, is an effective means of preventing excessive water in the storage.

Wet manure storage structures and basins are subject to high loading pressures on the sidewalls and banks. These pressures increase with manure depth. Wall failure allows the entire contents of the storage to escape and flow overland, which may cause environmental and property damage. Standards and specifications for manure storage construction have been developed by the NRCS to ensure design and construction procedures that will avoid structural failure. Although maintenance assistance is currently not a part of the NRCS plans, good maintenance is essential to a successful operation. For more information on earthen storages, see *Earthen Manure Storage Design Considerations,* NRAES–109.

Manure Storage Maintenance

Maintenance procedures for manure storage are highly dependent on the type of structure and the properties of the manure being held. Maintenance can be described as "efforts to ensure that dry manures are kept dry and wet manures are kept contained". Maintenance of solid manure storage systems includes preventing drinker water spillage, optimizing the effectiveness of house ventilation systems, securing outside stack plastic covers from the wind, repairing any damage from wind or machines to permanent structures, and monitoring stored manure temperatures with appropriate control measures employed in fire emergencies.

Maintenance of liquid and semi-solid manure storage systems includes periodic inspection of walls to identify cracks or buckling, periodic inspection of earthen banks to identify erosion or animal burrows, and removal of waste in a timely manner so as to keep the design freeboard and storm trapping volume available (Figure 7). Other maintenance efforts should be identified during the design stage and become part of a maintenance plan. Structural designs developed by the Natural Resources Conservation Service are accompanied by an operations and maintenance manual.

Manure Production and Characteristics

Manure properties are influenced by several factors:

- bird species,
- bird age,
- diet and nutrition,
- bird productivity, management, and
- housing, ventilation, drinker systems, nutrition, and other environmental factors.

What is "Manure"?

The waste management system will collect bedding or litter, water, soil, grit, feathers, nonutilized dietary minerals, and other materials that must be handled as "manure." The term "manure" generally includes raw feces, urine, waste feed, spilled water, absorptive bedding used in poultry houses, and any other waste material that is part of the waste stream from production houses.

Poultry Manure Production

Table 1 lists estimated manure production, as excreted by poultry. Data in this table represent averages from a wide database of published and unpublished information on poultry manure. Total manure production is presented per 1,000-bird capacity per day based on the weighted average daily live weight of the bird during its production cycle. Since manure production is generally based on the live weight of the bird, manure amounts may be increased or decreased proportional to the bird live weight. Each farm may have slightly different manure production rates due to factors already mentioned. Table 2 lists commonly used manure characteristics, as excreted.

Moisture Considerations

Table 3 lists manure and wastewater produced from various types of waste storage and treatment systems for commercial layer production facilities. Production amounts can be highly influenced by leaking waterers, rainfall surplus added to open storage pits and lagoons, and whether or not fresh water is used for flushing or cleaning manure collection alleys. Table 4 lists manure and wastewater characteristics, as removed from storage systems

Table 1. Manure production, as excreted

Bird type	Live weight (lbs)		Total manure production per 1,000 birds per day		
	Market	Average	(lbs)	(ft³)	(gallons)
Commercial layer					
Hen	4.0	4.0	260	4.2	32
Pullet	3.0	1.5	97	1.6	12
Broiler	4.5	2.25	177	2.8	21
Roaster	8.0	4.0	315	4.9	37
Cornish	2.5	1.25	99	1.5	12
Breeder	7.0	7.0	552	8.7	65
Turkey					
Poult	5.0	2.5	113	1.8	13
Grower hen	16.0	10.0	452	7.1	53
Grower tom, light	22.0	13.0	588	9.3	69
Grower tom, heavy	30.0	17.0	769	12.1	91
Breeder	20.0	20.0	905	14.3	107
Duck	6.0	3.0	328	5.3	39

NOTE: Data in this table represent averages from a wide database of published and unpublished information on poultry manure. Total manure production is presented per 1,000-bird capacity per day based on the weighted average daily live weight of the bird during its production cycle. Since manure production is generally based on the live weight of the bird, manure amounts may be increased or decreased proportional to the bird live weight. Each farm may have slightly different production rates.

Table 2. Typical manure characteristics, as excreted

Manure characteristics	Commercial layer	Broiler	Turkey	Duck
Density (lbs/ft³)	62	64	63	62
TS[a] (%)	25	26	25	27
VS[b] (%)	19	19	19	16
COD[c] (ppm)	176,000	197,000	236,000	169,000
		(lbs/ton)		
TKN[d]	27	26	28	28
NH_3N	6.6	6.7	8.1	7.4
P_2O_5	21	16	24	23
K_2O	12	12	12	17
Ca	41	10	27	29
Mg	4.3	3.5	3.1	4.1
S	4.3	2.0	3.3	3.6
Na	3.6	3.5	2.8	3.5
Cl	20	18	18	20
Fe	2.0	1.9	3.2	2.8
Mn	0.16	0.20	0.10	0.17
B	0.05	0.06	0.06	0.06
Zn	0.14	0.084	0.62	0.48
Cu	0.02	0.02	0.03	0.03

Sources: ASAE; Department of Biological and Agricultural Engineering, North Carolina State University
NOTE: All values are on a wet basis (as excreted)

[a]   TS = total solids (100 − TS = moisture or water content)

[b]   VS = volatile solids; the portion of total solids driven off as volatile (combustible) gases

[c]   COD = chemical oxygen demand; a measure of the oxygen-consuming capacity of inorganic and organic matter present in water or waste

[d]   TKN = sum or organic and ammonia(um) nitrogen, as measured by the laboratory Kjeldahl procedure

Table 3. Typical commercial layer waste production, as removed from storage

Bird type	Live weight (lbs)		Total waste production per 1,000 birds per day		
	Market	Average	(lbs)	(ft³)	(gallons)
			Undercage collection alley-scraped manure[a]		
Hen	4.0	4.0	155	2.5	19
Pullet	3.0	1.5	58	0.9	7
			High-rise, deep-pit stored manure[b]		
Hen	4.0	4.0	108	2.1	16
Pullet	3.0	1.5	41	0.8	6
			Liquid manure slurry[c]		
Hen	4.0	4.0	356	6.1	46
Pullet	3.0	1.5	134	2.3	17
			Anaerobic lagoon liquid[d]		
Hen	4.0	4.0	587	9.4	70
Pullet	3.0	1.5	220	3.5	26
			Anaerobic lagoon sludge[e]		
Hen	4.0	4.0	108	1.7	13
Pullet	3.0	1.5	41	0.65	5

Source: Department of Biological and Agricultural Engineering, North Carolina State University

NOTE: Production amounts can be highly influenced by leaking waterers, rainfall surplus added to open storage pits and lagoons, and whether or not fresh water is used for flushing or cleaning manure collection alleys.

[a] Manure scraped from paved alley and collected within two days

[b] Annual manure accumulation stored on unpaved surface

[c] Manure, excess water usage, storage surface rainfall surplus; does not include fresh water for flushing

[d] Manure, excess water usage, lagoon surface rainfall surplus; does not include fresh water for flushing

[e] No manure solids removal prior to lagoon treatment/storage

Manure tanks, pits, or earthen storage basins receive scraped manure, excess water spillage, and surface rainfall surplus for storage until spreading. Tanks and pits that can be covered usually receive less water, have a higher solids and nutrient content, and result in less slurry volume to handle than open earthen storage basins. Such storages are usually designed for four to twelve months= accumulation, depending on the amount of storage needed for an acceptable nutrient management plan.

Table 4. Typical commercial layer waste characteristics, as removed from storage

Manure characteristic	Undercage collection alley-scraped manure[a]	High-rise, deep-pit stored manure[b]	Liquid manure slurry[c]	Anaerobic lagoon liquid[d]	Anaerobic lagoon sludge[e]
Density (lbs/ft^3)	62	51	58	62	62
TS[f] (%)	35	53	11	0.49	17
VS[g] (%)	25	32	7.4	0.22	7.3
COD[h] (ppm)	270,000	286,000	106,000	2,950	28,600
			(lbs/ton)		
TKN[i]	28	34	57	6.6	21
NH$_3$N	14	12	37	5.6	6.5
P$_2$O$_5$	32	51	52	1.7	77
K$_2$O	20	26	33	10.3	9.8
Ca	41	76	33	1.1	47
Mg	5.5	5.7	4.0	0.34	12
S	7.1	4.8	4.0	0.61	7.1
Na	2.8	3.3	4.8	1.8	3.3
Cl	4.0	6.0	6.6	3.4	2.4
Fe	2.4	2.8	1.7	0.060	4.8
Mn	0.29	0.44	0.38	0.0069	1.6
B	0.022	0.036	0.030	0.0092	0.035
Zn	0.31	0.35	0.39	0.016	1.1
Cu	0.034	0.058	0.073	0.004	0.14

Source: Department of Biological and Agricultural Engineering, North Carolina State University
NOTE: All values are on a wet basis (as excreted)
a   Manure scraped from paved alley and collected within two days
b   Annual manure accumulation stored on unpaved surface
c   Manure, excess water usage, storage surface rainfall surplus; does not include fresh water for flushing
d   Manure, excess water usage, lagoon surface rainfall surplus; does not include fresh water for flushing
e   No manure solids removal prior to lagoon treatment/storage
f   TS = total solids (100 − TS = moisture or water content)
g   VS = volatile solids; the portion of total solids driven off as volatile (combustible) gases
h   COD = chemical oxygen demand; a measure of the oxygen-consuming capacity of inorganic and organic matter present in water or waste
i   TKN = sum or organic and ammonia(um) nitrogen, as measured by the laboratory Kjeldahl procedure

Grit is a poultry feed ration additive that aids digestion. When used excessively, grit will separate from manure. Dietary calcium can also separate from manure and reduce both the flow of manure to holding tanks and the capacity of holding tanks to store manure. This may present handling challenges for liquid storage and handling systems.

Anaerobic lagoons (those that provide treatment without oxygen) are designed for biological treatment. The relatively large surface areas of lagoons often collect more surplus rainfall than dry manure stacks or liquid manure storage tanks. As a result, relatively large liquid volumes with comparatively dilute nutrient contents result from lagoon treatment. Bottom sludges are usually handled with liquid slurry equipment and are removed infrequently. Some nutrients such as calcium, phosphorus, zinc, and copper may be concentrated in lagoon sludges.

Litter Considerations

Table 5 lists litter characteristics, as removed from production houses. Table 6 lists manure and litter production, as removed from various meat-type bird production facilities. Litter amounts are presented as tons per 1,000 birds. Bedding or litter materials (typically sawdust, wood shavings, peanut hulls, or rice hulls) are initially placed on floors to a depth of 2 inches or more (depending on the poultry integrator company) in production houses. The differences in management required by the different integrator companies, described earlier in this paper, will influence the litter nutrient values, as will changing the amount or type of litter used. The values supplied in Tables 5 and 6 are given as guides. Where litter will be used as fertilizer or for other purposes, it is wise to test the litter for nutrient content.

Moisture management in the production facilities can affect litter characteristics. As the litter becomes wetter (lower solids content), more ammonia will be released, and the nitrogen content of the litter will decrease. As litter dryness increases, increasingly dusty conditions will exist within the facility. A number of factors influence litter moisture. When fed in excess, certain dietary ingredients (especially salt) cause birds to consume and excrete large amounts of water, resulting in wet litter conditions. Some drugs stimulate excess water consumption and excretion. Environmental conditions, such as wet and humid weather or very cold temperatures, can cause wet litter if the house heating and ventilation system is not able to eliminate moisture effectively. Waterers, foggers, and evaporative cooling pads, if not managed and maintained carefully, can contribute greatly to wet litter problems. Ideally, litter moisture should be maintained at 20 to 25%.

Around waterers and feeders, additional manure and moisture tend to form crusted or caked areas of litter that have different handling and nutrient characteristics than the other house litter. This manure cake typically represents about 30 to 35% of the total litter. The manure cake around waterers is usually wetter and has a lower nutrient content than the total litter. But the area around feeders is often drier and has a higher nutrient content due to feed spillage. The cake may be removed with crusting equipment after each flock of birds.

Table 7 gives estimates of manure and litter volumes that have been removed from production facilities to an outside uncovered stockpile on an earthen surface for storage up to six months before field spreading. The litter mass removed from the stockpiles is, in most cases, only slightly less than that taken directly from the production facilities. Absorption of rainwater in the stockpiles is offset by a reduction in solids due to composting action. Covering the stockpile or storing the litter inside a roofed structure reduces losses and preserves a higher-quality litter. Covering the stockpile also reduces the potential for nutrient leaching or nutrient loss to the environment in runoff.

Table 5. Typical litter characteristics, as removed from production houses

Bird type	Density (lbs/ft³)	TS[a] (%)	VS[b] (%)	TKN[c]	NH₃N	P₂O₅	K₂O	Ca	Mg	S	Na	Cl	Fe	Mn	B	Zn	Cu
										(lbs/ton)							
Broiler																	
Whole litter[d]	32	79	63	71	12	69	47	43	8.8	12	13	13	1.2	0.79	0.057	0.71	0.53
Manure cake[e]	34	60	47	46	12	53	36	34	7.0	9.2	10	—[f]	1.2	0.69	0.044	0.60	0.41
Roaster																	
Whole litter[d]	29	76	59	69	16	70	47	41	8.4	14	13	—[f]	1.6	0.76	0.047	0.68	0.49
Cornish																	
Whole litter[d]	30	68	53	59	12	57	59	41	22	—[f]	—[f]	—[f]	—[f]	1.1	—[f]	0.92	0.61
Manure cake[e]	34	54	42	62	17	39	39	30	14	—[f]	—[f]	—[f]	—[f]	0.67	—[f]	0.50	0.46
Breeder																	
Whole litter[d]	50	69	29	37	8.0	58	35	83	8.2	7.8	8.3	—[f]	1.2	0.69	0.034	0.62	0.23
Turkey poult																	
Whole litter[d]	23	80	62	40	9.6	43	27	26	5.1	6.1	4.7	1.8	2.0	0.53	0.038	0.46	0.39
Grower																	
Whole litter[d]	32	73	53	55	12	63	40	38	7.4	8.5	7.6	12	1.4	0.80	0.052	0.66	0.60
Manure cake[e]	35	55	44	45	20	47	30	26	5.4	6.3	5.5	—[f]	1.2	0.56	0.038	0.47	0.48
Breeder																	
Whole litter[d]	50	78	34	35	7.6	47	18	72	4.6	7.4	4.3	—[f]	1.0	0.43	0.031	0.50	0.40
Duck																	
Whole litter[d]	50	37	24	17	3.6	21	13	22	3.3	3.0	3.0	—[f]	1.3	0.37	0.021	0.32	0.04

Source: Department of Biological and Agricultural Engineering, North Carolina State University

NOTE: All values are on a wet (as is) basis

[a] TS = total solids (100 − TS = moisture or water content)

[b] VS = volatile solids; the portion of total solids driven off as volatile (combustible) gases

[c] TKN = sum of organic and ammonia(um) nitrogen as measured by the laboratory Kjeldahl procedure

[d] Annual manure and litter accumulation; typical litter base is sawdust, wood shavings, or peanut hulls

[e] Surface manure cake removed after each flock

[f] Data not available

Table 6. Typical litter production, as removed from production houses

Bird type	Live weight (lbs)		Total litter production per 1,000 birds sold (tons)
	Market	Average	
Broiler			
Whole litter[a]	4.5	2.25	1.25
Manure cake[b]	4.5	2.25	0.4
Roaster			
Whole litter[a]	8.0	4.0	2.6
Cornish			
Whole litter[a]	2.5	1.25	0.625
Manure cake[b]	2.5	1.25	0.06
Breeder			
Whole litter[a]	7.0	7.0	24.0[c]
Turkey poult			
Whole litter[a]	5.0	2.5	1.0
Grower hen			
Whole litter[a]	16.0	10.0	8.0
Manure cake[b]	16.0	10.0	2.5
Grower tom, light			
Whole litter[a]	22.0	13.0	10.0
Manure cake[b]	22.0	13.0	3.3
Grower tom, heavy			
Whole litter[a]	30.0	17.0	14.0
Manure cake[b]	30.0	17.0	4.4
Breeder			
Whole litter[a]	20.0	20.0	50.0[c]
Duck			
Whole litter[a]	6.0	3.0	4.25

Sources: Department of Biological and Agricultural Engineering, North Carolina State University and Department of Agricultural Engineering, University of Delaware

[a]   Annual manure and litter accumulation; typical litter base is sawdust, wood shavings, or peanut hulls

[b]   Surface manure cake removed after each flock

[c]   Tons/1,000 birds/year

Table 7. Typical litter volume after open stockpiling

Bird type	Live weight (lbs)		Total litter production per 1,000 birds sold[a] (tons)
	Market	Average	
Broiler	4.5	2.25	1.0
Turkey grower	25.0	15.0	11
Duck	6.0	3.0	2.2

Source: Department of Biological and Agricultural Engineering, North Carolina State University

[a]   Annual house manure and litter accumulation removed to uncovered stockpile to be spread within six months; typical litter base is sawdust, wood shavings, or peanut hulls

Nutrient Considerations

Table 8 shows typical litter characteristics, as removed from open stockpiles. As can be seen from the table, phosphorus is conserved in open stockpiles and potassium levels are only slightly less than those of house litter. The nitrogen content, however, is about half that of broiler and turkey house litter due to the loss of ammonia caused by wetting and resulting biological activity. Again, storing litter in a dry structure will conserve nitrogen.

Table 9 estimates typical nitrogen losses between excretion and land application on a mass basis. Bedding and water dilute manure, resulting in less nutrient value per pound. Substantial nitrogen can be lost to the atmosphere as ammonia. The least nitrogen losses are associated with slurry storage pits, dry-house whole litter, and roofed storages. Deep-pit manure stacking and open stockpiled litter have moderate to high nitrogen losses. Lagoons have the highest loss.

Phosphorus and potassium losses are usually negligible, except with lagoons. Much of the phosphorus in lagoons concentrates in, and is recoverable with the bottom sludge. Moderate amounts of potassium may be lost from open uncovered stockpiles due to leaching.

Table 8. Typical litter characteristics, as removed from open stockpiles

Manure characteristic	Broiler	Turkey grower	Duck
Density (lbs/ft³)	33	24	50
TS[a] (% w.b.)	61	61	49
VS[b] (% w.b.)	43	44	32
	(lbs/ton)		
TKN[c]	33	32	22
NH_3N	6.9	5.5	4.8
P_2O_5	77	70	41
K_2O	32	30	22
Ca	63	45	34
Mg	8.2	7.1	5.2
S	10	7.4	4.5
Na	6.6	5.7	5.4
Cl	13	8.0	_[d]
Fe	1.8	2.1	1.5
Mn	0.70	0.76	0.56
B	0.039	0.042	0.031
Zn	0.63	0.63	0.50
Cu	0.29	0.42	0.05

Source: Department of Biological and Agricultural Engineering, North Carolina State University

NOTE: All values are on a wet basis (as is). Annual house manure and litter accumulation removed to uncovered stockpile to be spread within six months; typical litter base is sawdust, wood shavings, or peanut hulls

[a] TS = total solids (100 − TS = moisture or water content)

[b] VS = volatile solids; the portion of Total Solids driven off as volatile (combustible) gases

[c] TKN = sum of organic and ammonia(um) nitrogen as measured by the laboratory Kjeldahl procedure

[d] Data not available

Table 9. Typical nitrogen losses during handling, storage, and treatment

System	Nitrogen lost[a] (%)
Solid	
Paved collection alley, scraped[b]	30–40
Deep pit[c]	40–50
House litter[d]	25–35
Open stockpiled litter[e]	60–75
Liquid	
Slurry storage[f]	20–30
Lagoon[g]	75–85

Source: Department of Biological and Agricultural Engineering, North Carolina State University

[a] Nitrogen lost during handling, storage, and treatment compared to as-excreted manure nitrogen

[b] Manure scraped from layer undercage paved alley and collected within two days

[c] Annual layer manure accumulation stored on unpaved surface

[d] Annual manure and litter accumulation; typical litter base is sawdust, wood shavings, or peanut hulls

[e] Annual house manure and litter accumulation removed to uncovered stockpile to be spread within six months

[f] Manure, excess water usage, storage surface rainfall surplus; does not include fresh water for flushing

[g] Anaerobic lagoon liquid and sludge

Sampling and Testing

The values in Tables 1 through 8 are estimates based on averages from large databases. Actual farm-specific manure production and characteristics may vary considerably from the averages. As-excreted values may vary up to 30% from the average due to bird productivity, age, or diet. Variances of 25% for dry-house litter to as much as 60% for open liquid manure or lagoon systems are common because of differences in management or environmental factors. For these reasons, where possible, samples of the actual farm manure, litter, or wastewater should be collected and analyzed by local laboratories or testing facilities to provide more accurate information for planning, design, and land application.

The results from a regular sampling program should be entered into a farm-specific records database. Once actual farm averages have been developed, they should be useful in making management decisions.

Regional Differences in Production Practices

Commercial poultry production practices are inclined to be quite similar across the United States, with a few exceptions. Generally, differences tend to be climatically related, especially with regard to ventilation, and the appropriateness of spreading of wastes ("organic fertilizer") on fields.

Ventilation is a critical issue in producing poultry in houses. The challenge in hot weather is to provide sufficient air exchange in production houses to keep them from getting too warm. If this is accomplished, the secondary issue of providing enough fresh air will generally take care of itself.

The challenge in cold weather is to provide enough air exchange to remove ammonia, dust, and other air impurities without requiring an undue amount of supplemental heating, or chilling of birds in the house. This becomes relatively more difficult in production houses in colder climates than for those in warmer. Because of these climatic differences and critical nature of ventilation control, houses in cold climates will generally rely more on fan ventilation, especially in winter, and those in warmer climates will attempt to provide sufficient ventilation with moveable "curtain walls", and minimal fan boosted ventilation. As evaporative cooling has become more popular for hot weather use, housing systems have tended to become more standardized across the various production areas.

Perhaps greater regional differences in poultry housing are seen in manure handling systems. As farm nutrient management planning has come to be recognized as important to environmental quality issues, more states are limiting the time of year that nutrients can be applied to fields. Often limitations are related to the ability to sustain a growing crop on waste application fields, and whether or not application fields are frozen or covered in ice or snow.

Anaerobic and aerobic treatment systems that employ simple designs tend to work best in warm weather. Thus, these systems are more easily utilized in year-round warm climates.

Because of climatic influences, poultry housing systems that collect manure for long term holding (such as deep pit layer houses) or scraper systems in layer or breeder houses, with storage of manure in concrete or earthen pits, are more common in areas of the U.S. where winters are cold. This allows manure to be held until Spring or Fall when it can be applied to row crops or pastures when plant growth will occur and utilize waste nutrients according to a nutrient management plan.

Systems of housing that employ in-house flushing, or pit recharge systems for waste removal, tend to be found in warmer states such as those in the deep southern U.S. In these areas, anaerobic treatment lagoons work well year round for reducing waste strength, recycling effluent for waste removal, and the excess effluent can be applied (generally through irrigation) and utilized on growing pastures or fields for a longer period of the year than in northern locations. In addition, operational problems related to system/equipment freezing are not as significant as in cold climates.

Disposal of Poultry Mortalities

A by-product of even the most successful poultry production operations is dead birds. Despite improved health and production practices, intermittent mortality is to be expected in commercial flocks. Regardless of the cause of the mortality, proper disposal of carcasses is required to ensure biosecurity, protect the environment, and avoid offending others with nuisance conditions.

Two general categories of carcass disposal are: [1] normal mortality (typically about 0.1% per day, with fluctuation up to 0.25% not uncommon); and [2] disposal of large portions of the flock or a whole flock (usually associated with sacrifices due to contagious disease outbreak or death due to power outages or other catastrophes). Normal mortality is generally more easily handled because of the steady "flow" of material through the disposal system.

Methods for disposal of dead birds are burial, incineration, rendering, and composting. The method chosen must be compatible with the individual grower's situation and management capabilities and must comply with state laws. Fabricated pits for burial have been used in many areas, but questions have been raised about their impact on groundwater quality, and they have been prohibited in many states. Incineration is perhaps the most biosecure method of disposal, but it tends to cause odor and maintenance problems, and it tends to be slow and expensive. In addition, there may be regulatory issues in some states. For producers located close to a rendering plant or pickup route, this process can be an attractive and economical method for disposal of normal and catastrophic mortality. However, there are biosecurity risks associated with transporting carcasses to a rendering plant or with farm pickup systems. Also, some rendering plants may not be able to handle large quantities of carcasses from catastrophic events. Composting of dead birds is gaining popularity, but it requires investment in new facilities and has not received unconditional approval by regulatory authorities in all states. However, composting (Figure 8) has become the method of choice for disposal of normal mortality losses on many poultry farms.

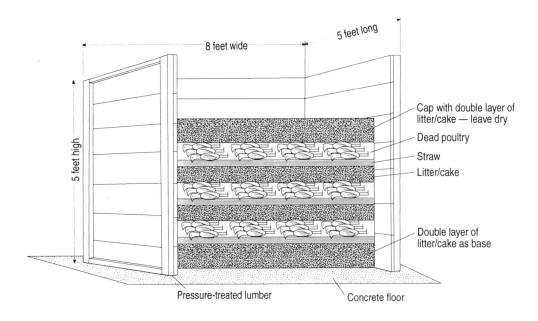

Figure 8. Layering arrangement for routine composting of normal mortalities

Summary

Poultry waste management systems are closely related to the type of housing used. Birds grown for meat are typically produced on floor litter systems, and produce a "dry" waste. Birds grown for breeder purposes, or to produce eggs are produced in housing systems that may produce a "dry" waste, or due to method of manure collection, removal, and treatment may be quite wet, or even "liquid". The waste handling (and housing) system chosen will depend upon grower preference, the character of waste involved, the level of waste moisture content, the soil type and other site conditions. The individual grower must weigh the advantages and disadvantages of each method and decide which one best fits into the overall poultry operation, and which best meets local, state, and federal regulations. Regional characteristics such as depth to groundwater, soil type, year-round climate, and other factors also influence methods of manure management chosen. Regardless of the system chosen, success will largely depend upon proper operation and maintenance.

Nutrients contained in poultry waste must be incorporated into a nutrient management plan. Tabular values for different categories of poultry waste are available, but wastes are highly variable from farm to farm. Tabular values are useful as estimates, but good nutrient management will require that samples of the actual farm manure, litter, or wastewater be collected and analyzed to provide more accurate information for planning, design, and land application plans.

Selected References

American Society of Agricultural Engineers. 1993. Standards: Manure Production and Characteristics. ASAE Standard D384.1. American Society of Agricultural Engineers. 530–532.

Collins, E. R., Jr. 1992. *Composting Dead Poultry.* Publication 442–037. Blacksburg, VA: Virginia Cooperative Extension, Virginia Polytechnic Institute and State University.

Collins, E. R., Jr., J. C. Barker, L. E. Carr, H. L. Brodie, and J. H. Martin, Jr. 1999. *Poultry Waste Management Handbook.* NRAES–132. Natural Resource, Agriculture, and Engineering Service, Cooperative Extension, Ithaca, NY 14853-5701.

Midwest Plan Service. 1993. *Livestock Waste Facilities Handbook.* MWPS–18. Ames, Iowa: Midwest Plan Service, Iowa State University.

United States Department of Agriculture. 1992. *Agricultural Waste Management Field Handbook.* AWMFH–1. Washington, DC: United States Department of Agriculture, Natural Resources Conservation Service.

Wright, P., W. Grajko, D. Lake, S. Perschke, J. Schenne, D. Sullivan, B. Tillapaugh, C. Timothy, and D. Weaver. 1999. *Earthen Manure Storage Design Considerations.* NRAES–109. Natural Resource, Agriculture, and Engineering Service, Cooperative Extension, Ithaca, NY 14853-5701.

Manure Management on Dairy Farms: Practices and Risks

Robert E. Graves, Ph.D.
Professor
Agricultural and Biological Engineering
The Pennsylvania State University

Biographies for most speakers are in alphabetical order after the last paper.

A dairy manure handling system is a cost item that must meet many requirements. The ultimate goals of a manure handling system should include:

- enhance management and worker productivity
- be cost effective
- be environmental friendly
- be safe for workers and visitors
- complement on farm biosecurity measures

Many observers would argue that on most farms the priority items are how the system works and how much it costs. Often it seems that too little attention is given to the affect of the system on neighbors or the environment; safety for workers or visitors; or biosecurity requirements for control of disease spread. A death or serious injury resulting from a manure handling system suddenly makes us aware of safety considerations. A visit from the regulatory community or subpoena from the court reminds us of our obligation to be a good neighbor. The spread of a disease through the herd, because of excessive cross contamination between manure and feed or water, emphasizes the relationship between manure handling and animal health.

A farmer or her teenage son going out to spread the daily load of manure may still be seen as bucolic and part of country life. A week of 12-hour days with two contract twin screw 3000 gallon tankers going up and down the road presents a different picture. Environmental protection requirements will continue to be high on the minds of the general public. As dairy facilities become larger and more complex the public will notice and the expectation for environmental protection will increase. Larger manure handing systems with more components, larger equipment, more hours of use, and more people involved will further highlight the importance of reducing hazards involved with manure systems. Larger herds increase the cost of an infectious disease and the complexity of controlling its spread throughout the herd.

The manure handling system is an integral part of any dairy facility design and management plan. Therefore, the best manure handling systems are usually developed as part of an overall plan for new construction, renovation or expansion. A well-planned system will be compatible with the type of housing and will include manure and wastewater from outside animal areas, the milking center, youngstock facilities, and silage effluent. It should also be compatible with anticipated changes in housing and management over the next 8-10 years.

A well-designed and installed manure handling system will include backup plans to allow continuous operation even if a key component malfunctions. Readily available spare parts or replacement equipment are required for dairies handling large amounts of manure or utilizing complex treatment or handling systems. Advanced thought should be given to how a large manure spill, resulting from a damaged storage or broken pipe, can be contained before large quantities of manure can run off the property or reach nearby streams or lakes.

Hazards

Manure systems present hazards from asphyxiation, poisoning, drowning, and machinery entanglement or entrapment. Pumps, pits and tanks can easily contain poisonous gases that will not be apparent until someone enters the tank and is overcome. Multiple deaths have occurred as a result of failure to follow appropriate procedures for working in these confined spaces. **Do not enter manure sumps, pits or storage tanks without appropriate safety apparatus and procedures.** No tool, pump part, or farm chore is worth a human life!

Another common hazard is failure to provide adequate guarding at manure tank openings and pushoff ramps to prevent entry by people, tractor scrapers, or cows. All open storages or openings into storages must have adequate fencing, guards, or covers to prevent visitors, including small children, from gaining entry. **Everyone has an obligation to design,**

supply, buy, operate, and maintain manure storage and handling systems that are safe for workers and visitors.

Complete Information

A brief paper in a proceeding cannot be considered a comprehensive resource on the variety and complexity of designing and managing a dairy cattle manure system. The reader is encouraged to make use of material available from a wide variety of education, government and private sources. These would include Natural Resource, Agriculture and Engineering Service (NRAES), Midwest Plan Service (MWPS), Natural Resources Conservation Service, Dairy Practices Council (DPC), American Society of Agricultural Engineers (ASAE), local cooperative extension offices, conservation districts, private consultants and equipment suppliers.

An excellent up-to-date reference on dairy cattle manure is *Guideline for Dairy Manure Management from Barn to Storage* (Weeks, 1998) available from NRAES or the DPC. A dated but comprehensive reference on manure treatment and handing systems is *Livestock Waste Facilities Handbook*, MWPS - 18.

Manure Characteristics

Dairy cattle manure as it comes from a milking cow offers a challenge for handling and storage. Its often said that its "too thick to pump and too thin to shovel."

The following material is abstracted from Weeks, 1998:
Manure is the feces and urine from farm livestock containing waste products from digestion and other bodily processes and is often described as:
- sticky (viscous)
- smelly (odorous)
- soupy, normal, stiff or dry.

Characteristics of manure are also affected by
- diet, development (age) and health of animals
- time (because of bacterial action)
- temperature (because of drying, bacterial action, freezing)
- added bedding materials and other materials
- added water and other liquids

Fresh dairy cattle manure is about the same density as water (about 62 pounds per cubic foot or 8.3 pounds per gallon). Addition of typical amounts of organic bedding materials or waste feed does not appreciably change this value. On the other hand sand laden manure from barns using sand bedding often weighs about 75 pounds per cubic foot or 10 pounds per gallon. Hauling and lifting equipment used with sand laden manure must account for this increase in density. Sand laden manure in a 5000 gallon tanker would weigh 25 tons. This is 4.25 tons more than 5000 gallons of liquid manure without sand bedding.
Sand laden dairy cattle manure presents a perplexing challenge to the managers, designers and suppliers of manure handling systems. Depending on moisture content, sand will settle

out in storages and pipes. The abrasive nature of sand reduces the service life of bearings, impellers, housings, pipes, scraper blades, cables, etc. Efforts to address the challenges of handling sand laden dairy cattle manure have included management changes; storages designed to allow clean out with both liquid pumps and front end loaders, back hoes or drag lines; and development of a specialized sand manure separation system based on a sand washing device. (Wedel and Bickert, 1994) (Stowell and Bickert, 1995)

Manure Production

Predicting quantity and quality of dairy manure is difficult and table values sometimes seem inadequate. In general modern high producing dairy cattle produce more manure than older tables suggest. Historically a thumb rule of manure production 8% of body weight was used for lactating dairy cattle. Recent studies would indicate that this value may be closer to 10%. (Weeks, 1998) Low producing cows, dry cows, growing heifer and calves will likely produce less. Use of table values must always be done with discretion and adjustments for local conditions. Table 1 provides an example of manure production from different categories of animals in a herd with 80 milking cows and representative dry cows and replacements.

Table 1 Manure production from all animals in atypical dairy herd with 80 milking cows plus dry cows and replacements. (Weeks, 1998)

Dairy Group	Number	Average Weight	Manure Factor (%)	Daily (lb)	Production (ft^3)
High milking cows	20	1,350	10.0	2,700	43.6
Mid milk production	40	1,350	8.5	4,590	74.0
Low milking cows	20	1,400	8.2	2,240	36.1
Dry cows	20	1,450	8.0	2,260	36.5
Bred Heifers	25	1,000	7.8	1,950	31.5
Young stock	45	500	7.5	1,690	27.3
Total	170	-	-	15,430	249.0

Moisture Content

Weeks(1998) describes dairy cattle manure by the following three categories:

Solid manure (16% or more solids) contains considerable fibrous bedding, easily travels up a gutter cleaner chute, and is easily handled with a front-end loader and a conventional or flail manure spreader. In most cases it can be stacked. Excess water (e.g. precipitation, from leaking waterers or runoff from roofs) must be kept out of the manure.

Semisolid manure (12%-16% solids) generally contains some bedding and can be handled with a front-end loader and a conventional or flail spreader. It will flow to some extent, but is too thick to agitate and pump from storage with liquid manure handling equipment. Increased amounts of bedding, or waste feed, make semisolid manure more solid. Precipitation or groundwater should be continuously drained away from the storage, otherwise semisolid manure becomes the consistency of liquid manure.

Liquid manure (12% or less solids) usually contains little or no bedding, and water may be added so that the manure can be agitated into a liquid consistency and handled with a liquid manure pump and a liquid manure spreader. If liquid manure is handled with irrigation equipment, considerable quantities of water must be added. Special high-pressure chopper pumps and large-orifice irrigation nozzles are also necessary.

The liquid manure category includes a wide variety of consistencies and sources. It can vary from semi-solid manure scraped from a freestall barn and mixed with milking center wastewater to more dilute water and manure mixtures from flush barns, to colored water that is washed off an outside lot during a rain storm. Unscreened dilute mixtures of manure water can still contain large particles that will plug pumps, screens and irrigation nozzles.

Sources of Manure
Manure on a dairy farm originates from four major areas:
Tie stall barns - cows spend considerable time restrained in bedded stalls with manure removed using mechanical gutter cleaners or gravity flow gutters. Manure will vary from solid to semi-solid depending on management of milking center washwater bedding and waste feed.

Freestall barns - cows rest in bedded freestalls and walk on concrete alleys to feed and water. Manure is deposited on concrete alleys as cows move about and is removed by scraping, flushing or falling through slotted floors. Manure will vary from semi-solid to liquid depending on bedding and waste feed management and manure removal method.

Bedded pack barns and bedded pens - animals are kept on a thick layer of organic bedding and manure removed periodically as a "solid type" material. In addition to the bedded manure there may also be manure produced in unbedded concrete alleys or outside yards where cows eat and drink.

Outside housing - can vary from rotationally grazed pastures, vegetative sacrifice or exercise lots, concrete feedlots, dirt exercise areas or large dirt corrals. Consistency of manure will vary from manure carried in water runoff in humid areas to dry dusty powder from corrals in arid areas.

Handling Systems
A complete manure handling and storage system allows for collection and removal of manure from animal housing areas, treatment if necessary, transport to storage system, short and long term holding or storage, transport to cropland, and land application.

Collection systems include gutter cleaners and gravity flow channels in tie stall dairy barns and tractor scrapers, automatic alley scrapers, flushing, and slotted floors in freestall dairy barns. Outside yards, lots, and feeding areas can be cleaned with scrapers or in some instances flushing. Bedded pack and pen areas are cleaned with tractor loaders.

Tractor scrapers with rubber edges or made from sections of large rubber tires provide less wear and polishing of concrete and tend to squeegee the floor. Metal blades or buckets with down pressure are more effective under freezing conditions. Manure may be pushed off an elevated lip directly into a spreader or pushed into a storage or collection gutter. In some cases it is pushed to an area with a buck wall to facilitate loading with a bucket loader.

Automatic freestall alley scrapers are labor savers and frequent operation provides cleaner alleys and cows. The cost and time required for maintenance of alley scrapers is often less than the total cost (labor, machinery, maintenance, injured animals) of daily tractor scraping. Unattended operation of alley scrapers where very small or new born calves could be dragged away by the slow moving blade is not recommended. Alley scrapers must discharge through a hole, over a collection channel, or off the edge of a storage. Locate and guard the drop off point for the manure to assure that people, animals or equipment will not inadvertently fall in.

Flush cleaning is a low labor method that allows for frequent cleaning and results in drier alleys and cleaner cows. Important components of flush systems are adequate water supply, water disposal system, elevations, slopes, pumps and pipes. Systems can successfully operate much of the year, even in cold climates, if adequate facilities are available to take care of storage of extra water. Access for tractor scraping is recommended for periods when the flushing system can not be used. The most common problems with flushing systems are the quantity of water required and separating solids for reusing water. First time flushing system users are often overwhelmed by the amount of water that must be handled and the need for more dilution water in recirculating systems than expected. Criteria for satisfactory flushing include alley slope, water volume per flush, flow rate, duration of flush, velocity of water, and depth of water. In general, a 3-inch depth of water and 5 feet per second velocity are recommended. A 3% alley slope is often considered ideal. Steeper slopes will require more water and a higher flow rate, shallower slopes will require a high rate of water to maintain velocity. Water can be supplied from tip tanks, reservoirs with large gates that open or delivered through large pipes from high volume pumps or elevated holding tanks or ponds.

Slotted floors provide a method for immediate removal of manure from the animal area. Once beneath the floor, manure may be stored in an underfloor tank or removed by an automatic scraper, flushing, or a gravity flow channel. Excellent ventilation at all times is critically important in slotted floor barns with underfloor manure holding. Manure stored under barns can result in gas, odor and moisture problems in barns that are not adequately ventilated. Under-ventilation often occurs during cold winter months as a result of efforts to raise interior temperatures to minimize manure freezing or for operator comfort. During extreme hot weather, increased microbiological activity in the stored manure can increase the release of gases. During agitation and clean-out, keep animals (especially young animals that are close to the floor) and people out of enclosed barns and provide maximum ventilation if a manure tank is located under the floor. Floors may be configured with long parallel slats and slots, or oblong holes in a so called waffle pattern. Given a choice of any type of slotted floor or a solid surface, animals will migrate to the solid surface. Slotted floors allow urine to drain quickly away and manure is pushed through the slots by animal traffic. The result is a drier environment for the cows' hoofs. In extremely cold situations manure will eventually freeze and not go through the slots. Provide access for a tractor scraper to remove manure during cold weather.

Removal systems move manure from the barn to the field for immediate application or to storage. For immediate field application the manure may be loaded directly into a manure spreader from a gutter cleaner discharge or push off lip, or loaded with a front end loader. More liquid manure, may be pumped from a collection channel, sump, or small tank that holds a day or two manure accumulation. If manure is to be pumped into a liquid spreader, wastewater from the milking center is usually included to make agitation and pumping easier. Manure can be conveyed to a storage located at the barn by the gutter cleaner, tractor scraper, large piston pump, centrifugal pump, gravity flow pipe or gravity flow channel. Manure can be transported to satellite storages located away from the barn by trucks, large spreaders or pipelines.

Manure storage systems

Manure can be stored at the barn or near the cropland where it will be utilized. If manure must be transported long distances, it is often more effective to provide satellite storages near the cropland. This allows the manure to be moved to the storage during low labor periods and makes for more efficient spreading. Liquid manure, especially if solids have been removed, can be pumped long distances through buried or temporary pipes to satellite storages. The storage must be compatible with the form manure is removed from the barn.

Heavily bedded manure can be easily stacked in three sided bunker type storages or on simple hard surfaced or packed crush stone or fly ash stabilized pads. Storages should be designed to prevent clean water from running into them and to direct any leachate or runoff water from the storage away from streams to vegetative filter areas or holding ponds.

Liquid manure can be stored in properly designed concrete or steel tanks; concrete, clay or membrane lined in-ground storages; and in some cases earth storages. It is critical that storages be located and constructed to assure that manure will not seep through storage walls or bottom to ground or surface water. Consult the USDA Natural Resource Conservation Service or a qualified soils engineer regarding location and design of in-ground manure storages. An annual inspection and maintenance program should be in place to assure continued safe operation of any type manure storage. **Liquid storages require appropriate signs, guards and fences to protect workers and visitors from unintended entry and possible loss of life.**

Treatment Systems

Various treatment systems have been proposed for use with dairy cattle manure. Typical reasons for treating manure include: ease of storage or transport, reducing odor potential, extracting energy, and concentrating, partitioning or removing nutrients. A comprehensive discussion on manure treatment *Manure Treatment and Handing Options* by Peter Wright is found later in this proceedings.

Solids separation will allow easier handling of liquid material, allows for recycling of water for flushing, and can provide a useful by-product. Separation systems can be categorized as gravity, screens, extruders, centrifuges, and cyclones. Various additives have been proposed to enhance the separation process. Settling tanks or basins use gravity and time to allow larger particles of liquid manure to settle or float out. Screens normally require some method to prevent particles from plugging or blinding the screen openings. This may be accomplished by sloping the face of the screen so material slides off or with mechanical scraping or vibration. Screens require a dilute material similar to that obtained with flushing systems. Extruders use screws, plungers, or belts to pack manure against a perforated cylinder, box, belt or plate. Liquid is forced out through the holes and the solids are discharged out the end. These devices tend to provide a drier solid and some will work with a less dilute, slurry consistency manure. Centrifuges may be compared to cream separators and use a rotating chamber to force solids to the outside for removal. Cyclone separators are similar to dust collectors. A very dilute flow of manure is introduced into a conical chamber that encourages large particles to separate from the liquid.

Anaerobic digestion or biogas production has been used by some farms as a method to extract energy from manure and reduce odor that results from long term liquid storage. The biogas process converts complex organic material such as manure into biogas and low odor effluent. A heated digester with a 15-25 day detention time is normally required. The primary constituents in biogas are methane (natural gas), carbon dioxide and trace gases. Originally the primary interest was in energy production, usually electricity. Economics tended to favor farms with more than 200-400 cows. Even at this size most farmers chose not to bother with the extra expense and management requirements. Recently there has been interest is biogas digesters as a method to solve odor and nuisance problems associated with storing and handling large quantities of dairy manure. The process does not significantly change the amount of manure or the nutrient content, but does alter the form of the nitrogen. The effluent will be more liquid and homogeneous as a result of the digestion process.

Composting is another form of manure treatment that dairy farmers consider to improve handling, enhance marketability, or reduce odor and nuisance problems. Composting is an aerobic process that requires a material with good porosity, 40-60% moisture content and proper carbon to nitrogen ratio. Some form of mixing and or aeration is required to provide satisfactory composting. The process generates its own heat, reaching temperatures of 120-140°F. When properly done, composting will eliminate most odors and result in a stable easily handled dry humus like material. Dairy manure or separated solids may require some form of dry carbonaceous material such as straw, sawdust or wood chips to provide porosity and maintain the desired moisture content and carbon nitrogen ratio. Some dairy farmers have found an additional income source by charging municipalities or industries to take materials such as paper, cardboard, tree trimmings, etc. This is then mixed with the manure and after composting either spread on land or sold as compost.

Land Application
Solid or semi-solid dairy cattle manure or separated solids are land applied with tractor pulled or truck mounted box or V bottom spreaders. Manure may be spread whenever it is removed from the barn or periodically throughout the year from manure storages. The manure is applied to the surface of crop or pasture land. Some managers choose to cover the manure immediately by plowing or disking to minimize nitrogen volatilization and odor losses. Spreaders are loaded directly from barns during cleanout using front end loaders, pushoff lips, gutter cleaner discharges, cross conveyors or gravity flow discharge pipes. Manure storages are unloaded into spreading equipment using front end loaders or gravity discharge pipes.

Liquid manure and wetter semi-solid dairy cattle manure is land applied using tractor pulled or truck mounted liquid tankers. The manure may be applied to the surface. To minimize odor and ammonia volatilization manure can be injected below the surface with special attachments. Manure tankers are loaded using centrifugal type liquid manure pumps, augurs, or gravity discharge pipes. Dairy manure storages can form thick floating crusts in addition to sludge layers of thicker manure. To agitate and homogenize the manure, special agitators or recirculating unloading pumps are required.

When liquid manure must be transported long distances it may be hauled to the field in large over the road trailers and then directly pumped into field spreaders or discharged into small holding containers for reloading.

Very liquid dairy cattle manure or liquids from solid separators can also be pumped through portable or fixed pipelines to fields for application to the surface using special large orifice irrigation sprinklers or drag hose systems. While a variety of irrigation systems can be used the most common is a traveling big gun type sprinkler. Liquid manure is supplied to the sprinkler with a large diameter soft or hard hose system. The sprinkler is attached to a wagon that is pulled across the field with a cable and winch or by the supply hose being wound up on a large powered reel. Problems with irrigation include, plugged nozzles, application of excess manure nutrients, runoff of liquid manure resolution from applying manure faster than the ground can absorb it and dispersion of odors off the farm. (Jarrett & Graves, 1999) (NRAES, 1994)

An alternative to irrigating liquid manure is to hook the delivery hose to a special tractor pulled device. The device will either inject the liquid manure directly into the soil or discharge the manure close to the ground in front of a tillage tool such as a disk harrow or field cultivator. The tractor is driven back and forth across the field in a pattern so that the hose is pulled along without being run over by the tractor.

Biosecurity

Manure can become a means for transfer of diseases around a farm. Diseases that can move between animals by the "fecal oral" route can be transmitted by the mismanagement of manure. Typical paths include sharing common equipment (front end loaders, scrapers, tractors) between manure removal and feeding chores. Manure scrapping and hauling routes that allow spillage from equipment, tires etc. directly onto feed floors, feed storage or where feed vehicles travel can be another path. Manure handing equipment or manure from infected animals that move through or among susceptible animals is another path of infection. The people responsible for herd health on a dairy farm must be included in decisions concerning manure handling and evaluate the likelihood of biosecurity problems.

Environmental Affects - land , air, water

There are various ways that manure handling systems affect the environment near and away from a dairy farm. The tendency is to focus on water pollution. However, dairy farms also can affect land and air quality.

Organic, nutrient and microbiologic constituents in manure can cause pollution if there is an uncontrolled release of dairy cattle manure into ground or surface waters. Obvious points of release include leakage from storages or pipes to ground water. Residuals from manure can also be carried to ground water as the result of excessive application of manure to cropland or accumulations of leachate from barnyards or load out areas.
Surface water pollution can result when there is over application of manure (typically from irrigation) and direct runoff to drainage systems. Rainfall events can flush manure from fields, outside animal yards or piles of manure and also cause liquid storages to over top. Catastrophic failure of a liquid storage structure is another potential source of water pollution.

Excessive quantities of manure resulting from heavy applications to crop fields or intensively populated dirt exercise areas can degrade soil quality. In arid climates, salt build up can occur.

Odors and ammonia emissions from barns, surfaces where there are manure accumulations, or manure storages can affect air quality. Spraying manure thorough the air, as in irrigation systems, can result in gases and particulates and aerosols being carried off the farm in the air. Before land applied manure is incorporated, odors, ammonia or other gases can also escape to the air.

Summary
The dairy farm of the future will have more cows kept in a more concentrated area. A larger percentage of feed nutrients will be hauled to the production unit from someplace else. Non-farm people and regulators will be more aware of, and more interested in dairy production systems. Odor problems and concern about the destination of "waste products" from the farm will intensify. The siren call of black boxes, quick fixes and magic powders will continue. Highly focused regulators and politicians looking at one small portion of this big picture will continue to look for the "one size fits all fix." Even after spending millions of dollars municipal sewage treatment plants end up putting more and more of the final product from their "treatment systems" on the land, where mother nature provides the ultimate disposal/utilization. "Everything has to be someplace!" A dairy/crop farm with adequate cropland would seem to be the ideal model for municipalities and industries to be following instead of the other way around. Animals eat crops to produce milk. Manure (and many of the nutrients) are discharged from the animals and subsequently used to grow crops to feed to the cows. If the manure is kept out of surface water resources there is no need to worry about the oxygen demand nature of this organic material. The same material that competes with fish for oxygen in water enhances soil structure when applied to cropland.

Manure handling problems resulting from more animals in one place and the decreasing distance between animals and people require more than technology to solve. Decisions and policy beyond the control of the farmers, as diverse as transportation subsidies and pricing and land use planning have resulted in major dislocation between crop production and milk production. Successful managers will have a more global view of their operations. They will understand the relationships between feeding strategies, housing strategies, manure handing strategies, feed production strategies and being a good neighbor. A well-planned manure handling system can result in a more efficient and environmentally friendly dairy

References
Dougherty, M., L. D. Geohring, and P. Wright. 1998. Liquid Manure Application Systems Design Manual. NRAES 89. Northeast Regional Agricultural Engineering Service, Cooperative Extension, 152 Riley-Robb Hall, Ithaca, NY 14853-5701. 168 pp.

Jarrett, A. J. & R. E. Graves. 1999. Irrigation of Liquid Manures. Penn State Agricultural and Biological Engineering Fact Sheet, 246 Agricultural Engineering, University Park, PA 16802. 11 pp.

MWPS. 1985. Livestock Waste Facilities Handbook. Midwest Plan Service, MWPS-18. Ames, IA. www.mwpshq.org 112 pp

NRAES 1994. Liquid Manure Application Systems, Design Management and Environment Assessment, NRAES 79., Northeast Regional Agricultural Engineering Service, Cooperative Extension, 152 Riley-Robb Hall, Ithaca, NY 14853-5701. 168 pp.

Stowell, R. R. and W. G. Bickert. 1995. Storing and Handling Sand-Laden Diary Manure. Extension Bulletin E-2561. Michigan State University, East Lansing, MI. 16 pp.

Wedel, A. W. and W. G. Bickert. 1994. Handling and Storage Systems for Sand-Laden Dairy Manure. IN Dairy Systems for the 21st Century Proceedings of the Third International Diary Housing Conference. ASAE-02-94. American Society of Agricultural Engineers, St. Joseph, MI. p. 715-723.

Weeks, S. A. 1998. Guideline for Dairy Manure Management from Barn to Storage. Dairy Practices Council, DPC 27. Keyport, NJ www.dairypc.org. Natural Resources, Agriculture and Engineering Service, NRAES-108. Ithaca NY www.nraes.org. 36 pp

Additional Information Sources

ASAE 2950 Niles Road, St. Joseph, MI 49085-9659. (616) 429-3852 http://asae.org

DPC - Dairy Practices Council, 51 E. Front Street, Suite 2, Keyport, J 07735. (732) 203-1947 www.dairyp.org

MWPS - Midwest Plan Service, 122 Davidson Hall, Ames IA 50011 (800) 562-3618 www.mwpshq.org

NRAES - Natural Resources Agriculture and Engineer Service, 152 Riley-Robb Hall, Ithaca, NY 14853-5701 (607) 255-7654 www.nraes.org

NRCS - Natural Resources Conservation Service, National Technical Information Service, US Department of Commerce, 5285 Port Royal Road, Springfield, VA 22161 703 487 4600 www.nrcs.usda.gov/

Manure Treatment and Handling Options

Peter Wright
Animal Waste Specialist
Agricultural and Biological Engineering Department
Cornell University

Biographies for most speakers are in alphabetical order after the last paper.

Agriculture is a vital component of any society. In North America agriculture provides wholesome cheap food to consumers. Farms are also important environmentally. Open space, wildlife habitats, and aquifer recharge can be important environmental benefits of farms. However, excess nitrates in the ground water, pathogens in the drinking water and excess nutrients, BOD, and sediment in surface water can have a negative effect on the environment. Farms can affect the environment through odors, as well as produce gases that contribute to the greenhouse effect and acid rain. Society has recognized some of these negative effects and is asking farms to improve. To maintain a competitive industry we need to be able to provide feasible alternative practices based on science and good engineering that allow productive agriculture while minimizing the effect on the environment.

There are a wide variety of farms. They vary in their resources and their environmental concerns. Some farms have access to more capital, skilled labor, management ability, land resources, water resources, and markets than other farms. Different manure treatment and handling methods will be needed to match the resources and needs of different farms.

Depending on the location and the management's personal values, each farm can have different environmental concerns. Those in a watershed that supplies drinking water may be more interested in controlling pathogens and phosphorus. Those upstream of a fresh water lake may be more concerned with sediment and phosphorus. Those with close or sensitive neighbors may be more concerned with odors. Those in a porous aquifer may be more concerned with nitrogen leaching and pathogens. Others may only be concerned about BOD loading that cause fish kills locally. Nutrient loading far downstream may be a concern to some farms. Manure treatment methods will be required to deal with each of these issues.

Some farms are interested in mass reduction to facilitate manure movement off their farm. Development of by-products that can be sold at a profit off the farm could help some farms maintain profitability while improving the environment.

There are many management issues that affect the choice of a manure treatment system Some of these issues include the desire to 1) minimize environmental damage, 2) maximize nutrient value, 3) minimize neighbor problems, 4) minimize damage to the land, 5) minimize cost and 6) minimize frustration. Although society may order these issues one through six, farmers may order them six through one. Manure management alternatives need to address these concerns.

There will need to be a variety of treatment methods that work with the variety of resources the farms may chose to allocate to them. Table 1 describes some manure management alternatives that either are being used or are proposed.

Table 1: Manure Management Alternatives

Manure Management Alternative	Advantages	Disadvantages
Daily Spreading is being practiced by many farms. Manure and other wastes are spread as they are produced throughout the year.	Capital costs are low. Environmental effects are hidden. Odor problems are minor. Labor and equipment use is steady.	Total costs may be high. Nutrient and pathogen losses during times of saturated soils may provide excessive delivery to waterbodies. Field accessibility may be a problem.
Storage to reduce spreading during periods of high nutrient loss and times when fields are inaccessible. Required in many areas, encouraged in all areas.	Nutrient management can be easier. Efficiencies in handling can be obtained to keep costs down. Manure can be spread when needed. Storage of solids is safer environmentally than liquid storage.	Odors are a big problem when spreading especially from a liquid storage. Large handling equipment needs to be available. Labor and equipment needs peak. Non-earthen storage can be very expensive. Dry storage may need an expensive roof or runoff controls. Catastrophic failure of liquid storage or heavy rainfalls right after spreading can cause peak pollutant discharges.

Manure Management Alternative	Advantages	Disadvantages
Odor Control of stored liquid manure is a major need. Chemical and biological treatments have been tried and proposed.	Would allow spreading of the manure during the growing season and eliminate neighbor complaints.	No technology has yet shown that it can significantly reduce odors without significant costs.
Solid Separation of the manure solids mechanically can produce a "solid" portion (15-30% DM) and a "liquid portion" (4-8% DM).	Liquids are easier to handle. Solids can be recovered for bedding, soil amendment, or exported off the farm.	High capital and operating costs. Maintenance of the equipment is a problem. Marketing of the solids may not be successful on all farms.
Composting dryer manure directly, or by adding bedding or an amendment to wetter manure, a biologically decomposed product can be produced successfully on some farms.	An excellent way to treat dryer manure. Odor reduction is an important advantage of well managed composting. Equipment for solids handling is available on most farms. Material may be marketed.	High moisture contents of some manure makes conventional composting difficult. Sales may depend on expensive specialized mixing equipment and good management. Composting outside on large areas can create runoff losses. Composting inside may take a large capital expense.
Biodrying of the manure by recycling dry compost as the amendment to composting, and using the heat generated in the aerobic decomposition to dry the manure/compost mix with forced air has been proposed.	Odor reduction, volume reduction, and weight reduction would occur. Equipment for solids handling is available on most farms. Storage of solids is safer environmentally than liquid storage. Material may be marketed.	Management of drying process will be critical. Costs of operation may be high. Material handling may be excessive. Additional amendment may be required. Winter operation may require closed buildings.
High Solids Anaerobic Digestion would produce a decomposed residual and produce methane gas. Heat from the gas or from an engine generator could be used to dry the material for recycling within the system. This system has been tried experimentally on dairy manure.	Odor reduction, volume reduction, and weight reduction would occur. Equipment for solids handling is available on most farms. Storage of solids is safer environmentally than liquid storage. Material may be marketed. Energy production would meet the needs of the farm and allow excess to be sold.	Management of digestion and drying process will be critical. Capital costs will be high. Electric utility connections may be difficult. Material handling may be excessive. Additional amendment may be required.

Manure Management Alternative	Advantages	Disadvantages
Anaerobic Digestion takes as produced manure and digests it, producing an odorless effluent that has reduced solids content while retaining the nutrients. Methane gas is recovered that can be used to run an engine generator.	Odor reduction and energy recovery will occur. Effluent is reduced in solids content and can be further reduced easily by mechanical solid separation. Demand for the anaerobically digested solids is greater than raw solids.	Management of digestion process will be critical. Capital costs will be high. Electric utility connections may be difficult. Diluted manure will require a large digester and more heating.
Lagoon Treatment of manure from the farms consists of diluting the manure, removing solids mechanically or allowing them to settle in large shallow pools then flow to a facultative lagoon to be recycled as flush water to dilute more manure. Liquids and solids are periodically removed from the system.	Odors are reduced and solids are separated. Works well with a flush system to remove manure from barns. Solids may be marketed. Liquids can be easily irrigated. Management is relatively easy.	Solid harvesting and dewatering can be difficult or expensive. Exposure of large surface areas may result in extra water volumes. Impermeable soils on moderately flat terrain are required to keep cost down. Aeration is sometimes required.
Sequencing Batch Reactors to reduce the COD, N, in the liquid effluent and concentrate the P from the manure have been proposed. A large tank(s) would alternately fill, react, settle and decant a treated liquid and concentrated sludge. Mechanical separation and dilution would precede the process.	Odors, Nutrients and COD would be reduced in the liquid effluent which could be spray irrigated at hydraulic loading rates on crop fields. High P solids could be exported so that a dairy would not be tied to a large land requirement based on manure disposal limits.	Capital and operating costs may be high.
Total Resource Recovery by combining the plug flow methane production process with solid separation, and hydroponically recovering the nutrients would eliminate the waste and maximize production of useful by products.	Odors would be controlled. Energy would be recovered. Nutrients would be recycled. There would be no waste.	Capital costs will be very high. Operating costs may not offset by-product sales and savings in a cheap energy, cheap nutrient situation.

Some of these existing and potential treatment methods have not yet been implemented on a farm. The time to implement in Table 2 gives relative estimates of when these systems could be available for farm use. Each system may be appropriate for some farms and not others. The size of the farm may determine the applicability of one system over another. The system and the specific pollutant(s) they will treat need to be balanced with the specific pollutant control needed on the farm, and the needs of the farm for an efficient manure handling system. The management skills of the farm as well as the closeness (and marketing skills of the operator) to a market will also have an influence on the choice of a system. Table 2 lists some of the characteristics of each system.

The descriptions above provide background and a basis for comparing these manure management alternatives. Without a regulatory incentive to control nutrients, pathogens, or odors, it will be hard for an economically rational farm manager to increase production costs to implement some of these alternatives.

Concentrating research on those alternatives that will provide odor control, P concentration, and pathogen control at a reasonable cost should be a priority. More documentation of the costs, pathogen removal, and phosphorus concentration are important for any alternative considered.

System Descriptions:

Daily Spreading

Description: Spreading manure and wastewater daily as it is produced is a low capital cost, low management option. Many farmers continue to daily spread most of their manure. Nutrient management to reduce fertilizer use can be difficult unless careful records and accurate spreading tactics are used. Spreading close to the source of the manure has caused soil buildup of excess phosphorus on many farms. Unfortunately, soils, particularly those close to the barns, may have already been saturated with phosphorus, contributing to the water quality problems. While odor issues are generally not a serious problem on daily spread sites, runoff and leaching losses during saturated conditions can add to the nutrient loading of a watershed.

Environmental Impact: There are a number of important questions which need to be answered in order to document the environmental effects of daily spreading. Critical among them is the transfer rate between soluble P and bound P in the soil. Manure applications must be managed in such a way that almost all of the P is adsorbed in the soil and bound so that it cannot readily escape. Field and laboratory research will need to determine the appropriate spreading rates and frequency on different soils.

A continual year round spreading on fields regardless of the conditions can incrementally deliver a major portion of the nutrients to the environment. Pathogen losses during saturated moisture conditions can occur. Losses from large runoff events may be less than from farms spreading stored manure depending of the timing of the spreading in relationship to the runoff event.

Table 2: Relative Characteristics for Manure Treatments

Scale: 1 = poor   10 = good

Characteristic	Daily spread	Liquid Storage	Odor Control	Solid Separation	Compost	Biodrying	High Solids Methane	Methane	Treatment Lagoons	SBR	Total Resource Recovery
Runoff and leaching	1	4	5	5	6	7	7	6	8	9	10
Odors	5	1	5-10	2	9	9	9	10	8	10	10
Small farm	5	3	?	2	5	7	6	3	6	4	5
Large farm	2	7	?	6	2	5	6	8	8	7	6
N reduced	5	4	n/a	5	4	4	4	6	7	10	10
P export	2	1	n/a	3	5	5	5	4	6	9	10
Pathogen control	1	3	?	3	7	7	7	6	5	8	10
Nutrients recycled	5	8	6	7	4	4	4	9	2	2	10
Compaction	1	5	6	6	8	8	8	9	9	10	10
Capital Costs	9	3-7	?	5	7	4	2	2	4	3	1
Operating Costs	5	7	?	5	3-8	?	3-8	3-8	5-7	2-4	?
Material sales	2	3	4	6	9	10	10	7	6	8	10
Time to implement	10	9	1-9	9	9	5	3	7	7	3	1
Simplicity	10	8	1-9	6	7	5	2	2	5	3	1

Manure Handling Options will have different relative values on different farms. Of course every farm is different both in their resources and their goals. This scale is an attempt to compare the systems with each other. For a specific farm it would have to be reevaluated to reflect actual conditions for that farm.

Economics: There are many misconceptions about the cost of daily spreading. Some studies have shown a wide range of costs. When no fertilizer credit is taken for the manure and inefficient methods are used to spread it the costs approach $200 per cow per year. Efficient spreading and using the fertilizer value of the manure can produce a positive value on some farms. The average farm in one study in western NY lost $75 per cow per year on their spreading operation.

Land Requirements: Daily spreading is a relatively land-intensive operation, especially if phosphorus limitations are placed on the spreading rate. Specialized application vehicles to move more manure further may be needed.

Management Strategy: This alternative can be implemented on individual farms, and in many cases can be accomplished using existing manure spreading equipment. Custom spreading may be appropriate for fields at considerable distances from livestock housing. Additional record keeping will be needed on most farms.

Acceptability: Because it is a traditional manure management practice, farmer and community acceptance of this alternative is likely to continue to be high. However, longer hauling distances, less efficient use of nitrogen with P balancing, and pathogen concerns may discourage this option.

Storage

Description: Storing manure and wastewater to spread it at an environmentally appropriate time can reduce the total loading of nutrients to a watershed. However, most farmers don't limit their spreading from a storage to times when the environmental effects will be minimized. Labor and equipment availability results in considerable application during the fall, winter and early spring. Nutrient management to reduce fertilizer use is easier when spreading from a storage. Records can be kept and spreading patterns in each field observed to be sure of more uniform coverage. There can be cost savings both in reduced fertilizer use as well as efficient use of equipment when spreading from a storage. However unroofed liquid storages may gain considerable precipitation that also needs to be spread. Roofed systems for dry manure can be expensive. Odor issues are a serious problem when spreading stored manure.

Environmental Impact: Management is the key to reducing the environmental impact of spreading stored manure. Odor problems create a continued pressure to avoid the warmer times of the year to spread. Catastrophic failures can result when liquid storages are breached, or when a large runoff event washes all the manure off the land after the manure storage was emptied on to many fields.

Economics: Storage can be an economic positive for the farm by allowing efficient spreading operations and the use of the manure as a fertilizer. Incorporating the manure to preserve the ammonia as a fertilizer is an option for all of the manure when spreading from a storage. Delays in planting or other operations due to the time it takes to complete the manure spreading operation can cause a large loss in net farm income. Weather variability makes spreading a year's worth of manure in the spring prior to planting a difficult task.

Land Requirements: Spreading from a storage can take even more land than daily spreading if it is incorporated, as the nitrogen can be utilized more completely. This will make it a relatively land-intensive operation, especially if phosphorus limitations are placed on the spreading rate. Less nitrogen will need to be added to a phosphorus balanced rate if it is incorporated. Specialized application equipment to move more manure further may be needed.

Management Strategy: This alternative can be implemented on individual farms, and in some cases can reduce spreading costs. Custom spreading may be appropriate on some farms. Management is very important when unloading a storage to prevent spills or overloading of fields. Odor control of the stored manure is an issue that each farm with stored manure needs to consider.

Acceptability: Because of the increase in foul odors the community may not accept this practice even when it can be shown to improve water quality. Farms that can reduce their spreading costs, or improve the ease of spreading operations may be willing to put up with some odor complaints. People will be less and less willing to accept the odors as the spreading is concentrated more in the warmer months and concentrated on larger farms.

Odor Control

Description: Many processes have been proposed to treat stored manure to reduce the objectionable odors. These include some of the treatment process contained in this document as manure management alternatives. Other biological, physical, and chemical proposals include: specific enzymes and bacteria, aeration, chemical reactants, heating, drying, raising the pH, chemical masking, magnetism, and electric currents. So far a process that is effective yet low cost has not been found. Research continues world wide to provide a solution.

Environmental Impact: Odor control is essential to allow farms to spread on land close to residences and during the summer. Potential leaching and runoff concerns during the cooler wetter times of the year are pushing those farms interested in water quality into a conflict with their neighbors as they attempt to spread manure during the summer. Spray irrigation on growing crops of an odorless manure would be the best method of manure application to provide nutrients to the crop while minimizing environmental losses.

Economics: The value of odor control is a hard quantity to define. Avoiding neighborhood conflicts, increased quality of life, and avoiding potential law suits does have a value. Regulations on odor emissions are a real possibility in the future. These regulations will make some form of odor control needed on many farms.

Management Strategy: Although there are some management techniques to reduce odors and avoid neighbor conflicts, the odors from stored manure without treatment will challenge even the best managers.

Acceptability: Any system that is low cost and easy to manage will be adopted rapidly by farms.

Mechanical Solids Separation

Description: Separating and hauling the separated solids from the manure produced on each farm could potentially allow for the export of approximately 20% of the phosphorus. There are a number of separators commercially available. Most require the addition of extra water. The screw press separator manufactured by Fan seems to work on slurry manure without additional water added and provides a fairly dry product. It will produce at least one cubic foot of 30% solid manure for every minute of operation on a dairy farm. That rate will handle about one cow's daily manure production per minute. This rate may be increased, depending on the size of the solids, the moisture content of the manure slurry, and the internal wear on the auger vanes. A truck mounted unit or units could go from farm to farm on a regular schedule separating the manure for export, or composting. A storage facility on each farm would need to handle the daily manure produced as well as the liquid remainder. This system is being used in Europe.

Environmental Impact: Since the amount of phosphorus in the separated manure solids is 20% of the total phosphorus in the manure, removal of the solid produced from the separator would only reduce the loading of phosphorus by 20%. It would have no effect on the pathogens. Nitrogen would also be removed by 20%.

Economics: The separation equipment costs about $30,000 for each machine. If permanently installed on a farm it would take another $25,000 for the building pipes and plumbing. One machine would be required for each 500 cows assuming an 8-hour day. The ownership costs which include depreciation, insurance, a 15-year life, and interest at 10% would be $3.50 per hour. The Fan separator uses 4 kW to turn the auger, and 0.15 kW to run a vibrator to keep the manure entering smoothly. Assuming a 10 hp manure pump motor uses 8 kW, the electrical use per 500 cows would be 8.3 hours times 12.15 kW or 102.5 kW hours per day. Maintenance on the separator is estimated at $2.50 per hour. The economics of truck mounted multiple units would have to be explored. Liquid manure handling equipment would need to be obtained for each farm left with the liquids.

Land Requirements: Two small three to four day storages for each farm would be needed to store the manure until it was separated and then to store the liquids until they could be spread. Vehicle access to these storages would be needed if portable separators were to be used.

Management Strategy: Farms that are not large enough to justify a dedicated separator could use a private operator to set up an operation to separate the manure, haul the solids to a central site for composting or high solids anaerobic digestion, and then market the product. The profit from the sales if any could be split between the farmer and the private operator. Large farms could use the outside expertise to market the solid by-product.

Acceptability: The farmers would need to convert to a liquid manure spreading system. If all the solid manure were to be moved off the farm, this would not be that big a burden. The storage and spreading of the liquid is a little easier to manage than handling a solid. More uniform applications could be expected as well as slightly more N retention since the liquid portion will infiltrate into the ground sooner. There may be a slight odor reduction because the liquid manure will infiltrate into the ground quicker.

Composting

Description: Composting is an aerobic decomposition process that is an established on-farm manure management method. The energy liberated during the decomposition process raises the compost temperature to accelerate decomposition and evaporate water, resulting in a dried, stabilized product in 3 to 6 months. A primary constraint on composting is the requirement for a relatively dry manure or a dry bulking amendment to create adequate porosity in high moisture manures. This requirement can be reduced or eliminated via the use of solid separation technology (see above) and/or recycling dried product in the mixture (see Biodrying).

Shared mobile composting equipment could make composting more feasible on small farms. Transport to a centralized facility for composting will depend on the tradeoff between equipment ownership and manure transportation costs. Product marketing would likely be a centralized function performed by a private contractor, who would also provide equipment maintenance and management services.

Environmental Impact: Export of the compost product would result in removal of phosphorous and nitrogen from the farm. To the extent compost was used on the farm, phosphorus would remain while nitrogen would be reduced through ammonia volatilization by 30 to 70%. In either case the high temperatures achieved during composting would greatly reduce or eliminate pathogens. The amount of manure composted could include only the excess intended for export, or it could include the entire manure stream as a way to treat manure before returning it to cropland. Application rates could be limited to the phosphorus required by the crop, so that nitrogen fertilizer would need to be imported. Odors will be controlled with proper management.

Economics: The economics of on-farm composting have been documented through a number of studies, including case studies from throughout New York State. Combined amortized capital and operation costs typically range from $5 to $20/ton. The value of the compost product will offset that cost to a limited degree, although marketing expenses will also need to be considered. Tradeoffs between material transport (bulking amendment and manure) and equipment transport will be analyzed for specific clusters of farms to determine the optimal scale of a facility. Marketing of compost could become a problem since there is and almost unlimited supply of raw materials.

Land Requirements: The composting site(s) could be outdoor windrow systems. These would have to be designed with adequate water quality protection measures. Retention ponds and pasture irrigation systems have been used successfully for runoff management at other farm composting facilities. Typical windrow composting operations require one acre for every 3000 yards of material. Including some space for runoff management, approximately 1.5 to 2 acres would be required for a 50-cow dairy. Some producers of higher solid manure have used in-channel composting effectively to compost manure in 21 days. These systems require a building where compost is treated in three-foot high concrete channels. Air is blown into the compost and automatic turners mix and move the material through the system.

Management Strategy: Several management options are possible for this technology. In the scenarios proposed, a subsidized private contractor could manage the manure for several farms, either moving equipment to individual on-farm sites or hauling manure to a centralized facility. In a decentralized processing scenario, equipment would need to visit each site regularly, with increased frequency during the summer and early fall to provide relatively dry finished compost for storage and sale the following spring.

Acceptability: Composting is an established manure management approach, and multi-farm implementations already exist in the Northeast. Several sites operate centralized facilities collecting manure from many farms, while one operation contracts compost turning and management services for several farms. Although these existing models demonstrate widespread acceptance of composting by farmers and private businesses, they also provide indications of the critical issues for community acceptability. Centralized facilities, because of their larger impact on the immediate neighborhood, are more likely to raise concerns about odor and traffic. If economic and transportation analysis suggests a centralized solution, these concerns must be carefully addressed.

Biodrying

Description: If managed carefully the heat generated by aerobic composting can provide the energy to reduce 12% DM manure to a 60% DM residual. Forced air composting, under a roof, with the air flow controlled carefully would optimize this process. Composting works best with an initial moisture content below 70%. Recent applications of composting operations that have used forced air to compost six foot high layers of manure in 21 days have shown the feasibility of this process. Recycled compost at 40% dry matter could be spread in the alleys of dairy farms about 3 inches thick to absorb one days production of 12% DM manure. The mixture could be scraped into a shed, piled 6-8 feet deep and aerated to produce 40% DM compost in 3 weeks. This recycle loop could be continued indefinitely. One third of the compost produced each day would not be needed to be recycled and would be stock piled for sale or land application on the farm. This process could potentially compost all of the manure produced with little additional amendment needed. The compost would be reduced one half in volume and one sixth in weight from the original manure due to water loss and solid conversion to gasses. Mechanical biodrying systems have been pilot tested for swine manure.

Environmental Impact: Pathogen control and odor control would be substantial. Heat produced during the compost process has been shown to reduce pathogen viability substantially. The aerobic nature of the composting process produces few odors if managed correctly. Storing and spreading a high solids product should reduce the runoff potential and eliminate the potential of a catastrophic failure from the storage system.

Economics: The capital cost for this system would consist of a three-sided composting shed with an aeration system installed in the floor. Most farms have the needed material handling equipment on the farm. The additional material handling, amendment if needed, and power for the aeration equipment would be the operating costs. These costs may be offset by sales of the product or by use of the compost as bedding.

Land Requirements: The composting shed would need to be large enough for 21 days storage of the compost manure mix piled 6-8 feet high. Additional storage for the excess compost could be provided on a pad with controls for rainwater runoff.

Management Strategy: Although the air control/temperature feedback system may need to be automated to optimize moisture removal, the rest of the system is well within the management capabilities of most dairy farm operators. Solid handling of odorless compost should make manure spreading much easier.

Acceptability: If successful this system would likely be adopted by many small and medium sized farms that have yet to adopt to liquid storage systems. Farmers and the community will enjoy the odor and pathogen reduction.

High Solids Anaerobic Digestion

Description: This system would take the manure produced on the farm, mix it with the dried by-product and send it through a high solids anaerobic digester. The system would produce a stabilized pathogen free by-product much like compost available to export off the farm. All the phosphorous from the manure would be in this by-product.

Manure at 12% solids would be mixed with 50% solid by-product to make a 28% solid material that would be placed in a closed tank and heated to thermophilic temperatures. This will produce methane which, when generated for electric use on the farm, will also produce the heat needed to both heat the digester and to help dry the effluent from the digester. The effluent would need to be dried in a roofed structure with aeration and waste heat from the generator. Most of the dried by-product would be reused in the process but ultimately a steady flow of 5 cubic yards per day per 100 cows of 50% solids would be produced to be marketed off the site.

This process has not been used on a farm but has worked in large-scale operations in previous research activity by Professor William Jewell at Cornell.

Environmental Impact: This process would package all the phosphorus produced on the farm in a stabilized, deodorized humus much like compost for export off the farm. The pathogens and weed seeds would probably be killed. The volatile nitrogen (NH_3) would be lost during the drying phase. The farmer would have a portion of the by-product available to use on the farm for fertilizer or organic matter amendments. Application rates would be limited to the P required by the crop, so that nitrogen fertilizer would need to be imported.

Economics: This system may be put on a single farm or at a central site. The advantages of placing it on each individual farm include: the ability to use the electricity generated on the farm to replace electricity being purchased; the equipment needed to handle the high solid materials are generally already on the farm; and transportation costs would be limited to those used to export the final by-product off the farm.

A central system would find some economies of scale in building the digester tanks and in the engine generator to use the methane. The expertise needed to run the system and market the by-product could be concentrated at one site.

The capital costs would be high for this system. They would include a closed vessel for the digestion, an engine generator, and a drying shed. These costs could be as much as $200,000 for a 150 Animal Unit farm.

The operating costs may provide a break even or better situation on a single farm resulting in savings from electricity generated and bedding produced. A centralized site may have difficulties selling the electricity and have added transportation costs to move the dried material back to farms for bedding.

Land Requirements: The area this system would take up would include a 20 foot high by 15 foot diameter vessel and a 70 x 100 drying shed for a 100 cow dairy. This should not be a significant constraint on most farms.

Management Strategy: This technology would be about the same complexity as existing anaerobic digestion systems. We can expect those farmers with an above average management ability being able to operate the system with ease, while those farms that are not managing well will find running the system to be a burden. Handling the manure as a solid would mean that most of the equipment to move the material would already be on the farm and the operator would be familiar with the operation of it. An outside service person could used to check the systems on a regularly scheduled basis if enough farms were using the system. At a central site a manager with the capabilities of a sewage plant operator would be required to run the operation efficiently.

Acceptability: There would be some disadvantages to this system on the farm. This system would potentially reduce the amount of recycled nitrogen and phosphorus to the land. The nitrogen deficit would have to be imported. It would add another enterprise that would have to be managed on the farm. The advantages include electricity and bedding cost savings and odor control. Good managers who were provided the capital cost should benefit from this system.

The community should accept this low odor processing. There would be an increase in truck traffic especially at a multi-farm site. This system would provide pathogen control, phosphorus export.

Anaerobic Digestion

Description: These biological treatment systems take manure as produced at 12% DM and heat it with waste heat from the engine generator, then anaerobically digest it in an enclosed insulated trough for about 20 days. The manure is continually being fed in and an odor reduced effluent at about 8 % DM is continuously released. Methane is produced that is used to power an engine which drives a generator. Electricity produced exceeds the average dairy farm consumption of electricity providing the possibility of power sales. The effluent with all of the nutrients still in it could be stored to apply to the land.

Environmental Impact: The anaerobic process does reduce most pathogens. The extent of this reduction depends on the temperature regime in which the digester is operated. Most are run mesothermically but thermophillic digesters are possible. Odor reduction is another benefit of this process. The slight liquification and the enhanced ability to separate the solids mechanically after digestion can make the effluent easier to pump and irrigate. With the odor reduced, spray irrigation on growing crops is a real possibility. This has the potential to reduce runoff and leaching losses.

Economics: There are economies of scale with this system. Farms with over 1000 Animal Units would be more viable and may make a profit with these systems. Smaller farms may not make a profit. Design modifications to reduce the capital costs will continue. Centralized systems must overcome transportation costs to be profitable. In cheap energy times the sales of power may not produce enough of a cash flow to warrant use of this system.

Land Requirements: The size of the digester needs to be large enough for 20 - 30 days of hydraulic retention time. The engine generator and solid separation system, if used, would require a building to house them. The total volume of manure is not significantly reduced with methane generation.

Management Strategy: Management of these systems is difficult for most farms. While the ordinary monitoring could be done with existing personnel on the farm, when problems develop like engine failure or reduced gas production in the digester, farm labor may not have the expertise or be available to fix them. A private design and maintenance organization should be able to provide this service for a group of farms.

Acceptability: Although these systems have been installed and demonstrated for over twenty years, few farms have continued operating them. Cheap energy and the high capital costs are economic disincentives. Maintenance demands also caused existing systems to be abandoned. The need for odor control at a low cost may make them more popular in the future.

Treatment Lagoons

Description: Lagoons provide treatment as well as storage. The treatment occurs by building a biomass of organisms that will decompose the manure either aerobically, anaerobically or faculatatively. These systems often use flushing to dilute the manure and recycle the biomass. Some systems use mechanical separation to remove the solids. Some systems use managed shallow pools to separate the manure solids into aquatically stabilized solids. The solids can then be harvested, dried, screened and sold as a soil amendment. The system recycles the biologically active liquid to move the manure through the lagoons.

Environmental Impact: Although the amount varies, the phosphorus can be concentrated in the solids so that up to 75% of it can be removed. Although there is no specific knowledge of pathogen reduction, this may be significant since the retention time in the system is long. There is no temperature increase above ambient in this system. Odors are much reduced when this system is operating correctly. Ammonium nitrogen is lost into the air from this system.

Economics: The lagoon system and recycling pump for flushing would be a relatively low capital cost on a favorable site. A flat site with low permeability soil would keep the costs of installation down. Steep sites that require an artificial liner would be much more expensive. The operating costs may be offset by the sale of the product. If the barns were set up for flushing, additional labor savings could be obtained as the recycled water would clean the barns. Retrofitting a flushing system into a flat barn would be very expensive. An existing 2% slope on the alley's would be ideal. A method to further treat or spread the extra wastewater would need to be provided.

Land Requirements: The land requirements for the lagoon treatment system would be high. A 150 Animal Unit operation would need about one acre of ponded area and storage for 1 million gallons of effluent.

Management Strategy: The operation could be managed by the farmer. In some cases, a company interested in marketing the solids may manage the system for a share of the solids. The capital costs for the installation and a management fee would be paid to this company and the profits from the sale of the solid by-product would be split between the company and the farmer.

Acceptability: This process should be acceptable to the farmer once the initial capital cost is taken care of. The effluent to be spread or treated in a wetland will need to be managed. The system includes over winter storage and does reduce odors. This system would be most efficient operated on the farm in conjunction with a flushed housing. By potentially removing a larger portion of the phosphorus and reducing odors, this system may become more popular with farms.

Sequencing Batch Reactor

Description: Certain microorganisms store phosphorus under alternating aerobic - anaerobic conditions. Biological phosphorus removal creates fluctuations in the oxygen content of a wastewater to encourage this excess P uptake by microbial biomass. During the anaerobic stage some of this phosphorus-enriched biomass is withdrawn as a sludge. This is an established treatment technology for municipal wastewater, but has not been applied often to agricultural wastes. In order to achieve adequate settling of the solids, manure collected as a semi-solid would need to be diluted prior to treatment. Much of the effluent from the system would be recycled as dilution water.

Environmental Impact: In this application the phosphorus enriched sludge would be dried into a marketable product. Export of the product would result in removal of phosphorous and nitrogen from the watershed. Effluent from the system (beyond that required for dilution) may be suitable for direct discharge, or might require spray irrigation fields.

Economics: This is an expensive option, with costs similar to those of a conventional wastewater treatment plant on a BOD basis. Professor Carlo Montemagno at Cornell is currently developing this system at a pilot scale, to establish feasibility and costs for agricultural manure applications

Land Requirements: Siting requirements would be similar to those of a wastewater treatment plant. If spray irrigation of the effluent were necessary, land requirements could increase significantly. However land needed to spread manure would not be required as the effluent would be the exportable solids and the very dilute liquid.

Management Strategy: As envisioned, manure in excess of crop phosphorus needs would be hauled off site perhaps under contract with a private operator. Management of these systems is difficult for most farms. While the ordinary monitoring might be done with existing personnel on the farm, monitoring to prevent problems from developing might best be done with a private design and maintenance organization. Cost could be low if they were able to provide this service for a group of farms.

Acceptability: On farms that need both odor control and phosphorus balancing, this system may be required to stay in business. Farms just meeting odor control may see the high cost as a disadvantage. If energy and nutrient prices increase this system will not be the preferred treatment system.

Total Resource Recovery

Description: This system in essence creates a high value recycling system for the manure nutrients. The liquid stream from a solids separator would be diluted and used to grow plant proteins hydroponically that could then be fed to the animals. On-farm production of a high nutrient feed would reduce the need to import extra feed onto a farm and into the watershed. The diluted manure could come from several different treatment systems. An anaerobic digestion system operated at thermophilic temperatures preceding the hydroponics system would provide pathogen and odor control. The hydroponics system would be contained in a greenhouse environment and potentially be used to grow bacteria, simple and complex plants.

Environmental Impact: By reducing the import of nutrients this intensive recycling program could improve the nutrient use efficiency on the farm significantly. This would eliminate the extra phosphorus loading onto the fields. Pathogens and odors would be completely controlled if the system was preceded by a thermophilic anaerobic digester. Caution should be exercised before using untreated diluted manure, which could potentially recycle pathogens to animals through the feed.

Economics: Depending on the operating costs, this system could be very feasible. A high protein feed, bedding or soil amendment from the fibers, and energy are produced. There would be no waste discharge. The capital cost would be high. A 150 Animal Unit farm could expect the initial costs to be on the order of $250,000. Professor Bill Jewell at Cornell is currently evaluating the feasibility of this system at pilot scale, and those results should provide the necessary information for a field scale pilot system.

Land Requirements: A 150 Animal Unit farm would need about 2 acres of greenhouse production. The anaerobic digester and the engine generator would also need a building space.

Management Strategy: This system would be very intensely managed. The green house would have to be managed by an expert. One scenario would be for the farmer to enter into a partnership with an individual or a corporation with this expertise. If successful, this strategy would provide a range of opportunities for local economic development, including services related to greenhouse production, digesters, and composting of separated solids.

A centralized site would provide the opportunity for economies of scale on the digester, and better management and production of the protein supplement. Transportation cost of the manure and the difficulty of selling the electricity produced would be the disadvantages.

Acceptability: This system could not be adopted by the farmer on his own. The hydroponics production would be a complex and new skill to learn. With the right partner, this system should be accepted by the farmers. The potential economic advantages are attractive.

This system should be seriously considered for the future. The capital costs will be high but it would achieve the goal of pathogen, odor, and phosphorus control. It should increase the economic viability of the area with the creation of jobs and cheap energy.

Conclusions

- There are advantages and disadvantages to each system that may be more or less important to each farm.

- The lagoon treatment system and the SBR would work very well with a flushing system to clean the barns.

- Gently sloping topography and relatively impermeable soils on farms using earthen storage will keep the initial costs low.

- Farms that don't need all the nutrients in the raw manure may benefit from the nutrient losses of the SBR, or lagoon treatment.

- The anaerobic digester system would be best for a farm that had high electric costs and could use the nutrients for crop production.

- Nutrient utilization and by-product sales are important in reducing the cost of a manure handling system. Marketing the separated solids or other by-products and fully utilizing the nutrients in the manure can help pay for odor treatment systems.

References:

Jewell, William J., P. E. Wright, N. P. Fleszar, G. Green, A. Safinski, A. Zucker, "Evaluation of Anaerobic Digestion Options for Groups of Dairy Farms In Upstate New York" Final Report 6/97 Department of Agricultural and Biological Engineering, College of Life Sciences, Cornell University, Ithaca, New York 14853

Wright, Peter and S. P. Perschke, "Anaerobic Digestion and Wetland Treatment Case Study: Comparing Two Manure Odor Control Systems for Dairy Farms" Presented at the 1998 ASAE Annual International Meeting Paper No. 984105 ASAE 2950 Niles Road St. Joseph, MI 49085-9659

Wright, Peter, Manure Spreading Costs and the Potential for Alternatives. Proceedings from Managing Manure In Harmony with the Environment and Society, 2/10-12/98 Ames, Iowa, West North Central Region of the Soil and Water Conservation Society.

The Bion Nutrient Management System: A Biology-Based Treatment Alternative

Jeffrey Poulsen
Bion Technologies, Inc.
Williamsville, New York

Introduction

The continued consolidation of the animal agricultural industry, along with a trend toward farm specialization, has outpaced the advancement of farming practices relating to the handling of waste products. Farmers continue to import more nutrients than they are capable of disposing of in a manner acceptable to either government regulators or the general public. The continued use of short-term manure storage and anaerobic lagoons combined with the use of sprayfields and field application is neither cost effective nor environmentally sound. In addition to environmental concerns the increasing quantity of manure produced on dairy and hog farms today leads to concerns about odor, poor public perception, increased trucking costs, increased production costs, potential legal liability.

To address these concerns, Bion Technologies, Inc. (Bion) developed a patented waste treatment technology which eliminates the problems associated with manure management in large scale agriculture operations in a simple and cost effective manner.

The Bion Nutrient Management System (NMS) is a biological treatment system derived from the activated-sludge technology practiced in the wastewater industry. While municipal wastewater treatment systems are designed to treat large volumes of low strength waste, the NMS system is capable of effectively treating very high strength waste. The premise behind a conventional wastewater treatment process is to build a microbial population that will live and thrive under a strict set of constraints. In order to treat the waste quickly, the majority of treatment facilities operate aerobically and incorporate a great deal of mechanical complexity. These systems tend to be very expensive to construct and operate which are undesirable qualities for a system designed for farm use. The NMS system relies on a mainly facultative microbial population derived from the bacteria naturally present in the waste stream to metabolize the biochemical oxygen demand (BOD) and assimilate the nutrients found in the organic material. Wastes which have been treated in the system include manure, silage leachate and milkhouse waste.

The NMS system can be divided into two distinct operational components; the BionSoil® Production Loop and the Polishing Ecoreactor (Figure 1).

Figure 1. Bion NMS System Schematic

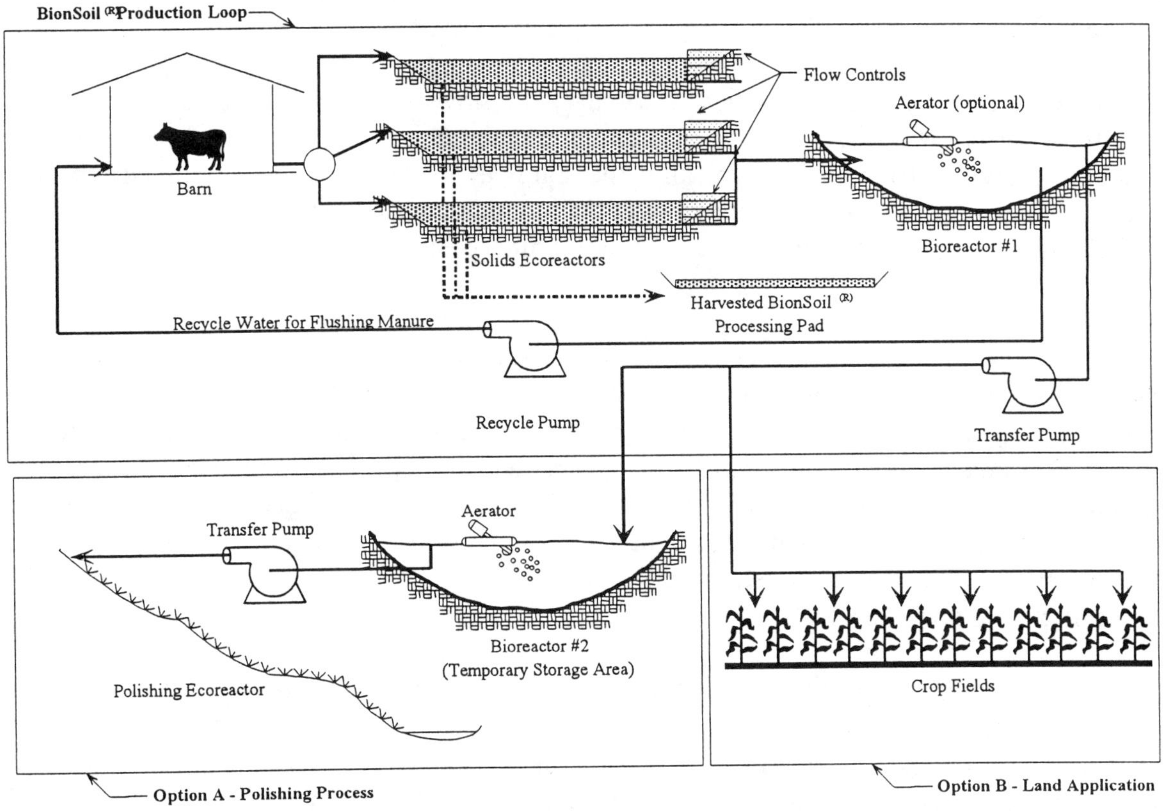

The BionSoil® Production Loop is designed to convert the waste into a stable product to be removed from the wastestream and is common to all of the farms where the NMS technology is used. The Polishing Ecoreactor is designed to treat the water to near discharge standards by bioconverting the remaining BOD and removing residual nutrients. The exact configuration and layout of the system components is arranged to meet the needs of the farm but the fundamentals of the system remain constant.

Solids Ecoreactors

To begin the treatment process, wastes are moved from the barns and collection areas to one of the Solids Ecoreactors (SEs). Solids ecoreactors are shallow basins designed to capture the solids in the wastestream. Typically, three SE cells are constructed to provide sufficient solids storage capacity within the system and to improve operational flexibility. In the SEs, the microbial population and the fiber from the wastestream settle or are filtered out by the previously accumulated solids in the cell. The SEs are designed with a length to width ratio that is conducive not only to promote removal of the solids, but also to cause channels to form in the deposited material which will allow fresh flush water to continue to circulate throughout the cell, thereby constantly exposing fresh microbes to the accumulated wastes. Once the biomass/manure solids mixture is deposited in the SE, the biomass will continue to grow, assimilate nutrients, consume the remaining BOD, and attack cellulosic materials in the manure as an energy source. This process is defined as curing. The available BOD and nutrients are used as the resulting solids become mature and stable. The rate limiting step in this process is the metabolism of BOD since nutrients are available in excess and there will not be sufficient time for cellulosic degradation to be a significant factor.

Once the first SE cell has been filled to capacity the flow will be switched to a second SE. The first cell is drained of free standing water and the solids are removed, resulting in a material known as BionSoil®. Typically, the solids ecoreactor cells are designed with the capacity to contain the solids generated over a 6 month period of time. Effluent from the SE is drained through the flow control device to the primary Bioreactor.

Bioreactor

Bion has developed a method used to size the bioreactor in a manner which will ensure that the biomass which develops which will be most suited to metabolizing the waste stream being presented to it. The system is managed in such a way as to provide a high degree of operational flexibility without loss of treatment effectiveness. The combination of aerobic, anaerobic and facultative bacterial populations in the system ensures that the biomass has the ability to rapidly adapt to any change in operating conditions. This combination also means that the system will convert wastes rapidly in the manner of an aerobic system without the associated high operating costs.

Rather than the bioreactor being periodically emptied as with a standard lagoon, a minimum process volume is left in the system to ensure adequate biomass concentration to effect treatment. In the majority of dairy systems the use of additional mechanical aeration is not necessary. Because hog waste has almost twice the BOD of dairy manure (NRCS), all of the hog manure based systems have been designed with some mechanical aeration in the system. Mechanical aerators have also been used in some dairy systems where there is a limited amount of space available or the farm has expanded to a size larger than planned for in the original design.

In addition to being constructed to contain the minimum process volume, the bioreactor is often sized to contain a sufficient storage volume to eliminate the need for spreading or irrigation of waste water during the winter months. When temporary waste storage volume is available, waste is transferred from the bioreactor to storage.

Recycle Water

In order to provide constant treatment, water from the bioreactor is used to flush the barn alleys which washes the manure into the solids ecoreactors. This flushing action helps to break the manure into smaller particles and exposes the fresh wastes to the microbe-laden bioreactor water. Farms without a treatment system often do not use, recycled water to flush the barns because of odor concerns and the high solids content in the water. Because of the efficiency at which the SE cells are capable of solids removal, farms with the NMS are able to use water from the bioreactor which contains less than 1 percent solids and is virtually odor free. Those farms which have chosen the NMS technology often retrofit their operations to include the use of a flushing system for the barns. When a farm does not chose to flush the barns, the recycled water is combined with the influent waste stream in a reception pit before being transferred to the solids ecoreactors.

The solids production loop is designed to divide the waste into a solids and a liquid stream. The solids fraction becomes BionSoil and the liquid fraction is suitable for field application or further treatment in the polishing ecoreactor.

BionSoil

When one of the SE cells becomes filled the flow is diverted to a second SE cell, and the free water is allowed to drain through a flow control structure to the bioreactor. The solids are then removed and stacked alongside the cell to continue curing. Typically, a system will produce 10 cubic yards of BionSoil for each milking cow and 1 cubic yard for each finishing hog. As part of their contract with the farmer, Bion removes the BionSoil from the farm removing a large percentage of the excess nutrients with it. There is a demand for BionSoil in the landscape, nursery, and the organic fertilizer markets. BionSoil has been proven to have characteristics which are beneficial for growing plants and for use as a fertilizer. BionSoil has been approved by the Northeast Organic Farming Association of New York (NOFA-NY) for use on certified organic farms. Testing by the Soil Foodweb Inc. in Oregon has shown that BionSoil contains large numbers of protozoa and a large active fungal biomass (Table 1). It is likely this diversity of living material in the soil which promotes good root growth and slow release nutrient value to the plants.

Table 1. Typical BionSoil Analytical Results

Moisture	66.1%
Total Nitrogen	7.50 lbs/cubic yard
Total Phosphorus	1.11 lbs/cubic yard
Potassium	1.15 lbs/cubic yard
Calcium	5.23 lbs/cubic yard
Magnesium	1.77 lbs/cubic yard
Iron	5.00 lbs/cubic yard
Zinc	0.11 lbs/cubic yard
Copper	0.03 lbs/cubic yard
Active Bacterial Biomass	69 µg/gram
Active Fungal Biomass	160 µg/gram
Flagellates	31,978 individuals/gram
Amoebae	39,949 individuals/gram
Ciliates	319 individuals/gram

A recent trial completed by the University of Georgia documented that BionSoil outperforms conventional 10-10-10 fertilizer in growth trials of peach orchards. The young peach trees were larger, greener and ready for intensive peach production sooner that those treated conventionally (Couvillon).

Growth trials conducted by the Plant and Soil Science Department at Utah State University (Newhall) show that the yields achieved by using BionSoil are comparable to those of conventional ammonium nitrate fertilizer. Grain corn grown with BionSoil produced 77% more bushels per acre than compost, and barley grown in BionSoil produced 12% more bushels per acre than the conventional fertilizer and 42% more than traditional manure compost.

Effluent Water
Excess water from the Bioreactor is eliminated from the system either through irrigation or further treatment in the Polishing Ecoreactor (PE). Increasingly, regulations require that waste only be applied to fields at levels balanced with the needs of the crop. The high levels of phosphorus in manure mean that wastes are having to be applied to fields further from the farm than is economically viable. The ability of a waste treatment system to reduce the amount of nutrient in the waste stream as well as the volume of waste which needs to be applied will allow the farm to utilize the nearest acreage for waste disposal. This will not only reduce the cost of trucking waste to the distant fields but also save the cost of labor and liability.

In addition to lower nutrient content, the NMS system provides the farmer with a liquid effluent containing fewer solids, allowing more efficient pumping to satellite lagoons or an expanded variety of irrigation options such as center pivots or hard hose reels. When compared to a standard storage lagoon, the water from the NMS contains only a fraction of the nitrogen, phosphorus and potassium loads (Table 2). The effluent can therefore be applied to fewer acres than untreated waste and often using hydraulic rather than nutrient based application rates.

Table 2. Typical Waste Nutrient Analysis

	As Excreted [1] (mg/L)	From Storage[2] (mg/L)	NMS Effluent (mg/L)	Reduction from "As Excreted" (%)
Nitrogen	5,578	2,169	839	85
Phosphorus	855	723	156	82
Potassium	3,217	2,289	699	78

(1) NRCS
(2) Wright

Polishing Ecoreactor
Following primary treatment in the BionSoil production loop, waste effluent from the primary bioreactor can be further treated in the Polishing Ecoreactor (PE). The PE has been designed to reduce the amount of liquid which needs to be irrigated from the farm as well as

further reducing the nutrient load in the effluent. The polishing process consists of a second stage Bioreactor (B2) and a constructed wetland area. B2 is an aerated basin designed to further the treatment process by stimulating microbial growth and has a longer retention time than the solids production loop. Water in B2 is pumped into the constructed wetland area during the active growing season. The wetland area is a flooded vegetated area in which nutrients are removed from the waste stream by means of a vegetative-microbial complex. This stage of the treatment process is designed to contain all of the effluent and has both wet and dry areas to promote nitrification/denitrification. The wetland area has the appearance of a native wetland, which provides habitat for wildlife, and in general presents an attractive environmental image.

At present vegetation within the system is not mowed or removed from the wetland. Further testing is needed to determine the rate of phosphorus buildup within the system. In the future it may be necessary to harvest a crop with a high rate phosphorus uptake or to periodically remove the top layer of soil to eliminate excess nutrients.

Pathogens
In the fall of 1999 in cooperation with the University of North Carolina at Chapel Hill, Bion began a research program to determine the effect of the NMS treatment system on pathogens with Dr. Mark Sobsey and doctoral candidate Vince Hill. They had previously published data on the levels of microbial indicator organisms found in the sprayfield irrigation water of anaerobic swine lagoons (Hill and Sobsey, 1998). Preliminary results from the Bion NMS swine system indicate that the indicator organisms were 96-99% lower in Bion's sprayfield irrigation water than those published for anaerobic lagoons. While the available data is directly related to hog systems, similar results are expected from the dairy systems.

Analysis of BionSoil for parameters required by the EPA Part 503 regulations has shown that there are no detectable populations of *Helminth* ova, enteric viruses, E.coli or *Salmonella* in the dairy or hog based BionSoil.

CASE STUDY: DREAM MAKER DAIRY
Dream Maker Dairy (DMD) currently milks approximately 325 cows plus replacements (500 Animal Units). DMD owns approximately 100 acres adjacent to the barns, approximately 60 acres of this were existing crop fields (hay and corn). DMD, unlike most dairy operations, does not grow its own feed for the herd. Instead, DMD purchases all of its feed from local farms near the dairy. Prior to installing the NMS system, DMD paid an outside contractor to haul and spread manure onto nearby crop fields. DMD evaluated many manure management options and decided to install the Bion NMS process because the system is designed to remove nutrients from the waste stream and requires minimal daily maintenance.

The Bion NMS was installed in phases. The first phase was initiated in the spring of 1996, when one SE cell and one bioreactor was constructed. Recycled water from Bioreactor 1 (B1) was pumped to a distribution channel to flush the manure into the system. The topography of the site was sufficient to allowed the waste material to flow by gravity into the system.

The second phase of the project began in the summer of 1997 and was completed in Spring 1998. The second phase consisted of constructing a center berm, which divided the original SE cell into two cells. A new B1 was constructed as well as a second bioreactor/temporary storage area, and containment berms for the wetland area. The new B1 was positioned so that the SE cells could completely drain by gravity. The barns were equipped with flush tanks, which allowed the barn alleys to be cleaned with recycled water from B1. The containment berms for the wetland area were constructed during the fall 1997 so that plants would have time to become established prior to the system being charged with water treated in the BionSoil production loop. Construction was completed and the system was started in June 1998.

A sampling program was initiated in June 1998 in cooperation with the State University of New York at Buffalo to track system performance. Water samples were collected from key locations throughout the system and were analyzed for a number of parameters including BOD, Total Kjeldahl Nitrogen and Total Phosphorus. A summary of the sampling results is presented in Table 3 below. Results indicate a dramatic decrease in nutrient loading as the waste stream moves through the treatment process. A comparison of "As-Excreted" data and samples from the wetland portion of the Bion NMS system shows that 99% of the BOD, 98% of the total nitrogen, and 98% of the total phosphorus are removed from the wastestream. Sampling continues and will provide additional insight on the mechanisms involved in the removal of nutrients in the Bion NMS process.

Table 3. Dream Maker Dairy Analytical Data

Sample Location	BOD		Nitrogen		Phosphorus	
	(mg/L)	(lbs/1000 gal)	(mg/L)	(lbs/1000 gal)	(mg/L)	(lbs/1000 gal)
As Excreted[1]	19,784	165	5578	46.3	855	7.1
Bioreactor #1	2,353	19.6	617	5.1	119	1.0
Recycle/Flush Water	193	1.6	48	0.4	10	0.1
Bioreactor #2	1,095	9.1	250	2.1	77	0.6
Wetland #1	130	1.1	56	0.5	15	0.1
Wetland #2	118	1.0	64	0.5	15	0.1
Wetland #3	98	0.8	44	0.4	9	0.1

(1) NRCS

Conclusions

As of January 2000 Bion is managing operations of 9 dairy systems and 7 swine NMS systems in New York, North Carolina, Illinois, Florida, and Utah. Three additional systems are scheduled to be constructed in New York and Vermont in the summer of 2000.

Bion is continuing to develop operational data from the systems and is aggressively pursuing the development of the NMS technology on both hog and dairy operations. Currently, the technology is being promoted to producers with more than 2000 sow equivalents or 300 milking cows. As the operational characteristics of the system are more clearly defined and more is learned about the biological treatment mechanisms involved, the systems will be able to be designed with increased efficiencies and reduced construction costs.

University evaluation of BionSoil's plant growth capabilities and the fate of pathogens in both the BionSoil and the treatment system will expand tremendously over the next year.

References

Couvillon, Dr. Gary. University of Georgia, College of Agriculture and Environmental Sciences. personnel communication with Edward Lamb, Bion Technologies Inc. October 1999.

Hill and Sobsey. 1998. Microbial Indicator Reductions in Alternative Treatment Systems for Swine Wastewater. Water Science Technology Vol. 38, No. 12, pp. 119-122.

Newhall, Dr. Robert. Utah State University, Plant and Soil Science Department. Personal communication with Dr. Cindy Frick, Bion Technologies, Inc.

NRCS, National Engineering Handbook-Part 561 Agricultural Waste Management Field Handbook Agricultural Waste Characteristics, Natural Resources Conservation Service, Washington. DC June 1999

Wright, Peter. "Survey of Manure Spreading Costs Around York, New York" Presented at the 1997 ASAE Annual International Meeting Paper No 972040. ASAE 2950 Niles Road, St. Joseph, MI 49085-9659 USA

Feed Management to Reduce Excess Nutrients

Feeding Poultry
to Minimize Manure Phosphorus

Roselina Angel, Ph.D.
Assistant Professor
Department of Animal and Avian Sciences
University of Maryland, College Park, Maryland 20742

Biographies for most speakers are in alphabetical order after the last paper.

INTRODUCTION

Recent events in certain areas of the US have brought the issue of phosphorus (P) content in poultry litter to the forefront. In the Delmarva area new legislation will limit the use of excess P application to soil (based, in part, on soil P content). Given limited land for litter application in certain areas of the U.S.A. where the greatest concentration of poultry production exists today, the poultry industry and poultry producers must find strategies that reduce P in litter. The challenge of minimizing excreta P is one, that sooner or latter, will have to be faced by the poultry industry nationwide. Different strategies exist to reduce excreta P but the most effective is a multi-strategy approach.

Changes made to the feed can have a great impact on the amount of P that is excreted by broilers and accumulated in litter. Several feed and management related strategies will be discussed that have the potential for decreasing excreta P significantly. 1) The first strategy is formulating feeds and feeding birds closer to their P requirements. To do this, several issues must be addressed which would include: establishing P and calcium (Ca) requirements under commercial conditions and management systems; use of phase feeding systems with a maximum number of phases taking into consideration practical and economic constraints; use of ingredient knowledge on nutrient content and its variability within ingredient; use of rapid analytical techniques at feed mills to determine actual P and Ca content in ingredients; etc. 2) The second strategy is the use of feed additives that maximize the availability of P for broilers. These feed additives would include enzymes, such as phytase, and/or enzyme "cocktails" and vitamin D_3 metabolites. Feed formulated with enzyme(s) addition must take into account the increased availability of nutrients when feed additives are used. 3) The third strategy is to use new ingredients that are low in phytate P (PP) such as the high available P (HAP) corns currently being developed and tested.

FEED FORMULATION

In formulation of diets for commercial use, several factors have to be considered. Among these factors are: ingredient variability; availability of nutrients within an ingredient and changes in that availability due to processing, growing season, soil where ingredients were grown or where ingredients were mined and processed; specific plant genotype; bird strain; requirements that change with physiological factors such as sex and age; environmental factors or stressors such as heat or density; and the mixing accuracy within a specific system in a feed mill. Formulators need to have these factors in mind when determining the nutrient levels to use in diets. Together, these factors can lead to errors that have dictated the use of safety margins in formulation that are in place to prevent deficiencies from occurring.

Minimizing formulation safety margins will lead to feed formulation that is closer to requirements. For this to happen formulators need increased knowledge about ingredient nutrient content and its variability; use of near infrared reflectance (NIR) technology or other rapid analysis techniques at the feed mill for rapid analysis of ingredients for real time formulation; minimizing mixing "errors" through changes in ingredient delivery systems; formulation of diets for specific strains and physiological phases that should be as narrowly defined as possible; and possibly seasonal formulation. Cost of implementation, availability of information and technology, and previous lack of economic or legislative incentives to overcome implementation costs have resulted in very limited use of new technologies to minimize safety margins.

Ingredient Selection and Variability

Before getting into ingredient selection issues and how they affect P retention, it is important to clarify terms (related to P) before one goes any further. Book values, such as those found in NRC (1994) for nonphytate P (nPP) levels in plant ingredients are often referred to as being available P (aP) values. This misuse of the aP term has led to confusion as to the meaning of terms. Available P refers to the P that is absorbed from the diet into the animal. Retained P refers to the P that stays in the body (i.e., feed P minus excreta P).

Ingredient selection can plays an important role in the potential to decrease excess levels of most nutrients. This is the case with P. In most formulation systems inorganic sources of P are generally assumed to be 100% available by poultry. This belief is not correct since inorganic sources of P are clearly not 100% available. Inorganic sources of P vary in greatly in availability (Weibel *et al.*, 1984; Potchanakorn and Potter, 1987; Potter *et al.*, 1995; De Groote and Huyghebaert, 1996; Van Der Klis and Versteegh, 1996). Monocalcium phosphate has a relatively higher bioavailability than dicalcium phosphate, with the lowest bioavailability being seen with deflourinated phosphate, regardless of publication. This is consistent among experimental trials done in the same research unit (Weibel et al., 1984) as well as among researchers (Weibel *et al.*, 1984, Potchanakorn and Potter, 1987; Potter *et al.*, 1987; De Groote and Huyghebaert, 1996) and bioavailability assays (Potchanakorn and Potter, 1987; Potter *et al.*, 1995). Researchers have found that the experimental conditions under which P availabilities are determined affect absolute P availability results (De Groote and Huyghebaert, 1996) and thus commercial application of these data must be done carefully. P availability also appears to change due to the physical form of the diet (Sandberg et al., 1987), with pelleted diets having lower P availability. This is possibly, in part due, to decreases in endogenous phytase activity brought about by enzyme inactivation due to the heat associated with pelleting. The extensive use of absolute P bioavailabilty data (CVB, 1994) in commercial feed formulation in Europe should be questioned and perhaps a relative bioavailability system should be applied instead (De Goote and Huyghebeart, 1996). Data presented by Van Der Klis and Versteegh (1996) demonstrates that nonphytate P and available P are not synonymous . These authors found that of the total P in corn 24% was nPP but 29% was available to broilers. Similarly, of the total P in SBM, 39% was nPP but 61% was available to broilers.

Actual P and phytate P content in different ingredients varies somewhat between different published papers and publications (NRC, 1994; Van Der Klis and Versteegh, 1996; Nelson *et al.*, 1968). Data are still limited (Nelson *et al.*, 1968) as to the variability in phytate P content within an ingredient and how soil and environmental factors may affect this content (Cossa *et al.*, 1997). Work done by Cossa *et al.* (1997) showed, in 54 corn samples, a total P content of 3.11g/kg on a dry matter basis and reported a standard deviation (SD) of 0.28 with low and high values of 2.55 and 3.83 g/kg, respectively. Average phytate P was 2.66 mg/kg (SD of 0.34) with low and high values of 1.92 and 3.54 g/kg DM, respectively. These researchers found no apparent differences between locations and early, medium and late varieties of corn on the phytate P content of the corn. There is also limited information on potential variability in the availability of phytate P (Van Der Klis and Versteegh, 1996; Cossa *et al.*, 1997) (Table 2) within an ingredient and on how diet manufacturing process may affect this availability (De Goote and Huyghebeart, 1996).

Another strategy to maximize P retention from feeds is the selection of plant based ingredients. New plant genotypes are being developed that contain lower levels of phytic acid and thus of phytate P, as is the case in the new high available phosphorus (HAP) corn (Stillborn, 1997; Stillborn, 1998). This new genotype contains the same level of total P as normal corn varieties. In HAP corn only 35% of the total P is phytate P versus 75 to 80% in other corn varieties. Chick studies have shown that the P in HAP corn is indeed more available (Ertl *et al.*, 1998; Kersey *et al.*, 1998; Huff *et al.*, 1998). Other key ingredients are currently being selected for high availability of P. Soybean phytic acid content could be reduced (13, Raboy and Dickinson, 1993) with a concomitant decrease in phytate P from 70% to 24% of total P through breeding efforts (Raboy *et al.*, 1985). Another strategy being implemented is the incorporation of fungal phytase gene(s) into plants such that phytase is expressed in the seed at high levels (Stillborn, 1998). Work has been done in chicks with transgenic canola meal (Ledoux *et al.*, 1998) and soybean meal (Denbow *et al.*, 1998). Results from these chick trials showed that similar levels of added endogenous phytase and added exogenous phytase were effective in improving PP utilization. Processing is still a concern in terms of inactivation of phytase regardless of how it is added to the diet. Post expansion and/or pelleting application of exogenous phytase to feed can be done (Aicher, 1998) thus avoiding heat inactivation of the enzyme. This would not be possible with transgenically incorporated phytase.

From a practical standpoint, the use of new ingredients in commercial diets poses some challenges. New ingredients must be identified from planting to actual incorporation into diets. The logistics and economics of accomplishing this are still being worked out. The simplest solution so far is for feed manufacturers to contract fields for planting specific genotypes. This solution still leaves some of the logistical and economic challenges unanswered. In a feed mill, bin space for ingredients is always at a premium and thus new ingredients would, in most cases, displace other ingredients.

NIR technology for quick determination of protein, fat, and fiber exists and has been in place in some commercial mills in the U.S.A. for several years. Application of NIR for amino acid and digestible amino acid predictions has been developed (van Kempen and Simmins, 1997) but so far it is not used much in commercial mills in the United States. Other potential applications for NIR are determinations of organically bound Ca and P. This technology has the potential to allow for feed formulation based on real time ingredient nutrient content beyond protein, fat, and fiber. This would allow for much closer formulation, under commercial mills situations, to actual requirements and would allow for smaller formulation safety margins.

Phosphorus Requirements

There is limited information as to the P requirements of broilers (NRC, 1994; Van Der Klis and Versteegh, 1996; Gillis *et al.*, 1949; Fritz *et al.*, 1969; Twining *et al.*, 1965; Lillie *et al.*, 1964; Van Der Klis and Versteegh, 1997a) but no conclusive values have been established apart from those published by NRC (1994). The values proposed by NRC (1994) are recommended levels and reflect only information published through the peer review process. Thus, it does not include levels used with success commercially. NRC (1994) recommendations for nPP from hatch to 3 weeks of age appear to be well supported both under controlled experimental conditions as well as under commercial conditions. It is in the grower and finisher phases that NRC (1994) recommended levels for Ca and nPP exceed those used successfully in the field and shown to be adequate under experimental conditions (Skinner and Waldroup, 1992; Waldroup, 1998). Although NRC (1994) recommends a nPP level of 0.30% from 42 to 56 days of age, commercial use levels during this age period (42 to 50 days of age) can be 0.17% or lower. Establishing minimum adequate levels for nPP under defined conditions is a necessity if one is to maximize the effect of feed additives in decreasing excreta P.

It is imperative that certain factors be defined when requirements are being determined. The factors that most affect P requirements are dietary Ca level (and Ca:P ratio), level and type of vitamin D in the diet, and the presence and amount of plant P in the diet (Fritz *et al.*, 1969; Twining *et al.*, 1965; Van Der Klis and Versteegh, 1997a; Van Der Klis and Versteegh, 1997b; Davies *et al.*, 1970). Plant P in the diet should be at least defined by analysis as total P, and nPP. Age (Lillie *et al.*, 1964; Van Der Klis and Versteegh, 1997b) and strain (Lillie *et al.*, 1964) of the birds also have an effect on P requirements. The more closely diets can be formulated for poultry of specific ages, the lower the P excretion will be. To accomplish this, phase feeding systems should be implemented. These systems should maximize, within economic logistical constrains, the number of feeding phases.

Preliminary Results on Phosphorus Requirements in a Four Phase Feeding System

Several trials have been done to determine more accurately what the available phosphorus needs of broilers grown in a four phase feeding system are (Angel, Applegate, Dhandu, Ling, unpublished data). The four phases studied were: starter, hatch to 18 days of age; grower, 18 to 32 days of age; finisher, 32 to 42 days of age; and withdrawal, 42 to 49 days of age. Two more studies are needed to finalize this work.

So far this research has shown that nPP can be reduce (versus average commercial usage levels) by 5% in the grower diet and by 15% in the finisher diet without affecting bone strength or performance. Withdrawal phase requirements are currently being determined. Requirements for all phases need to be confirmed in a study that closely mimics commercial conditions. A study currently underway will focus on processing plant condemnations and bone breakage data to allow us to determine if the levels we have found to be adequate in floor pen studies are "really" adequate when birds go through the processing plant.

Having more accurate available phosphorus requirement information has profound consequences in terms of phosphorus nutrient management. Given the results up to date, we can potentially see a reduction of at least 10% in the amount of available phosphorus we feed broilers. This would mean at least a 10% decrease in litter phosphorus. Having more accurate phosphorus requirement information will also allow us to more fully use feed additives, such as phytase, to decrease phosphorus in poultry litter.

FEED ADDITIVES THAT MAXIMIZE PHOSPHORUS RETENTION

Poultry diets usually contain plant-based ingredients. A high proportion of P from plants occurs in the seed as phytate P (PP). Phytate chelates other minerals and it binds to proteins and starches making them unavailable to birds. Extensive information is available on phytate and phytase (Kornegay, 1998; Parsons, 1998; Nelson, 1967; Harland and Morris,

1995). Phytase and factors affecting its activity and efficiency have been extensively discusses (see Nelson *et al.*, 1971; Simons *et al.*, 1990; Biehl *et al.*, 1995; Harland and Morris, 1995; Ravindran *et al.*, 1995; Van Der Klis *et al.*, 1997; Mitchell and Edwards, 1996; Quian *et al.*, 1996; Van Der Klis *et al.*, 1997d; Gordon and Roland, 1998; Kornegay, 1998; Parsons, 1998) and thus, the focus here will be on the potential use of several feed additives ("cocktails") together.

Work done by Zyla *et al.* (Zyla *et al.*, 1995a; Zyla *et al.*, 1995b; Zyla *et al.*, 1996; Zyla *et al.*, 1997) demonstrated that, under *in vitro* conditions simulating turkey intestinal conditions, the use of an enzymatic "cocktails" could release 100% of the PP contained in a corn-soy diet. The enzymatic "cocktail" contained a microbial phytase, acid phosphatase, acid protease, citric acid, and *A. niger* pectinase. From their work, it was clear that phytase alone could not release 100% of the PP present in a corn-soy diet. These researchers (Zyla *et al.*, 1995a, Zyla *et al.*, 1995b) found that phytase preparations are not pure phytase. These phytase preparations (both commercial and laboratory derived sources) generally contain, in varying amounts depending on the phytase preparation, acid phosphatases, acid proteases, and pectinase. These researchers found a negative correlation between purity of the phytate preparation and its capacity to release PP. Only when the appropriate balance between the different components of the "cocktail" was obtained did 100% release of PP from the corn-soy diet occur.

To determine whether the enzymatic "cocktail" developed *in vitro* would work as effectively *in vivo* an experiment was done with 7 to 21 day-old turkeys (Zyla *et al.*, 1996). These researchers fed a corn-soy-meat meal diet with a Ca level of 1.2% and an aP level of 0.6% which met NRC (1994) recommendations, a positive control diet containing 0.42% aP and 0.84% Ca (positive control), and diets containing 0.84% Ca and 0.16% aP to which enzyme preparations (phytase (1000 u/kg of diet), an enzyme cocktail, or *A. niger* mycelium) were added . They found that P retention from 31.0% in the NRC (1994) based diet, 42.8% in the positive control diet, 66.8% in the diet with phytase, 77.0% in the diet with the enzyme "cocktail", and 79.5% in the diet with the *A. niger* mycelium. Addition of acid phosphatase, pectinase, and citric acid to phytase (enzyme cocktail) also increased P retention (P<.05).

Questions remain as to why these researchers (Zyla *et al.*, 1995a; Zyla *et al.*, 1995b; Zyla *et al.*, 1996; Zyla *et al.*, 1997) obtained 100% release of PP *in vitro* but only 77% *in vivo*. Procedural problems may be part of the answer since analyzed TP levels in the diets where enzymes were added were higher than formulated. The authors speculated that the feed additives themselves could have been a source of P that was not accounted for. Another factor is potential differences in the time allowed for digestion *in vitro* versus actual residence time of digesta in the proventriculus/gizzard and small intestine. The *in vitro* assay consisted of two 30 minute periods simulating the action of the proventriculus and gizzard, then incubation for 240 minutes after the addition of enzymes or "cocktails" being tested. Passage rate in the small intestines of turkeys and broilers is 94.9 and 73.3 minutes, respectively (Hurwitz *et al.*, 1979). From these data, it would seem that specially in broilers, increasing the speed of the reaction could increase the release of PP *in vivo*. This could be done by looking at other enzyme combinations and increasing the accessibility of the substrate to the enzymes by decreasing particle size and the use of structural carbohdrases may help in maximizing PP release.

Other feed "additives" that need to be considered are vitamin D_3 and its metabolites. Not only does vitamin D stimulate P transport mechanisms in the intestine (Harrison and Harrison, 1961; Mohammed *et al.*, 1991; Biehl and Baker, 1997) but it also appears to enhance phytase activity (41, 42). Vitamin D as well as its metabolites, 25-hydroxycholecalciferol and 1,25-dehydroxycholecalciferol ($1,25(OH)_2D_3$) (43,44), have been shown to enhance phytase activity. It appears that $1,25(OH)_2D_3$ and phytase act in an additive manner rather than a synergistic one (Mitchell and Edwards, 1996; Biehl *et al.*, 1995). Mitchell and Edwards (1996) found that the addition of $1,25(OH)_2D_3$ and phytase could replace 0.2% of the inorganic P addition in the diet in 21 day-old chicks. Phytase and $1,25(OH)_2D_3$ alone could each only substitute for close to 0.1% of added inorganic P. Vitamin D metabolites have a clear role in improving P retention and their use in conjunction with other feed additives (phytase and/or enzyme "cocktails") is indicated.

Preliminary Results on the used of a *Lactobacilus*-Based Pro-Biotic in Broiler Feed on Phosphorus and Nitrogen Content of Litter

Some *lactobacillus*-based pro-biotics have been shown to improve growth and feed conversion in poultry. The objective of this research was to determine if broilers fed low P, Ca, and protein in a diet containing a *lactobacillus*-based pro-biotic would perform similarly to broilers fed a control (commercial levels of P, Ca, and protein) diet. If performance was not negatively effected, we wanted to determine if litter P and nitrogen (N) would decrease. To do this a series of two studies were done. The data from the first study has been summarized and presented (Angel, *et al.*, 1999a; and Angel *et al.*, 1999b), while the retention data from the second study is still being summarized. Broilers were fed a grower control diet (19.3% protein, 0.37% aP) and a low nutrient diet from 18 to 28 days of age. A finisher control diet (17% protein, 0.30% aP) and a low nutrient finisher diet were fed from 28 to 42 days of age Each diet was fed with and without the *lactobacillus*-based pro-biotic. The low nutrient diet was reduced by 12% in protein and by 18% in aP.

In both the grower and finisher phase, broilers fed the low nutrient diet had significantly lower weight gain, and feed conversion. Bones (tibia) from these birds broke easier than bones from control birds. When the pro-biotic was added to the low nutrient diet, broilers performed as well as those fed the control diet and bone breakage was also similar to that of controls. At 42 days of age birds fed the control diet weighed 5.8 lb and birds fed the low nutrient plus pro-biotic diet weighed 6.1 lb. Performance data from the second study has been summarized and supports the results of the first study.

Nutrient retention was improved when the pro-biotic was added to the diet. P retention was 22% higher and N retention was 10% higher in birds fed the low nutrient plus pro-biotic diet than in the birds fed the control diet. The addition of the pro-biotic to the low nutrient diet allowed broilers to grow as well as those fed a control diet in part because they were more efficient in retaining nutrients. Feeding a low nutrient diet with pro-biotic decreased excreta phosphorus by 33% without adversely affecting performance or bone strength. This decrease would also be seen in litter phosphorus content. Nutrient retention data from the second study is currently being summarized and is needed to corroborate data from the first study.

It is important to note that additions made to a diet that improve the availability of PP do not necessarily translate to higher P retention from the diet. The higher P availability resulting from the addition of phytase, enzyme "cocktails", and/or D_3 metabolites has to be associated with reduced dietary inorganic P levels if increased P retention is to occur.

SUMMARY

Implementation of strategies to maximize P retention and thus minimize P in litter needs to be multifaceted. Numerous changes in different areas must be implemented simultaneously if P retention is to be maximized. Changes that must be considered if P retention is to be maximized are: formulation choices based on actual P levels in ingredients and on values for availability of the P in those ingredients; ingredient selection, including the use of new ingredients with low PP content; dietary nutrient balance (vitamin D_3, Ca:P ratios, etc.) that maximizes P retention without adversely affecting economics; use of rapid analytical methods to formulate diets based on real time nutrient content of ingredients; use of several feed additives (enzyme "cocktails" that include phytase, and D_3 metabolites) that together maximize PP release and absorption of P; decreased inorganic P levels in diets when feed additives that increase P availability are used; and management changes that reduce stress. Only through multifactorial changes can P retention be maximized.

REFERENCES

Aicher, E. 1998. Post pelleting liquid systems for enzymes. In: Use of Natuphos Phytase in Broiler Nutrition and Waste management. BASF Technical Symposium, Atlanta, January 19, 1999. pp 35-46. Published by BASF Corporation,Mount Olive, NJ, USA.

Angel, C. R., R. A. Dalloul, N. M. Tamim, T. A. Shellem, and J. A. Doerr, 1999a. Performance and nutrient use in broilers fed a lactobacillus-based pro-biotic. Poultry Science 78 (Supplement 1): 98.

Angel, C. R., P. Melvin, R. A. Dalloul, N. M. Tamim, T. A. Shellem, and J. A. Doerr, 1999b. Performance and nutrient retention in broilers fed a lactobacillus-based pro-biotic. Poultry Science 78 (Supplement 1): 58.

Belay, T., K. E. Bartels, C. J. Wiernusz, and R. G. Teeter, 1993. Poultry Science 72:106-115.

Biehl, R. R. and D. H. Baker, 1997. J. Animal Science 75:2986-2993.

Biehl, R. R., D. H. Baker, and H. F. De Luca, 1995. 1 alpha-hydroxylated cholecalciferol compounds act additively with microbial phytase to improve phosphorus, zinc, and manganese utilization in chicks fed soy-based diets. J. Nutrition 125:2407-2416.

Coon, C. and K. Leske, 1998. Retainable phosphorus requirements for broilers. Maryland Nutrition Conference. pp 18-31.

Cossa, J., K. Oloffs, H. Kluge, and H. Jeroch, 1997. Investigation into the total phosphorus and phytate phosphorus content in different varietes of grain maize. 11th European Symposium on Poultry Nutrition, Proceedings of the World's Poultry Science Association, Faaberg, Denmark. pp 444-446.

CVB, 1994. Voorlopig systeem opneembaar fosfor pluimvee. (Interim system of available phosphorus for poultry). Centraal Veevoeder Bureau, Lelystad, The Netherlands, CVB-reeks nr. 16:1-37.

Davies, M. I., G. M. Ritcey, and I. Motzok, 1970. Intestinal phytase and alkaline phosphatase of chicks: Influence of dietary calcium, inorganic and phytate phosphorus and vitamin D. Poultry Science 49:1280-1286.

De Goote, G. and G. Huyghebaert, 1996. The bio-availability of phosphorus from feed phosphates for broilers as influenced by bio-aasay method, dietary calcium level, and feed form. Animal Feed Science and Technology 69:329-340.

Denbow, D. M., E. A. Grabau, G. H. Lacy, E. T. Kornegay, D. R. Russell, and P. F. Umbeck, 1998. Soybeans transformed with a fungal phytase gene improve phosphorus availbility for broilers. Poultry Science 77:878-881.

Edwards, H. M., Jr., 1993. Dietary 1,25-dihidroxycholecaciferol supplementation increases natural phytate phosphorus utilization in chickens. J. Nutrition 123:567-577.

Ertl, D. S., V. Raboy, and K. A. Young, 1996. Plant genetic approaches to phosphorus management in agricultural production. J. Environmental Quality 27:299-304.

Fritz, J. C., T. Roberts, J. W. Boehne, and E. L. Hove, 1969. Factors affecting the chick's requirement for phosphorus. Poultry Science 48:307-320.

Gillis, M. B., L. C. Norris, and G. F. Heuser, 1949. The effect of phytin on the phosphorus requirement of the chick. Poultry Science 28:283-288.

Gordon, R. W., and D. A. Roland, 1998. Influence of supplemental phytase on calcium and phosphorus utilization in laying hens. Poultry Scienece. 77:290-294.

Harland, B. F. and E. R. Morris, 1995. Phytate: A good or a bad food component. Nutrition Research 15:733-754.

Harrison, H.E. and Harrison, H.C. 1961. Intestinal transport of phosphate: action of vitamin D, calcium and potassium. Am. J. Physiol. 201:1007-1012. Mitchell, R. D. and H. M. Edwards, Jr., 1996. Effect of phytase and 1,25-dihydroxycholecalciferol on phytate utilization and the quatitative requirement for calcium and phosphorus in young broiler chickens. Poultry Science 75:95-110.

Huff, W. E., P. A. Moore, Jr., G. R. Huff, J. M. Balog, N. C. Rath, P. W. Waldroup, A. L. Waldroup, T. C. Daniel, and V. Raboy, 1998. Reducing inorganic phosphorus in broiler diets with high available phosphorus corn and phytase without affecting broiler performance or health. Poultry Science 77 (Suppl.):116.

Hurwitz, S., U. Eisner, D. Dubrov, D. Sklan., G. Risenfeld,and A. Bar, 1979. Protein, fatty acids, calcium and phosphate absorption along the gastrointestinal tract of the young turkey. Comp. Biochem. Physiol. 62A:847-850.

Kersey, J. H., E. A. Saleh, H. L. Stilborn, R. C. Crum, Jr., V. Raboy, and P. W. Waldroup, 1998. Effects of dietary phosphorus level, high available phosphorus corn, and microbial phytase on performance and fecal phosphorus content. 1. Broilers grown to 21 days in battery pens. Poultry Science 77 (Suppl.):1.

Kornegay, E. T. 1998. A review of phosphorus digestion and excretion as influenced by microbial phytase in poultry. In: The Use of Natuphos Phytase in Broiler Nutrition and Waste management. BASF Technical Symposium, Atlanta. pp 69-81.

McGillivray, J. J., 1978. Biological availability of phosphorus sources. In: 1st Annual International Mineral Conference, IMC, St. Petersburg Beach, pp 73-85.

Mitchell, R. D. and H. M. Edwards, Jr., 1996. Effect of phytase and 1,25-dihydroxycholecalciferol on phytate utilization and the quatitative requirement for calcium and phosphorus in young broiler chickens. Poultry Science 75:95-110.

Mohammed, A., M. J. Gibney, and T. C. Taylor, 1991. The effect of dietary levels of inorganic phosphorus, calcium, and cholecalciferol on the digestability of phytate phosphorus by the chick. Br. J. Nutr. 66:251-259. Mitchell, R. D. and H. M. Edwards, Jr., 1996. Effect of phytase and 1,25-dihydroxycholecalciferol on phytate utilization and the quatitative requirement for calcium and phosphorus in young broiler chickens. Poultry Science 75:95-110.

National Research Council, 1994. Nutrient requirements of poultry. 9th revised edition. Natinal Academy Press, Washington, DC.

Nelson, T. S. 1967. The utilization of phytate phosphorus by poultry – a review. Poultry Science 46:862-871.

Nelson, T.S., L. W. Ferrara, and N. L. Storer, 1968. Phytate content of feed ingredients derived from plants. Poultry Science 67:1372-1374.

Nelson, T. S., T. R. Sheih, R. J. Wodjinski and J. H. Ware, 1971. Effect of supplemental phytase on the utilization of phytate phosphorus by chicks. J. of Nutrition 101:1289-1293.

Parsons, C. M., 1998. The effects of dietary available phosphorus and phytase level on long-term performance of laying hens. In: The Use of Natuphos Phytase in Layer Nutrition and Waste management. BASF Technical Symposium, Atlanta. pp 24-35.

Potchanakorn, M. and L. M. Potter, 1987. Biological values of phosphorus from various sources for young turkeys. Poultry Science 66:505-513.

Potter, L.M., M. Potchanakorn, V. Ravindran, and E. T. Kornegay, 1995. Bioavailability of phosphorus in different sources using body weight and toe ash as response criteria. Poultry Science 74:813-820.

Quian, H., E. T. Kornegay, and D. M. Denbow, 1996. Phosphorus equivalence of microbial phyatse in turkey diets as influenced na Ca:P ratios and P levels. Poultry Science 75:69-81.

Raboy, V. and D. B. Dickinson, 1993. Phytic acid levels in seed of Glycine max and G. soja as influenced by phosphorus status. Crop Science 33:1300-1305.

Raboy, V., S. J. Hudson, and D. B. Dickinson, 1985. Reduced phytic acid content does not have an adverse effect on germination of soybean seeds. Plant Physiology 79:323-325.

Ravindran, V., W. L. Bryden, and E. T. Kornegay, 1995. Phytate: occurrence, bioavailability and implications in poultry nutrition. Poultry and Avian Biol. Rev. 6:125-143.

Sandbrg, A. S., H. Andersson, N. G. Carlsson, and B. Sandstrom, 1987. Degradation products of bran phytate formed during digestion in the human small intestine: Effect extrusion cooking on digestibility. J. Nutrition 117:2061-2065.

Simons P. C. M., H. A. J. Bersteegh, A. W. Jongbloed, P. A. Kemme, P. Slump, K. D. Bos, M. G. E. Wolters, R. F. Beudeker, and G. J. Vershoor, 1990. Improvement of phosphorus availability by microbial phyatse in broilers and pigs. British J. of Nutrition 64:525-540.

Skinner, J. T., and P. W., Waldroup, 1992. Effects of calcium and phosphorus levels fed in starter and grower diets on broilers during the finishing phase. J. Applied Poultry Research 1:273-279.

Stilborn, H. L., 1997. Progress in development of corn for nutrient potential. Maryland Nutrition Conference. pp 1-8.

Stillborn, H. L., 1998. Cultivars: Utilizing plant genetics to reduce nutrient loading in the environment. In: Proceedings of the National Poultry Waste Management Symposium. pp 154-159.

Twining, P. F., R. J. Lillie, E. J. Robel, and C. A. Denton, 1965. Calcium and phosphorus requirements of broiler chickens. Poultry Science 44:283-296.

Van Der Klis, J.D. and H. A. J. Versteegh, 1996. Phosphorus nutrition in broilers. Pages 71-83 in: Recent Advances in Animal Nutrition. Ed.: P. C. Garnsworty, J. Wiseman, and W. Haresign. Nottingham University press, Nottingham, UK.

Van Der Klis, J. D. and H. A. J. Versteegh, 1997a. The degradation of inositol phosphate in broilers. 1. The effect of dietary calcium and absorbable phosphorus content. 11th European Symp. on Poult. Nutr., World's Poult. Sci. Assoc. p 465-467.

Van Der Klis, J. D. and H. A. J. Versteegh,. 1997b. The degradation of inositol phosphate in broilers. 2. The effect of age. 11th European Symp. on Poult. Nutr., World's Poult. Sci. Assoc. p 468-470.

Van Der Klis, J. D. and H. A. J. Versteegh, 1997c. The role of 1.25-dihydroxycholecalciferol in the degradation of inositol phosphates in broilers. 11th European Symp. on Poult. Nutr., World's Poult. Sci. Assoc. p 471-473.

Van Der Klis J. D., H. A. J. Versteegh, P. C. M. Simons and A. K. Kies, 1997d. The efficacy of phyase in corn-soybean meal-based diets for laying hens. Poultry Science 76:1535-1542.

Van Kempen, T. A. T. G. and P. H. Simmins, 1997. Near –infrared reflectace spectroscopy in precision feed formulation. J. Applied Poulttry Research 6:471-477.

Waldroup, P. W., 1998. Nutritional aproaches to reducing phosphorus excretion in broilers. In: Proceedings of the Multi-State Poultry Meeting, Indianapolis, May 19-21, 1998.

Weibel, P. E., N. A. Nohorniak, H. E. Dzuik, N. M. Walser, and W. G. Olson, 1984. Bioavailability of phosphorus in commercial phosphate supplements for turkeys. Poultry Science 63: 730-737.

Zyla, K., D. R. Ledoux, A. Garcia, and T. L. Veum. 1995a. An in vitro procedure for studying enzymatic dephosphorylation of phytate in maize-soyabean feeds for turkey poults. British J. Nutrition 74:3-17.

Zyla, K., D. R. Ledoux, and T. L. Veum, 1995b. Complete enzymatic dephosphorylation of corn-soybean meal feed under simulated intestinal conditions of the turkey. J. Agric. Food Chem. 43:288-294.

Zyla, K., D. R. Ledoux, M. Kujawski, and T. L. Veum, 1996. The efficacy of an enzymatic cocktail and a fungal mycelium in dephosphorylating corn-soybean meal-based feeds fed to growing turkeys. Poultry Science 75:381-387.

Zyla, K., J. Koreleski., and S. Swiatkiewicz, 1997. Simultaneous application of phytase and xylanase to broiler feeds based on wheat. Preliminary in vitro studies and feeding trial. 11th European Symposium on Poultry Nutrition, World's Poultry Science Association. pp 474-476.

●●●●●●●●●●●●●●●●●●●●●●●●●●●●●●●●●●●

Managing Swine Feeding to Minimize Manure Nutrients

Allen F. Harper, Ph.D.
Extension Animal Scientist-Swine
Virginia Polytechnic Institute and State University
Tidewater Agricultural Research and Extension Center
6321 Holland Road, Suffolk, Virginia 23437

Biographies for most speakers are in alphabetical order after the last paper.

●●

Swine manure, as collected in liquid or slurry form on commercial hog farms, consists of feces, urine and varying amounts of wasted drinking and wash water. Contained within the manure are significant quantities of biologically essential nutrients, mostly notably, nitrogen (N) and phosphorus (P). Other nutrients of biological importance present in swine manure include potassium, calcium, copper, zinc, manganese and other minor elements. The majority of N and P in manure originates in feedstuffs and represents N and P not retained in the pig for maintenance, growth and reproduction. The proportion of consumed N and P excreted by the pig is high. Estimated excretion rates as a percentage of intake range from 60 to 80 % (Jongbloed and Lenis, 1992; Table 1).

Table 1. Intake, excretion and retention of nitrogen (N) and phosphorus (P) in swine.

		Intake, lbs.	Excretion, lbs.	Retention, %
Starter pig (20-55 lbs.)	N:	2.07	1.23	40%
	P:	.46	.29	37%
Grower-finisher pig (55-235 lbs.)	N:	13.93	9.35	33%
	P:	2.69	1.81	33%
Breeding sow (19.6 piglets/year)	N:	61.78	49.43	20%
	P:	14.46	11.95	17%

(Adapted from Jongbloed and Lenis, 1992)

As essential plant nutrients, waste N and P from swine feeding can and should be considered an asset to crop and forage production. However, the nutrients that are excreted in manure must be recycled onto cropland in a manner that is not damaging to the environment. Furthermore, limitations on the amount of available manure storage capacity, and the amount of crop acres available for application, necessitate that the levels of excreted nutrients be managed and controlled. This is also important from an economic standpoint because excess excretion of N and P is not cost effective for the swine feeding enterprise. Therefore, the goal of swine feeding management is to optimize the levels of N and P (and other nutrients) retained in the pigs being fed. In doing so, the amount of waste nutrients are kept at a manageable level such that they can be used for beneficial purposes without excess available for potential contamination of ground or surface waters. This concept has been referred to as *environmental nutrition* (Kornegay and Harper, 1997). The Following discussion offers nutrition and feeding strategies that can assist in reducing the level of nutrients excreted into swine manure.

Minimize Feed Waste

Excessive feed waste on commercial hog farms is a problem that can occur when operators or managers fail to recognize the importance of monitoring and preventing feed waste on a regular basis. An obvious negative impact of excessive feed waste from self-feeders and other feeding equipment is a substantial increase in feed costs per unit of pork sold. For example, in a swine finishing operation, a five percent level of feed waste may result in income loss of approximately $1.80 per hog (Harper, 1994; Table 2).

Table 2. Feed waste impacts nutrient management.

Percent feed waste	Feed loss per pig, lb.	Income loss per pig, $	Feed N waste per pig, lb.	Feed P waste per pig, lb.
1%	6	.36	.14	.04
3%	18	1.08	.43	.11
5%	30	1.80	.72	.18
7%	42	2.52	1.01	.25

Based on growing-finishing pigs from 50 to 250 lb. body weight, 3:1 feed:gain ratio, 2.4% N and .60% P in the diet and $.06/lb. diet cost.

Another important problem with excess feed waste is the unnecessary addition of N, P, and other nutrients into the swine waste storage system. At a 5 % level of feed waste in a swine finishing operation, an additional .72 lbs. of N and .18 lbs. of P per hog may be added to the manure storage and distribution system (Table 2). This may be equivalent to 10 to 15 % of the total N and P in the system. Use of proper feeder designs and regular maintenance and adjustment of feeder equipment is essential to prevention of excess feed waste. Well managed swine farms make prevention of feed waste a routine objective for both economic and environmental reasons.

Optimize Feed Conversion Efficiency

The amount of feed consumed per unit of pig growth (feed/gain ratio or feed efficiency) has a direct impact on economic returns for swine producers. Even for producers that are feeding swine under contract with no direct costs for feed have economic incentives to optimize feed efficiency. This is because contractors typically offer bonus payments for improved feed efficiency with each group of pigs fed. In growing-finishing pigs being fed from 50 to 250 lbs., feed/gain ratio may range from very inefficient levels as high as 3.5 to very efficient levels as low as 2.5.

Improved feed efficiency has direct implications for nutrient excretion as well. As the amount of feed consumed per unit of growth is reduced the quantity of nutrients excreted is also reduced. Coffey (1992) has reported that reducing feed/gain ratio in finishing pigs by .25 units (3.00 vs. 3.25) would reduce overall N excretion by 5 to 10%. A similar report by Henry and Dourmad (1992) indicates that for each .1 unit reduction in feed/gain ratio, N output is reduced by 3%.

A number of factors contribute to optimal feed efficiency. The most predominant is genetic capacity. Selection and use of breed crosses or genetic lines of pigs with high potential for good feed conversion will have a measurable effect on feed efficiency. Environmental factors such as maintaining barn temperature within or near the "thermoneutral" zone and ventilating the barn for good air quality will also contribute to better feed efficiency. Feeding high quality diets and certain feed processing methods also have impact. For example pelleting swine diets has been shown to improve feed efficiency as compared to feeding diets in meal form. Use of growth promoting feed additives and performance modifiers may also improve feed efficiency. But use of these agents is strictly regulated and may not be feasible in all situations.

Avoid Over-formulation of Diets

The goal of swine nutrition is to formulate diets to contain the correct concentration of energy and essential nutrients for optimal growth or reproductive performance and carcass quality. Traditionally nutritionists have included slightly elevated concentrations of certain essential nutrients to insure that nutrient needs were met. This practice of slight "over-formulation" of diets beyond the expected nutrient requirement of the pig was regarded as a safety factor for optimal performance. A major disadvantage of this practice is that feeding dietary nutrient concentrations that are beyond the pig's true requirement results in a higher rate of nutrient excretion. For example, Dourmad and Corlouer (1993) fed grower-finisher pigs three dietary protein concentration regimes, all of which were nutritionally adequate and all of which produced similar pig growth performance. However, pigs fed the diets that were excessive in dietary protein concentration produced as much as 29% more manure N and 32% more N emissions than pigs fed the lower but more nutritionally accurate protein level (Figure 1).

A similar situation exists with P concentration of the diet. Latimer and Pointillart (1993) compared performance and P excretion of growing-finishing pigs fed diets with either

.4%, .5%, or .6% total P. The .4% dietary P level was determined to be nutritionally inadequate because pig performance was reduced. Pigs fed the .5% or .6% dietary P levels performed equally well. But, those fed the .6% P diet excreted 33% more P than those fed the .5% P diet (Figure 2).

Figure 1. Dietary protein effects N excretion. (Dourmad and Corlouer, 1993).

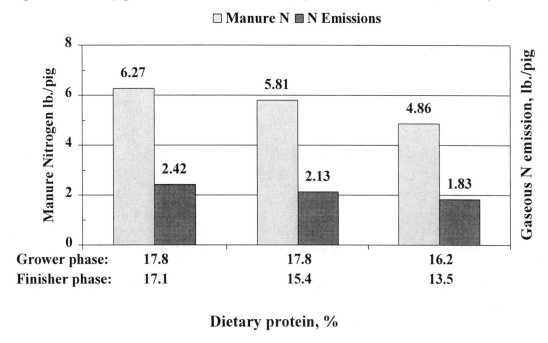

| Grower phase: | 17.8 | 17.8 | 16.2 |
| Finisher phase: | 17.1 | 15.4 | 13.5 |

Figure 2. Dietary P level effects P excretion. (Latimer and Pointillart, 1993).

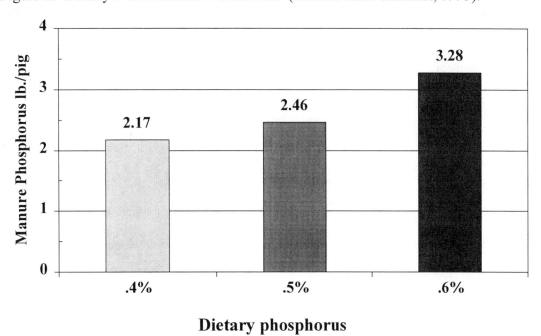

Directly related to prevention of over-formulation is accurate knowledge of pig nutrient requirements and accurate estimates of the nutrient content of feedstuffs. As the practice of swine nutrition becomes more refined, professionals formulating diets will use the best information available to optimize animal performance and minimize nutrient excretion. For example, the recently revised "Nutrient Requirements of Swine" (NRC, 1998) now includes a computer model that allows the formulator to predict nutrient requirements based on the size, sex and genetic capacity of the pigs being fed. The model also allows for adjustments with different environmental conditions in swine facilities.

Feeding for Optimum Rather than Maximum Performance

Professionals involved in swine feeding enterprises have begun to realize that feeding for maximum animal performance is not necessarily the best objective. Optimal growth and performance would be a better objective. In this case diets are formulated with the intent that the pig will closely approach, but not fully attain maximum performance. Under this objective, economic return over feed cost may actually be higher. And, just as important, nutrient excretion may be reduced.

The basis for feeding nutrient concentrations at levels slightly below that necessary for maximum performance response may best be explained with the concept of diminishing returns (Kornegay and Harper, 1997). As the dietary concentration of N (protein and amino acids), P or other essential nutrients increases from a deficient level to an adequate level, the pig responds with improved growth performance. However, as performance approaches a maximal level, each incremental increase in dietary nutrient concentration results in progressively smaller improvements in performance (Figure 3). At some point, increased dietary nutrient concentration cannot be justified because the magnitude of response in pig performance is too small to be economically justified. Furthermore, an additional problem will be the unnecessary increase in nutrient excretion. This concept must be balanced with the understanding that optimization of different performance traits may actually have different nutrient requirements. For example, a higher dietary protein and amino acid concentration (the significant N containing part of the diet) is required for optimal carcass leanness than is needed for optimal growth rate. This is important because market hog prices at packing plants are adjusted according to percentage of carcass lean.

Figure 3. Example of diminishing returns for nutrient inputs as the level of nutrient fed increases. At point A, one unit of input produces 0.27 units of gain, whereas, at point B, one unit of input produces 0.05 units of gain.

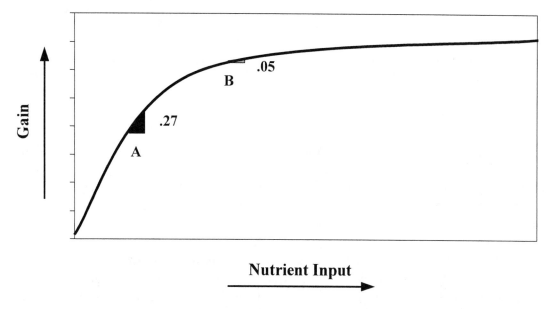

Phase Feeding and Separate Sex Feeding

Nutrient requirements as a percentage of the total diet change as pigs grow heavier (NRC, 1998). The challenge for nutritionists and swine producers is to determine how frequently or how many diet changes will be made from the nursery period (12 to 45 lbs.) and throughout the growing-finishing period up to market weight (about 250 lbs.). Historically, nutritionists and producers have chosen a less complicated approach with only a limited number of diet "phases" during grow-out. A disadvantage of this approach is that there may be extended periods during which the pigs are receiving higher concentrations of nutrients than actually required and nutrient excretion would be increased.

More frequent diet formulation changes during grow-out allow for more precision in meeting nutrient requirements of the pig. Henry and Dourmad (1992) have demonstrated that a 3-phase feeding regime can reduce N excretion in growing-finishing pigs as compared to a 2-phase or single-phase regime. In this work the 3-phase program provided dietary crude protein levels of 17% (55 to 110 lbs. pig weight), 15% (110 to 165 lbs. pig weight) and 13% (165 lbs. to market weight). Pigs fed in this manner excreted 8% less N than pigs fed a 17% protein (55 to 120 lbs. pig weight) followed by 15% protein (120 lbs. to market) (Table 3).

Table 3. Effect of feeding strategy during the growing-finishing period (55 to 230 lb) on N output.*

Item	Crude Protein, %		
	Single-feed 17%	Two-feeds[a] 17-15%	Three-feeds[b] 17-15-13%
N output, lb/day	0.070	0.064	0.059
% of two-feed strategy	110	100	92

[a]Crude protein changed at 120 lb.
[b]Crude protein changed at 110 and 165 lb.
*Adapted from Henry and Dourmad (Feed Mix, May 1992).

It has been recognized for some time that gilts have a different body composition than barrows during the finishing phase and at market weight. Gilts have slightly reduced feed consumption, deposit less body fat and deposit more lean tissue during growth resulting in a higher percent lean content at slaughter. Consequently, gilts have higher nutrient density requirements than barrows (NRC, 1998).

Many intensively managed hog farms are now taking advantage of these sex differences in finishing pigs. Typically pigs are separated into uniform weight groups of gilt only pens and barrow only pens during the grower-finisher period. The benefits of separate-sex feeding are realized during the finisher phases but the most practical time to segregate the pigs by sex is going into or coming out of the nursery. Under these conditions, gilts are fed specific diets to meet their nutrient needs while barrows are fed a more economical, lower nutrient density diet that specifically meets the needs for their body composition. As with phase feeding, this allows more precise diet formulation for the specific type of pig being fed. In addition there is a reduction in overfeeding certain nutrients and a reduction in nutrient excretion.

Use of High Quality Protein Sources and Crystalline Amino Acids

The protein component of swine diets consists of amino acids, all of which contain N. The pig does not have a requirement for protein per se, but for the essential amino acids that are the components of protein in feedstuffs. High quality protein supplements such as soybean meal or fish meal have an abundant and well balanced supply of amino acids for the pig. Poorer quality protein supplements such as peanut meal or cottonseed meal may be used in limited quantities as protein supplements, but if excessive quantities are used, N excretion can be increased. This may be due to the fact that excessively high total protein levels may have to be fed with these ingredients in order to meet the limiting amino acid requirement of the pig. Or, protein digestibility may be poorer in some lower quality protein supplements.

Even with high quality protein supplements like soybean meal, some excess feeding of total protein and N may occur. This is because enough soybean meal must be included in the diet to meet the pig's requirement for the first limiting essential amino acid, lysine. Consequently, when meeting the pig's requirement for lysine by adding soybean meal to the diet, some excess of total protein and N is fed. The development of synthetic lysine and other crystalline amino acids has given swine nutritionists and producers a highly effective means to reduce total protein content of pig diets while still maintaining good growth performance and carcass leanness. With typical corn-soybean meal based diets, total crude protein level can be reduced up to 2 percentage units if crystalline lysine is added to the diet to balance the pig's lysine requirement with no loss in performance (Cromwell and Coffey, 1994; Table 4). Furthermore such a practice can reduce N excretion from growing pigs by up to 22% (Table 5). Even greater reductions in dietary N content and excretion may become possible as crystalline sources of other limiting amino acids such as threonine, tryptophan and methionine become more economical to use.

Table 4. Performance of finisher pigs fed low protein, lysine supplemented diets. (Cromwell and Coffey, 1994).

Diet	Daily gain, lb.	Feed/Gain
14% C.P.	1.54	3.16
12% C.P	1.36	3.52
12% C.P. + .15% Synthetic lysine	1.52	3.18

Table 5. Theoretical model of the effects of reducing dietary protein and supplementing with amino acids on N excretion by 200 lb. finishing pigs.[a]

N balance	14% CP	12% CP + Lys	10% CP + Lys + Thr + Trp + Met
N intake, g/d	67	58	50
N digested and absorbed, g/d	60	51	43
N excreted in feces, g/d	7	7	7
N retained, g/d	26	26	26
N excreted in urine, g/d	34	25	17
N excreted, in total, g/d	41	32	24
Reduction in N excretion, %	-	22	41

[a]Assumes intake of 3,000 g/d, a growth rate of 900 g/d, a carcass lean tissue gain of 400 g/d, a carcass protein gain of 100 g/d (or 16 g of N/d), and that carcass N retention represents 60% of the total N retention. Adapted from Cromwell (1994).

Improve the Availability of Phosphorus in Feeds with Phytase

Much of the P in grains and oilseed meals is chemically bound in molecules called phytate. Phytate is the mineral salt of phytic acid and serves as the storage form of P within plant seeds. Approximately 66% of the P in corn and 61% of the P in soybean meal is bound in phytate form. Certain rumen bacteria in cattle and sheep produce an enzyme called phytase that liberates P from phytate making it available to the animal. Nonruminant animals such as pigs do not produce intestinal phytase; therefore, phytate bound P in feedstuffs is essentially unavailable to pigs.

To meet the pig's total dietary requirement for P, inorganic P sources, such as dicalcium phosphate or deflourinated phosphate, are added to the diet. The amount of P excreted in the manure and urine of pigs is equal to the amount of total P in the diet less the amount utilized and retained by the pig. Total P excretion would be reduced if phytate bound P could be made nutritionally available to the pig and additions of inorganic phosphorus to the diet were reduced.

The concept of using microbially derived phytase to improve phytate phosphorus utilization in non-ruminants was put forth many years ago (Nelson et al., 1971). However, the low yield and slow process of producing phytase from microbes such as *Aspergillus niger* made the technology too costly. More recently, through recombinant DNA technology, strains of *Aspergillus niger* have been developed that allow production of large, economical quantities of phytase for use in the feed industry. Commercial sources of feed grade phytase have recently been approved by the Food and Drug Administration and are available to the swine and poultry feeding industries.

Animal nutritionists in Holland and other northern European countries became very interested in investigating phytase after government mandates were implemented to reduce phosphorus (P) addition to soils from animal manure. This earlier work demonstrated that addition of phytase to low-P broiler or pig diets greatly enhanced phosphorus digestibility and substantially reduced P levels in the feces (Simons et al., 1990; Jongbloed et al., 1992). The percentage reduction in manure P content ranged from 25 to 45 % when compared to adequate P diets without phytase addition.

Recent studies have demonstrated that phytase supplementation of low-P pig diets can be effective under U.S. feeding conditions as well. For example, addition of phytase over a range of 0 to 500 U/kg to low P diets produced positive dose responses in growth performance, P digestibility, bone ash and bone strength in weanling pigs (Radcliffe and Kornegay, 1998; Table 6) and growing-finishing pigs (Harper et al., 1997; Table 7). Responses to these same traits were effected in a similar manner by increasing inorganic P level in diets without phytase. At the 500 U/kg phytase dose, growth and bone traits were restored to a level similar to the positive control diet containing elevated P. Excretion of P in the manure was reduced by a factor of 20% with the 500 U/kg phytase dose.

Table 6. Phytase and inorganic-P supplementation of weanling pig diets.[a]

Diet		Response trait		
P[b] (%)	Phytase (U/kg)[b]	Weights gain (lb./day)	P digestibility (%)	Rib shear strength (N)
.35	0	.80	21.5	442
.35	167	.90	28.3	488
.35	333	.92	34.3	500
.35	500	.96	35.4	580
.35	0	.80	21.5	442
.40	0	.88	32.0	523
.45	0	.97	39.1	627
.50	0	.98	37.1	702

[a] Data from Radcliffe and Kornegay, 1998.
[b] Significant linear effects of increasing phytase or phosphorus (P<.01) in all traits shown

Table 7. Phytase and inorganic-P supplementation of growing-finishing pig diets.[a]

Diet		Response trait		
P[b,c] (%)	Phytase[b] (U/kg)	Weight gain (lb./day)	P digestibility (%)	Rib shear strength (kg)
.38G/.33F	0	1.93	28.7	1495
.38G/.33F	167	2.02	32.6	1580
.38G/.33F	333	2.05	37.4	2000
.38G/.33F	500	2.06	41.3	1990
.38G/.33F	0	1.93	28.7	14.95
.42G/.37F	0	2.04	33.9	1620
.46G/.41F	0	2.06	39.5	1880

[a] Data from Harper et al., 1997.
[b] Linear effects of phytase (P<.01) and phosphorus (P<.01 to .08) on all traits shown.
[c] Phosphorus levels indicated for grower (G) and finisher (F) phases.

Because the dietary Ca:P ratio can have a significant impact on P availability, researchers have assessed impact of Ca: P ratio on efficacy of phytase. In finishing pigs, Liu et al. (1996) demonstrated that 500 U/kg of supplemental phytase was more effective in restoring growth and bone strength when Ca level was reduced to provide a Ca: P ratio of 1:1 than when the Ca: P ratio was 1.5:1. The results indicate that that maintaining the dietary Ca: total P ratio within a narrow range or near 1:1 is necessary for optimal phytase responses in pigs fed low-P diets. Subsequent studies indicate that phytase also releases some phytate bound calcium. Excessive addition of inorganic calcium to phytase supplemented diets may actually interfere with P utilization due to an excessively "wide" (greater than 1.3:1) dietary calcium to P ratio.

A final note on phytase concerns temperature stability. Like most enzymes, phytase is susceptible to heat degradation. Pelleting or other feed processing methods that involve temperatures above 60 degrees Celsius is likely to result in reduced phytase activity. Fortunately, technology and equipment that allows post-pellet application of phytase enzyme is available and effective. In these systems the phytase preparation is sprayed onto the feed pellets after they have been manufactured and cooled. Several commercial feed mills in the U.S. have installed spray-on enzyme application equipment for this purpose. For diets that are to be fed in meal or mash form with no heat treatment, phytase may simply be added at the mixer.

Summary

A significant portion of the N, P and other nutrients fed to pigs are excreted into the manure storage and handling system. Waste nutrients in swine manure storage can be considered an asset as nutrients for crop or forage production. However, keeping the total quantity of N, P and other manure nutrients at manageable levels is important so that storage capacity and land application area remains adequate for manure nutrient application plans.

Many of the current nutrition and feeding technologies intended to optimize pig performance and economic return over feed costs also have a positive impact on reducing excessive excretion of N, P and other nutrients into swine manure storage systems. These include minimization of feed waste, optimizing feed conversion efficiency, avoiding nutrient over-formulation, feeding for optimum rather than maximum performance, phase feeding and separate sex feeding, use of high quality protein supplements and crystalline amino acids, and finally, improvement of dietary P availability with supplementation of phytase. Some of these methods will have greater individual impact than other methods. When these methods are used collectively, a substantial reduction in nutrient excretion into the swine waste stream may be accomplished.

References

Coffey, M. T. 1992. An industry perspective on environmental and waste management issues: Challenge for the feed industry. Georgia Nutrition Conference Prodeedings, p. 144, Athens, Georgia.

Cromwell, G. L. 1994. Feeding strategies urged as techniques to decrease pollution from hog manure. Feedstuffs, July 25, p. 9.

Cromwell, G. L. and R. D. Coffey. 1994. Nutritional methods to reduce nitrogen and phosphorus levels in swine waste. Presentation Paper, Southern Section of the American Society of Animal Science Meetings, Feb. 8, 1994, Nashville, Tennessee.

Dourmad, J., and A. Corlouer. 1993. Effects of three protein feeding strategies for growing-finishing pigs on growth performance and nitrogen output in slurry. 25th French Swine Days Report 25: 54.

Harper, A. F. 1994. Feeding technologies to reduce excess nutrients in swine waste. In Proceedings: Meeting the Challenge of Environmental Management on Hog Farms. Second Annual Virginia Tech Swine Producers Seminar. August 4, p. 44, Carson, Virginia.

Harper, A. F., E. T. Kornegay, and T. C. Schell. 1997. Phytase supplemetation of low phosphorus growing-finishing pig diets improves performance, phosphorus digestibility and bone mineralization, and reduces phosphorus excretion. J. of Animal Science 75: 3174-3186.

Henry, Y., and J. Y. Dourmad. 1992. Protein nutrition and nitrogen pollution. Feed Mix (May issue): 25.

Jongbloed, A. W., and N. P. Lenis. 1992. Alteration of nutrition as a means to reduce environmental pollution by pigs. Livestock Production Science 31:75-94.

Jongbloed, A. W., Z. Mroz, and P. A. Kemme. 1992. The effect of supplemental *Aspergillus Niger* phytase in diets for pigs on concentration and apparent digestibility of dry matter, total phosphorus, and phytic acid in different sections of the alimentary tract. J. of Animal Science 70:1159-1168.

Kornegay, E. T., and A. F. Harper. 1997. Environmental Nutrition: Nutrient management strategies to reduce nutrient excretion of swine. The Professional Animal Scientist 13:99-111.

Latimer, P., and A. Pointillart. 1993. Effects of three levels of dietary phophorus (.4- .5- .6% P) on performance, carcass traits, bone mineralization and excreted phophorus of growing-finishing swine. 25th French Swine Days Report. 25:52.

Liu, J., D. W. Bollinger, D. R. Ledoux, and T. L. Veum. 1996. Effects of calcium concentration on performance and bone characteristics of growing-finishing pigs fed low phosphorus corn-soybean meal diets supplemented with microbial phytase. J. of Animal Science 74(Suppl. 1):180(Abstr.).

NRC. 1998. Nutrient requirements of swine (10th edition). National Academy Press, Washington D. C.

Nelson, T. S., R. R. Shieh, R. J. Wodzinski, and J. H. Ware. 1971. Effect of supplemental phytase on the utilization of phytate phosphorus by chicks. J. of Nutrition 101:1289-1294.

Radcliffe, J. S., and E. T. Kornegay. 1998. Phosphorus equivalency value of microbial phytase in weanling pigs fed a maize-soyabean meal based diet. J. of Animal and Feed Sciences 7:197-211.

Simons, P. C. M., H. A. J. Versteegh, A. W. Jongbloed, P. A. Kemme, P. Slump, K. D. Bos, M. G. E. Wolters, R. F. Beudeker, and G. J. Verschoor. 1990. Improvement of phosphorus availability by microbial phytase in broilers and pigs. British J. of Nutrition 64:525-540.

Managing the Dairy Feeding System to Minimize Manure Nutrients

T. P. Tylutki
Department of Animal Science
Cornell University, Ithaca, New York

D. G. Fox
Department of Animal Science
Cornell University, Ithaca, New York

Biographies for most speakers are in alphabetical order after the last paper.

Introduction

Applying manure nutrients in excess of crop requirements increases the risk of nutrients leaking into surface and groundwater (Fox et al., 1998; Wang et al., 1999). Reducing nutrient excretion in dairy cattle involves two main areas: lowering the levels of nutrients fed (especially purchased nutrients) above animal requirements and increasing the overall farm management. Rations have typically included safety factors for many nutrients. These safety factors are included to minimize production risk due to not being able accurately predict animal requirements, and variation in composition of the ration delivered to each group of animals in each unique production situation. Increased knowledge about the supply and requirements for nutrients in ruminants coupled with improved analytical methodology for feeds and management of critical control points in delivering the formulated diet to the intended group of animals is allowing us to decrease these safety factors.

A major factor limiting the adoption of very low safety factors is the management level required on the farm. Management's role in decreasing nutrient excretion covers many areas. High levels of management are required throughout the entire feeding system. This includes silo management, adequate feed analysis to describe the feeds fed, feeding accuracy, feed bunk management, and other areas One of the most critical variables affecting nutrient accumulation on the farm and profitability is the quality and quantity of homegrown feeds

(Wang , 1999). Homegrown feed quality includes: harvesting at the correct stage of maturity and dry matter, storage management, and minimizing variation in the feed. Tylutki et. al. (1999) simulated the impact of observed dry matter and fiber variation with corn silage at harvest. Variation observed in this feed resulted in variations in income over feed costs greater than $40,000 and nitrogen excretion of 240 pounds per 100 cows annually.

In this paper, we summarize the prediction of the supply and requirements of phosphorus and nitrogen, and feeding management required to decrease nutrient excretion. We end with an example of how to implement recommended management practices, using a case-study farm that has been implementing them.

Accurately meeting requirements for phosphorus

The first step is using accurate feed composition values in ration formulation. *Typical* phosphorus levels and normal ranges for many feedstuffs analyzed by DairyOne are found in Table 1. These levels can be quite different from 'book' values. As an example, NRC (1989) reports a P level of .22% for alfalfa hay early bloom. The average analyzed value is .26%, or 18% higher than the NRC value. This demonstrates the need for laboratory analysis of feeds used in ration formulation.

One form of phosphorus is bound to phytate. In ruminants, this is not a concern as the rumen produces high levels of the enzyme phytase. Estimates of phytate digestibility in the rumen are in excess of 99% (Morse et. al., 1992). Inorganic sources of P vary in their availability. They can be ranked (from highest to lowest availability) as: sodium phosphate, phosphoric acid, monocalcium phosphate, dicalcium phosphate, defluorinated phosphate, bone meal, and soft phosphates (NRC, 1989). As can be seen in Table 1, the high protein feeds (e.g. soybeans) are high in phosphorus.

The next step is to accurately determine P required. Nutrient requirements are often expressed as dietary percentages. However, this only represents the concentration of a nutrient needed when the assumed amount of dry matter intake is consumed by the animal. As we move towards decreasing nutrient excretion, diets need to be formulated and evaluated based upon the grams of nutrient fed compared to the grams required. As an example, differences in diets containing .41% P versus .40% P may appear unimportant. When this difference is computed on an annual basis per 100 cows, this difference becomes 161 pounds of additional excreted P to manage. In addition, the concentration of a diet may appear higher or lower than expected due to differences in dry matter intake. As an example, a requirement of 100 grams of P for a cow consuming 55 pounds of dry matter is a .40% concentration required. At 40 pounds of dry matter intake, the concentration required increases to .55%.

Approximately 86% of the phosphorus in dairy cattle is in the skeleton and teeth (NRC, 1989). It is a key mineral in energy metabolism, and is an essential component of blood and other body fluid buffering. Phosphorus is absorbed in the small intestine. Absorption is dependant on the P source, level of intake, intestinal pH, animal age, and the amount of other minerals in the diet. If P is fed in adequate amounts, the calcium to phosphorus ratio does not seem to be a concern. There is little published information regarding the absorption efficiency of different feedstuffs, and phosphorus recycling increases the difficulty in obtaining these numbers (NRC, 1989).

Table 1. Average Phosphorus content of feeds (adapted from Chase, 1999).

Feed	Mean	SD	Normal Range
Legume Hay	.26	.06	.21 - .32
Legume Silage	.32	.06	.27 - .38
Grass Hay	.24	.08	.16 - .32
Grass Silage	.31	.07	.24 - .38
Corn Silage	.23	.03	.2 - .36
Bakery byproduct	.40	.08	.32 - .49
Barley grain	.28	.16	.12 - .44
Beet pulp	.10	.03	.06 - .13
Blood meal	.20	.16	.05 - .39
Brewers grain	.62	.06	.56 - .68
Canola meal	1.14	.16	.98 - 1.29
Corn, ear	.30	.05	.26 - .35
Corn, shelled	.32	.11	.2 - .43
Corn germ meal	.71	.54	.17 - 1.25
Corn gluten feed	.90	.21	.68 - 1.11
Corn gluten meal	.77	.41	.37 - 1.18
Cottonseed hulls	.21	.08	.13 - .29
Cottonseed meal	.97	.28	.69 - 1.24
Cottonseed, whole	.66	.11	.55 - .78
Distillers grains	.82	.12	.71 - .94
Feather meal	.28	.06	.22 - .33
Fish meal	3.39	1.14	2.25 - 4.53
Hominy feed	.56	.21	.36 - .77
Linseed meal	.92	.11	.81 - 1.03
Meat meal	4.35		
Meat and bone meal	3.05	.98	2.07 - 4.04
Molasses	.68	1.20	up to 1.88
Oats	.39	.06	.32 - .45
Soyhulls	.17	.12	.05 - .29
Soybeans	.66	.11	.55 - .76
Soybean meal, 48	.68	.11	.57 - .79
Wheat	.47	.23	.24 - .69
Wheat bran	1.03	.31	.72 - 1.34
Wheat midds	.88	.21	.67 - 1.08

Post-absorption, large amounts of P are recycled through the salvia (NRC, 1989). Excess P is then excreted in the feces (Very little P is excreted through the urine.) (INRA, 1989). Preliminary results from INRA suggests that as P levels increase with increasing levels of concentrate feeding, P begins spilling into the urine (Agabriel, personnel communication, 1999).

Much research over the last several years has focused on the impact of phosphorus on reproductive efficiency of lactating cows (Satter and Wu, 1999). Satter and Wu (1999) summarized 13 trials where P levels were varied from .32 to .61% of the diet. No differences were found in any of the trials in days to first estrus, days open, services per conception, days to first breeding, or pregnancy rate. Satter and Wu (1999) also summarized the data from several trials that varied the P level to determine differences in milk production. Again, as long as the grams of P required daily were met, no differences were seen in milk production.

In an attempt to improve accuracy in predicting dietary requirements, the Cornell Net Carbohydrate and Protein System version 4.0 (CNCPSv4.0) calculates the phosphorus (and the other macro-minerals) requirements for cattle using the INRA (1989) system. Table 2 lists the equations used for various classes of cattle and Table 3 shows example calculations based on these equations. The INRA system was chosen for macro-mineral calculations due to its factorial approach. This system describes net macro-mineral requirements by physiological function (maintenance, lactation, growth, and pregnancy). The maintenance requirements are further partitioned into endogenous fecal and urinary losses. Varying transfer coefficients (based upon body weight or physiological state) are then applied to the net requirements to calculate dietary requirements. The INRA system utilizes a Total Absorption Coefficient (TAC) to convert net P required to dietary P required. The TAC is a combination of absorption efficiency as well as P digestibility.

The mineral section of the CNCPSv4.0 can be used to evaluate mineral balances, calculate macro-mineral excretion, and optimize mineral utilization within groups. At the herd level, the mineral section predicts herd phosphorous and potassium excretion, efficiency of mineral use (product/input), and a mass nutrient balance for the feeding program.

Table 2. Equations used to calculate Phosphorus requirements (gms/d) for dairy cattle (INRA, 1989)[1].

		Heifer	Dry cow	Lactating Cow
Maintenance				
	Fecal	(23 * SBW) / 1000	(23 * SBW) / 1000	((22 + (0.2 * Milk)) * SBW) / 1000
	Urinary	(2 * SBW) / 1000	(2 * SBW) / 1000	(2 * SBW) / 1000
Growth		(7 * SWG)	(7 * ADG)	(7 * ADG)
Pregnancy		If Days Pregnant > 187 Then Pregnancy Requirement = 4		
Lactation				(0.9 * Milk)
Total		SBW < 150, 80%	57.5%	57.5%
Absorption		SBW < 250, 75%		
Coefficient		SBW < 350, 65%		
		SBW > 350, 55%		

[1]Where: SBW = shrunk body weight, kg   Milk = milk production, kg/d
          SWG = shrunk weight gain, kg/d   ADG = average daily gain, kg

Table 3. Phosphorus requirements (gms/d) of various classes of cattle calculated using the equations in Table 2 as applied in the CNCPS v 4.0.

	Heifer	Dry Cow	Lactating Cows		
Body weight, lb	750	1400	1350	1350	1350
Milk production, lb/d	0	0	60	80	100
Gain, lb/d	2	0	0	0	0
Days pregnant	0	200	0	0	0
Expected dry matter intake, lb	15	28	42	48	54
Maintenance requirement, g/d					
Fecal	7.8	14.6	16.8	17.9	19.0
Urinary	0.7	1.3	1.2	1.2	1.2
Growth requirement, g/d	6.4	0.0	0.0	0.0	0.0
Pregnancy requirement, g/d	0.0	4.0	0.0	0.0	0.0
Lactation requirement, g/d	0.0	0.0	24.5	32.7	40.9
Total Absorption Coefficient	65.0	57.5	57.5	57.5	57.5
Total requirement, g/d	22.9	34.6	74.0	90.2	106.3
Dietary concentration, % of DM	0.34	0.27	0.39	0.41	0.43

Satter and Wu (1999) report survey data showing the average lactating dairy cow is fed a diet containing .48% P. As seen in Table 3, this level is in excess of that required to produce 100 pounds of milk. Figure 1 represents the relationship between three P dietary concentrations to daily milk production and the percent of the 1989 Dairy NRC recommendations. It is evident in this Figure that a diet containing .55% P results in severe P over-feeding over this range of milk production. The .45% level also results in excesses over most of this range.

Meeting requirements for Nitrogen

Ration formulation/evaluation systems using the CNCPS model (CNCPSv4, CPM Dairy, DALEX) to more accurately match sources of N with animal requirements partition protein supply into five pools:
1. A fraction: rapidly available non-protein nitrogen
2. B1: rapidly available true protein
3. B2: intermediate ruminal degradation rate
4. B3: slowly degradable
5. C: indigestible, bound to lignin.

These pools are calculated from feed analysis as shown in Table 4.

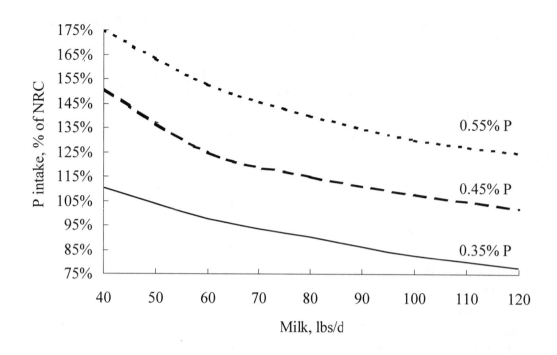

Figure 1. The relationship between daily phosphorus requirement and three levels of intake to milk production (adapted from Chase, 1999).

Table 4. CNCPS protein pools from feed analysis (Sniffen et. al., 1992).

Analytical result and the pool it contains	Protein Pool				
	A[1]	B1[2]	B2[3]	B3[4]	C[5]
Crude Protein	X	X	X	X	X
Soluble Protein	X	X			
N-bound to NDF				X	X
N-bound to lignin					X

[1] A protein = Soluble Protein - NPN
[2] B1 protein = Soluble Protein – A protein
[3] B2 protein = 100 – A protein – B1 protein – B3 protein – C protein
[4] B3 protein = N-bound to NDF (NDICP) – C protein
[5] C protein = N-bound to lignin

The purpose of these pools is to predict dietary N required to maximize rumen microbial growth, and the amount needed to supplement microbial protein to meet animal metabolizable protein requirements. This is accomplished by predicting microbial growth from ruminally-degraded carbohydrates, based on their digestion and passage rates. Then metabolizable protein and amino acids are predicted from intestinally available microbial protein. In order to meet the amino acid demand of high producing cows, feeds are included as needed that have a low ruminal protein degradability to supply feed amino acids to the small intestine to supplement the bacterial amino acids.

Nitrogen requirements need to be viewed as bacterial requirements and animal requirements. In ruminant nutrition, the objective is to maximize rumen microbial growth to supply the cow with energy and protein and then supplement the microbial supply with feed. Properly matching microbial requirements with animal requirements allows for less total protein to be fed resulting in lower nitrogen excretion.

Microbial nitrogen requirements are dependant upon the type of carbohydrate being fermented. Two primary pools of bacteria ferment feed in the rumen: those that ferment NSC, and those that ferment fiber. There is another pool that ferments amino acids; however they represent a small proportion of the total microbial population. Microbes that ferment NSC prefer peptides (B1 and some B2 protein) as their nitrogen source. Adequate peptide levels act as a growth stimulant. In the absence of peptides, they can meet their nitrogen requirements with ammonia (A protein). Fiber fermenting microbes rely strictly on ammonia as their nitrogen source. CNCPS v4 predicts that inadequate ruminal ammonia decreases microbial protein and reduces fiber digestibility (Tedeschi et al., 2000). Excess ruminal ammonia is absorbed through the rumen wall. Some is recycled back to the rumen; the remaining is excreted in urine and milk.

The animal requires protein for maintenance (tissue turnover, scurf, and metabolic fecal), pregnancy, growth, and lactation. Protein supply in excess of requirement is excreted primarily in the urine. This excretion requires energy to convert the excess ammonia to urea, resulting in decreased animal performance (growth or lactation).

Nitrogen excretion in CNCPSv4.0 is predicted by partitioning N excretion from the predicted N balance into feces and urine:
1. Fecal nitrogen (gms/d) = (FFN + BFN + MFN)
2. Urinary nitrogen (gms/d) = (BEN + BNA + NEU + TN)
   Where:
   FFN = fecal nitrogen from indigestible feed;
   BFN = bacterial fecal nitrogen, primarily bacterial cell wall;
   MFN = metabolic fecal nitrogen;
   BEN = excess bacterial nitrogen;
   BNA = bacterial nucleic acids;
   NEU = metabolizable nitrogen supply – net nitrogen use (i.e., inefficiency of use); and
   TN = degraded tissue nitrogen.

Minimizing nutrient excretion through ration and management strategies

General principles

Methods that can be used to minimize nutrient excretion include short-term (can be implemented within days or weeks) and longer-term (require one or more crop years to implement). Implementation of these changes must be done so that milk production, growth, reproduction, and animal health are not compromised. These methods revolve around two areas: 1) decreasing N and P inputs brought on the farm by more accurately formulating rations, and 2) improving the efficiency of nutrient utilization through improved feed and crop management strategies.

Short-term methods

1. Use more accurate ration formulation to decrease P fed to NRC or INRA requirements where possible. This will decrease P excretion and ration cost, as P is an expensive nutrient. Even though P levels are decreased to recommended levels, many groups will be overfed P due to the P levels in the forages and concentrates fed to meet energy and protein requirements.
2. Use more accurate ration formulation to decrease N fed to rumen and animal requirements. To accomplish this, feed carbohydrate and protein fractions must be known and combined optimally to maximize rumen microbial growth. Programs using the CNCPS such as CPM-Dairy, DALEX, and CNCPSv4.0 can be used to accomplish this.
3. Modify grouping strategies to improve accuracy of ration formulation. Logical alternative grouping strategies need to be investigated for each farm. Through proper grouping, it may be possible to reduce N and P and ration cost while maintaining milk production and body condition replenishment goals. As Figures 1 and 2 show, the nutritional needs of lower producing cows can be met with lower ration N (figure 2) and P (figure 1). A cow producing 120 pounds of milk may need an 18% crude protein diet while a cow producing 60 pounds only requires 14% crude protein. These values may even be lower if ration formulation maximizes ruminal microbial production and N supplementation by matching feed carbohydrate and protein fractions.

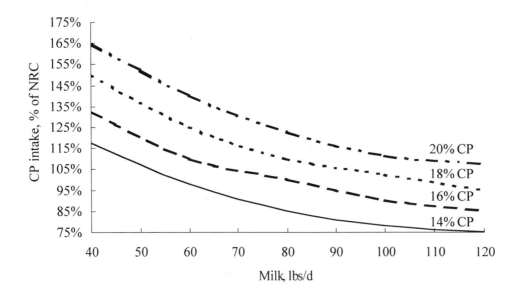

Figure 2. The relationship between daily crude protein requirement and three levels of intake to milk production (adapted from Chase, 1999).

216

4. Obtain Feed analysis as needed to accurately represent the feeds being fed. In order to decrease nutrient excretion, ration safety-factors need to be reduced in addition to matching protein and carbohydrate feed fractions. To accomplish this, a routine feed analysis protocol needs to be followed. A farm specific feed database should be developed that includes forages as well as concentrates. As is seen in Table 1, the laboratory-analyzed values for P vary considerably. Similarly, protein and NDF in forages (Tylutki et. al., 1999) and concentrates (Kertz, 1998) can vary greatly.

5. Determine dry matters as needed to account for individual feed variation. Tylutki et. al. (1999) simulated the impact of NDF and dry matter variation in corn silage using the average values and standard deviations as sampled on a 500-cow farm. The impact of improper forage analysis and lack of control over the dry matters at feeding resulted in a large annual variation in nutrient excretion (242 pounds N excretion and 64 pounds of P excretion), feed inventory required (61 tons of corn silage), and income over feed costs ($21,792) per 100 cows annually. The majority of variation was due to changes in the corn silage dry matter. Our current recommendation is to determine dry matters of all silages at least twice weekly (more often if wide fluctuations in intakes are observed)., then adjust as fed formulas as needed.

6. Improve feeding accuracy. Most farms assume that what is being mixed and fed is what is supposed to be fed. In many cases, this is not a valid assumption. Tylutki et. al. (1999) evaluated the impact of varying feeding accuracy ± 3%. The addition of feeding error increased annual variation in P excretion (18 pounds), corn silage inventory (9 tons), and income over feed costs ($19,148) per 100 cows annually. Feeding accuracy needs to be tracked to identify sources of variation, as well as to manage inventory. Commercial software and hardware is available that can be linked to the mixer scales to track this information.

7. Monitor dry matter intake to improve accuracy of ration formulation and animal performance.
    a. Track intakes. Proper ration formulation relies on many inputs from the farm, including animal body weight, feed inventory, and actual dry matter intakes. To decrease nutrient excretion, actual dry matter intakes must be known in order to ensure adequate grams of nutrient are provided. The data can also be used as a diagnostic tool. For example "our close-up dry cows are consuming 21 pounds of dry matter, and we are experiencing high levels of post-calving metabolic disease". Are they related? If so, why are we only achieving 21 pounds of intake?"
    b. Improve feed-bunk management to increase intake, and consistency of animal performance. This includes: daily cleaning, pushing feed up several times daily, and all other good management practices. More consistent performance allows the ration to be more accurately formulated for milk production level.
    c. Make ration changes to improve accuracy. By increasing the dry matter intake 5%, ration nutrient concentrations can be lowered. Chase (1999) calculated that by increasing intake 5%, it is possible to decrease diet crude protein about one percentage unit to achieve the same pounds of crude protein intake. This would result in higher inclusion rates of homegrown feeds, thus decreasing purchased nutrients.

8. Control the level of refusals. Most farms' feed refusals from the lactating herd are fed to replacement heifers. From a bio-security viewpoint, this is not a good practice. From a nutrient excretion viewpoint, this is an expensive practice. Mineral and protein levels that are adequate for lactating cows do not fit most replacement heifer groups. The amount of refusals must be at a level that is consistent with farm management to achieve maximum dry matter intake, however extremely high levels need to be avoided.

9. Use the proper 'tools' to track the impact of changes in ration formulation and feeding management. These 'tools' include:
    a. Milk production,
    b. Milk components,
    c. MUN's,
    d. Manure analysis. Manure needs to be analyzed two ways. The first is to determine what is not being digested by the cow. If large fiber particles or corn grain is evident, rations and feeding management need to be addressed. The second is to analyze manure that is being spread. As N and P levels are decreased in the rations, the levels found in manure will decrease as well. Most of the change in nitrogen will be found in the ammonia N pool. Tracking this analysis over time will provide an index of how consistent nutrient excretion is.

Long-term methods

1. Improve silo management. Silo capacity and management can play a significant role in decreasing nutrient excretion.
    a. Have adequate capacity to store separately different crop types and quality. Many farms in the Northeast have varying soil types that are best suited for different crops. The storage system must be able to handle each crop type individually (e.g., corn silage, grass silage, alfalfa silage, and different qualities of each). This will allow the nutritionist to better match protein and carbohydrate pools with specific groups of animals. An example of this would be to feed high quality alfalfa silage and corn silage to the high producing cows and feed the grass silage to lower producing cows and heifers.
    b. Minimize storage losses. During expansion, most farms will over-fill bunk silos for several years until additional capacity can be built. This over-filling results in poorer management of the bunk silo. Tylutki et. al. (1997) and Kilcer (1997) calculated the feed requirements, storage capacity, and storage losses for a 500-cow farm. They found that when the height of the corn silage pile was increased, dry matter storage losses increased (losses were calculated to be in excess of 35%) because of reduced ability to properly pack the silo during filling. By decreasing storage losses, inventory would have been high enough to allow higher home grown forage levels to be fed, thus decreasing purchased nutrients.

2. Match cows/crops/soils. Alfalfa and corn are not always the best choices for dairy producers due to soil constraints. The farms nutritionist and field crops consultant

need to work together to determine what is the best mix of crops to grow and how they can be fed to maximize production and minimize nutrient excretion.

3. Increase the amount of homegrown feeds in the ration. Increasing the amount of homegrown feeds in the ration decreases the amount of purchased nutrients. To accomplish this, homegrown feeds need to be of high quality to maintain or improve production and animal health.

   a. Impact of Forage quality. To increase the amount of forages in the rations, forage quality must be high. Maximum intake from forages can be expected when alfalfa is 40% NDF, grasses are 55%, and corn silage is 40-45% (Chase, personnel communication). A cow is limited in forage NDF intake to (1 to 1.1% of bodyweight (Mertens, 1994). As an example, a 1400 pound cow at 1.1% NDF capacity can consume 28 pounds of dry matter from grass at 55% NDF but only 24 pounds at 65% NDF. This four pounds of dry matter difference would have to be made up with purchased feeds.

   b. Impact of Grains. Homegrown grains (protein sources) decrease the amount of purchased nutrients. Most dairy farms do not have an adequate land base to produce their own grain; therefore, they should maximize forage quality and choose purchased concentrates that accurately supplement their forages.

4. Add more land or export nutrients. After all of these areas are addressed, nutrient excretion will still be in excess of crop requirements on most dairy farms. Increasing the land base to increase homegrown feed production and being able to utilize the manure N and P for crop production will be required. Chase (personnel communication, 1999) calculated the required land base for the 500-cow farm described by Tylutki et. al. (1997) to spread manure based on P recommendations. The resulting required land base was three times the current land base.

Case study

McMahon's EZ Acres is a 500-cow dairy farm located in Homer, NY. It is owned by two brothers in partnership. Four years ago, the herd was moved into a new facility from four old tie-stall barns. Since then, milk production has increased (milked 2x with no rBST) and herd health has improved. This change has been the result of a step-wise consolidation. In 1992, a bunk silo complex with a commodity building was built. This eliminated the use of numerous tower silos. Barns were then setup as production groups and a TMR was delivered twice daily. In 1996, a transition calf barn and a heifer barn were added at the site of the new complex. In 1998, the farm added milk metering and cow identification to the parlor. The farm has been a cooperator with this project since 1997 (Tylutki and Fox, 1998).

In 1997, baseline data was collected and an initial analysis of the herd and cropping system was conducted (Tylutki and Fox, 1997; Bannon and Klausner, 1997; Kilcer, 1997). It was concluded from this analysis that increased bunk silo capacity and improved bunk silo management were required. In addition, an increase in acreage of intensively managed grasses would allow for a higher proportion of homegrown feeds in the rations. The farm has been adopting these recommendations since then.

Since April 1999, rations for all groups have been formulated with CNCPSv4.0 by the farm's herd nutrition consultant. Most of the short and long-term strategies to lower N and P excretion described in this paper have been implemented by the consultant and farm management. These include:

Short-term methods implemented

1. Phosphorus levels in all groups have been decreased to CNCPSv4.0 computed requirements. Lactating cows currently are consuming a diet with .41% P (1 gram in excess). Non-lactating animals range in P excess from 2 to 10 grams with low levels of supplemental phosphorus used.
2. Protein levels have been lowered and are matched with carbohydrate feed fractions to optimize rumen microbial growth as computed by the CNCPS model. Lactating cow diets currently contain only 40 grams of metabolizable protein in excess of requirement (1% excess).
3. Lactating Cows are currently grouped by level of production. Further refinements in the grouping strategies are being explored.
4. Intensive feed analysis is being conducted as part of our research project on the farm. The project is designed to describe the variation in homegrown and purchased feeds and then how to account for this variation in ration formulation and daily feeding management practices. Daily samples of all forages are being collected and analyzed for DM, NDF, and crude protein. Weekly composites are analyzed for all protein and carbohydrate and protein fractions. Results of the weekly composites are used for ration formulation. Tables 5 and 6 show the averages and standard deviations of feed analysis by feed type from the case-study farm. These samples have been collected over the past 18 months. The SD column is the standard deviation and is a statistical measure of the variability around the average. The higher the standard deviation, the greater the variation. Variation in all feeds has been higher than anticipated. Another measure of variation is the coefficient of variation (CV). It is calculated as the standard deviation divided by the average. Calculating the CV for several of the feeds in Table 5, we find a range of 6.5 to 25.0% for phosphorus with homegrown feeds showing the highest variation and 2.7 to 33.2% for crude protein. Methods to account for this variation in ration formulation are being examined.
5. Dry matters of silages are determined at least twice weekly. If a large change is observed either in refusals or in bunk appearance, dry matters will be determined daily until they are consistent. Dry matters are charted by the feeder. The chart is used to look for patterns in changes. If a sample is greater than five units different from the last sample, another sample is taken and analyzed that day. The feeds in Table 6 with "at feeding" as part of their name illustrate the wide range in dry matters observed and the need to track dry matters on a regular basis. The corn silage standard deviation shows that the dry matter ranged from 18.2 to 43.4% as it was fed from the bunk silo (the average $\pm$ 2 standard deviations).

Table 5. Selected feed analysis averages and standard deviations from the case study farm (1999 data).

	Dry matter		Crude Protein		Soluble Protein		NDF		Fat		Phos.		
	Avg.	SD	Avg.	SD	Avg.	SD	Avg.	SD	Avg.	SD	Avg.	SD	n
Corn meal	89.0%	1.5%	8.8%	.6%	21.7%	14.5%	11.6%	1.7%	4.2%	.2%	.28%	.02%	17
Hi Moist Shell Corn	71.1%	5.7%	8.2%	1.0%	26.9%	9.6%	10.4%	1.2%	4.1%	.3%	.31%	.02%	7
Corn Silage[a]			7.6%	.5%	58.7%	4.6%	50.0%	2.8%	3.8%	1.1%	.20%	.03%	27
Grass hay	89.3%	8.4%	7.5%	2.5%	23.0%	9.4%	68.3%	3.8%	2.5%	.4%	.20%	.05%	15
Grass Silage[a]			20.5%	3.7%	61.7%	9.1%	55.0%	8.4%	6.9%	1.2%	.39%	.06%	3
MMG silage[a]			17.4%	3.1%	55.6%	11.8%	53.4%	7.7%	5.2%	.7%	.39%	.05%	23
MML Silage[a]			20.5%	2.9%	58.7%	13.1%	51.9%	6.4%	4.4%	1.2%	.37%	.04%	6
Whole cotton	87.3%	3.7%	24.6%	2.0%	24.9%	6.9%	54.7%	6.2%	19.7%	3.4%	.65%	.08%	17
Soy 48	88.9%	.7%	53.5%	1.5%	25.3%	9.4%	11.7%	2.0%	3.9%	8.1%	.74%	.06%	15

[a]The forage results are from the composites of daily samples. Within each sample, there are three to seven individual samples. Dry matter values for home grown forages are summarized in table 6.

Table 6. Homegrown forage dry matter and NDF averages and standard deviations from the case-study farm.

Feed Type	N	Dry Matter		NDF	
		Avg.	SD	Avg.	SD
Corn silage at harvest[a]	1057	26.2%	3.5%	50.0%	16.0%
Corn silage at feeding[a]	376	30.8%	6.3%	49.6%	7.6%
Grass haycrop at harvest[b]	29	26.0%	4.2%	58.4%	7.9%
Grass haycrop at feeding[a]	141	33.4%	12.7%	57.5%	6.3%
Legume haycrop at harvest[b]	9	21.8%	8.3%	49.8%	19.4%
Legume haycrop at feeding[a]	63	46.7%	13.1%	42.0%	7.7%
Mixed haycrop at feeding[a]	77	41.7%	27.2%	44.1%	14.8%

[a]1998 forages.
[b]1999 forages.

6. In 1998, the farm began using EZ Feed. EZ Feed is a commercial software package that interacts with the scale head on the mixer. It records the actual pounds of each feed added to the mix and time spent loading and unloading each feed/batch. Daily reports can be printed that list for each feed the formulated and actual pounds fed by batch. The use of EZ Feed resulted in the farm changing feeders to improve accuracy. It was calculated that greater than $20,000 was being lost due to over-feeding. The current feeder typically is within .3% of expected amounts (<150 pounds total over-feeding with approximately 60,000 pounds fed daily).

7. EZ Feed is also used to track dry matter intakes of each pen. Feed refusals from the lactating and dry cows are loaded into the mixer by pen and recorded.

   a. Intakes for the lactating cows are charted by the feeder daily. Charts are used to compare expected intakes with observed and intake versus milk production.

Intakes of dry cows are analyzed weekly. It was discovered in April that far-off and close-up dry cows were consuming below expected levels. Rations were adjusted for both groups and the far-off dry cows were moved to another location. Intakes of both groups now are 5% greater than expected values.

b. A feeder checklist was developed. The feeder walks each feed bunk each morning prior to feeding and scores the bunk. Feeding level is adjusted based on changes in cow numbers and this feed bunk score. The farm feeds once daily and pushes feed up every two hours.

c. Ration formulas are reviewed and adjustments made as needed. Time between ration formula adjustments ranges from weekly to monthly depending on inventory and trends in dry matter intake. As an example, weather changes resulted in observed intakes that were 12-15% greater than trend for two weeks. During this time, ration formulas were adjusted based on the higher intakes. When intakes returned to near expected levels, ration formulas were again adjusted based on intake.

8. Given the accuracy of the current feeder, refusal goals have been lowered. A current goal for lactating and close-up dry cows is 3-5% refusal.

9. In August 1999, a milk sampling method was implemented allowing for group composites to be taken weekly. These weekly samples are analyzed for fat, protein, SCC, and MUN. Fat is charted weekly by group. Each is evaluated weekly for changes compared to the trend. Bulk tank milk production is calculated daily and evaluated for trends, and 150-day milk by lactation group is calculated and charted.

10. Since December 1, 1999, weekly manure samples have been collected and analyzed from the lactating cows. Total nitrogen has averaged 40 pounds per 1000 gallons and phosphate equivalent has ranged from 9 to 11 pounds per 1000 gallons.

Long-term methods

1. During the 1999 cropping season, several changes in crop harvest and storage were made. Hay crop silage was stored with grass in one bunk silo, and alfalfa in another; in previous years, they were stored in either bunk by cutting: all first cutting in one bunk, second in another, etc. regardless of hay type. This allowed each forage to be harvested at desired dry matter levels. To accomplish this, a driveway and apron had to be constructed behind the bunks. An unexpected benefit of this was discovered during corn harvest. There are no back walls on the bunks and historically, a steep slope was made while filling. This resulted in poor packing and an impossible slope to keep covered. With the new apron, the corn bunk could be extended and a slope maintained for packing that has allowed for adequate covering. Four 12-foot bags were also filled with corn silage in order to decrease bunk silo height. In 1998, the bunk was measured at 26 feet tall; in 1999, it is 13 feet tall.

2. The farm has been working with their agronomist to increase grass production on the poorer, more erodable hillside soils and maximize corn silage and alfalfa yield on the valley soils. The farm has gone from zero grass acres in 1996 to greater than 225 in 1999.

3. In November 1999, the farm began moving towards higher levels of forage in the diet. The highest level achieved in the lactating cow diets was 48%, a level never achieved on

this farm before. Further, increases in these levels were planned; however given the current low inventory of hay silages they were decreased.

4. The land base has increased greater than 15% since 1997. As more land becomes available, it will either be rented or purchased.

Conclusions

Nutrient excretion is affected primarily by four factors: feed quality (homegrown and purchased), quantities of homegrown feeds, ration formulation, and ration delivery. Farm management directly controls three out of these four with some control over ration formulation. Homegrown feeds (quantity and quality) are the most important factors. Increasing the quantity and quality of homegrown feeds allows for higher inclusion levels during ration formulation. In cases such as phosphorus, there is a gram for gram replacement opportunity (increase homegrown P one gram, reduce purchased P one gram).

Many of the steps discussed in this paper revolve around decreasing the safety factors used in rations. Removing these safety factors requires high levels of farm management to decrease the risk of production fluctuations. If large levels of variation are present in forages, large variations in milk production will be observed. Avoiding these fluctuations requires that a forage sampling and dry matter determination protocol be developed and followed.

Accomplishing a reduction in nutrient excretion requires a team effort including the farm's nutritionist, crop consultant, management, and employees. This requires a Total Quality Management approach where all concerned share a vision for the farm. This includes sharing the farms financial and environmental goals with all parties so that the farm can meet its goals and is sustainable.

References

Bannon, C. D. and S.D. Klausner. 1997. Application of the Cornell Nutrient Management Planning System: predicting crop requirements and optimum manure management. Proc. Cornell Nutr. Conf. p. 36.

Chase, L.E. 1999. Animal management strategies—how will they change with environmental regulations? Proc. Cornell Nutr. Conf. p. 65.

Fox, D.G., C.J. Sniffen, J.C. O'Connor, J.B. Russell, and P.J. Van Soest. 1992. A net carbohydrate and protein system for evaluating cattle diets. III. Cattle requirements and diet adequacy. J. Animal Sci. 70:3578.

Fox, D.G. and T.P. Tylutki. 1998. Dairy Farming and Water Quality I: Problems and Solutions. Proceedings NRAES Dairy Feeding Systems Conference: 116:313. NE Regional Ag. Eng. Serv. Riley Robb Hall, Ithaca, NY 14853.

Kertz, A.F. 1998. Variability in delivery of nutrients to lactating dairy cows. J. Dairy Sci. 81:3075.

Kilcer, T. 1997. Application of the Cornell Nutrient Management Planning System: Optimizing Crop rotations. Proc. Cornell Nutr. Conf. P. 45.

Institut National de la Recherche Agronomique. 1989. Ruminant Nutrition. John Libbey Erurtext, Montrouge, France.

Mertens, D.R. 1994. Regulation of forage intake. Page 450 in Forage quality, evaluation, and utilization. G.C. Fahey, Jr., ed. Am. Soc. Agron., Madison, WI.

Morse, D.H., H.H. Head, and C.J. Willcox. 1992 Disappearance of phosphorus in phytate from concentrates in vitro and from rations fed to lactating dairy cows. J. Dairy Sci. 75:1979.

National Research Council. 1989. Nutrient Requirements of Dairy Cattle. National Academy Press, Washington, DC.

Satter, L.D., and Z. Wu. 1999 Phosphorus nutrition of dairy cattle—what's new? Proc. Cornell Nutr. Conf. p. 72.

Sniffen, C. J., J. D. O'Connor, P. J. Van Soest, D. G. Fox, and J. B. Russell. 1992. A net carbohydrate and protein system for evaluating cattle diets. II. Carbohydrate and protein availability. J. Animal Sci. 70:3562-3577.

Tedeschi, L.O., D.G. Fox, and J.B. Russell. 2000. Accounting for the effects of a ruminal nitrogen deficiency within the structure of the Cornell Net Carbohydrate and Protein System. J. Animal Sci. (in press).

Tylutki, T. P., and D.G. Fox 1997. Application of the Cornell Nutrient Management Planning System: optimizing herd nutrition. Proc. Cornell Nutrition Conf. p. 54.

Tylutki, T. P., and D.G. Fox. 1998. Dairy Farming and Water Quality II: Whole Farm Nutrient Management Planning. NRAES Dairy Feeding Systems Conference: 116:345. NE Regional Ag. Eng. Serv. Riley Robb Hall, Ithaca, NY 14853.

Tylutki, T.P., D.G. Fox, M. McMahon, and P. McMahon. 1999. Using the Cornell Net Carbohydrate and Protein System Model to evaluate the effects of variation in corn silage quality on a dairy farm. Proc. Vth Inter. Workshop on Modeling Nut. Util. in Farm Animals. In Press.

Wang, S.J. 1999. Optimizing feed resources to improve dairy farm sustainability. PhD thesis, Cornell University, Ithaca, NY.

Wang, S.J., D.G. Fox, D.J.R. Cherney, S.D. Klausner and D. Bouldin. 1999. Impact of dairy farming on well water nitrate level and soil content of phosphorus and potassium. J. Dairy Sci. 82:2164.

Session 7

Fate of Land-Applied Nutrients and Pathogens

Nutrient and Pathogen Transport in the Watershed

William J. Gburek
Hydrologist
Pasture Systems and Watershed Management Research Laboratory
USDA-ARS
University Park, Pennsylvania

Biographies for most speakers are in alphabetical order after the last paper.

Introduction

Many of the issues associated with phosphorus (P), nitrogen (N), and pathogens in the environment – potential sources, problems, impacts, and management – have been addressed in the previous papers. However, a major issue facing watershed planners today is how land use at the management scale (i.e., fields or farms) relates to the quantity and quality of watershed outflow. In fact, the *transport* processes within a watershed are what tie management to watershed outflow. Under humid-climate conditions, the dominant hydrologic transport processes result in relatively well-defined areas of the watershed, sometimes of limited areal extent, contributing much of the nonpoint-source sediment, P, N, and possibly pathogens, to watershed outflow. So from the perspective of management, it is critical that we develop concepts, sampling protocols, and modeling tools to define, delineate, and assess the impacts of what we will term *critical source-areas* (CSAs). These are the watershed areas having the highest priority for control, treatment, or remediation.

Background

The importance of the source-area concept in developing effective approaches to water quality management has generally been only recently accepted. There are, however, a limited number of earlier references to this need. For instance, Dunne (1978), one of the original field-based investigators of the variable-source-area runoff generation process, stated, "The transport of fertilizers, herbicides, or animal wastes, for example, can be highly dependent upon where the material is placed in relation to the runoff source areas." – note his specific use of the terms *transport* and *source areas*. In a study in Vermont, Kunkle (as referenced in Betson and Ardis, 1978) concluded that, "…because of the runoff processes involved, upland contributions of bacteria to streams were small compared to contributions from land surfaces near channels, the channel itself, or direct inputs." Here, there are more indirect, but still obvious, references to *transport* and *source areas*. Nonetheless, this has not historically been a prevalent view.

Relevant to a phrase within the topic of this meeting, *Managing Nutrients…,* Shuyler (1994) stated, "Nutrient management should reduce soluble nutrient transport and (yet) provide enough nutrients to produce a realistic crop yield." When concerned with the *…Animal Agriculture* part of the meeting title though, as well as the session in which this paper appears, *Fate of Land-Applied Nutrients and Pathogens*, we must recognize that nutrient management (as well as that of pathogens) sometimes takes on aspects of nutrient disposal (via animal wastes), when, because of large animal numbers, manure must be applied to the land surface such that nutrients contained therein are in excess of crop needs. Under these conditions, it is critical that we have the capability to predict nutrient and pathogen *transport* within and from a watershed, because the transport processes are what move these components through the watershed to their point of impact, or *fate*. Only when we consider a particular point of impact does the subject of transport become important – if there is no defined point of impact, or concern with *fate*, the sources of N, P, or pathogens might be considered to be relatively benign.

Development of a comprehensive nutrient and pathogen management strategy – one that encompasses the beneficial use of both nutrients (to meet crop needs), as well as a disposal component (when nutrients are in excess of crop needs or when pathogens occur on the land surface) – must address downgradient water quality impacts. A management strategy that addresses both use and disposal should be capable of integrating effects at the local scale where the particular management is implemented (i.e., the field), with the scale of the logical management unit (i.e., the farm), and finally with the larger scale at which results of the strategy are sampled and evaluated (i.e., the watershed). A management strategy developed in this way must accommodate the concept of transport, as well as its interactions with and controls on CSAs.

Pathways of transport

The singlemost important factor necessary to define transport pathways and CSAs is knowledge of water movement within the watershed, i.e., the hydrologic cycle. This is because moving water can translocate contaminants from their source-areas of application (typically on or within the soil zone), to or through zones of reaction and sinks within the watershed (in either the surface or subsurface), and finally to positions where they are removed from the watershed (generally streamflow but possibly via ground water). To determine flow components of the hydrologic cycle relevant to the problem of nutrient and pathogen management, we must first define the dominant interactions between these components and the natural flow system. The important nutrients from animal agriculture that are transported by a watershed's flow system to affect downgradient water quality are N and P. And finally, transport of pathogens from animal wastes applied to the land surface is a related concern.

Hydrologic controls on the transport of N and P within a watershed are different – in some ways they are opposed. In the case of P (and also pathogens), surface runoff, with its associated water-borne sediment, is the primary flow component of concern. With reference to P movement, there is also some concern with potential for subsurface transport. However, this route is considered to be of importance only in sandy or well-drained (highly permeable) soils having high water tables, or under conditions when soils are artificially drained. There is also a concern with soils having preferential flow pathways (i.e., cracks, wormholes, etc.), but again, there must be a high water table or artificial drainage involved to provide a connection to the stream. In the case of N transport, the concern is usually with nitrate (NO_3), a soluble species. Here, the subsurface flow system is of primary importance since NO_3 is generally moved into the subsurface by infiltrating water and ultimately exported from the watershed via subsurface flow.

To confuse the issue further though, all flow components within the hydrologic cycle are closely connected under the humid-climate upland watershed conditions being considered here. Rainfall, soil moisture, evapotranspiration, ground water recharge, surface runoff, and ground water discharge to the stream all respond at the same time scales, both event-based and seasonal. Finally, both surface and subsurface waters are important within and from these watersheds. The rural population of the Northeast relies almost entirely on ground water for water supply, so effects of nutrients (especially NO_3) on ground water quality is of concern. This same ground water is also the dominant source of streamflow, up to 70-80% of annual flow (Gburek et al., 1986). So the subsurface-derived flow, combined with surface runoff, provides the water for the larger downgradient rivers, impoundments, and estuaries important to fisheries, recreation, and water supply. Thus, development of management strategies to protect surface and ground water quality implies quantification of disparate but interactive hydrologic flow components and corresponding delineation of CSAs controlling losses of both nutrients and pathogens.

Watershed hydrology – Response verses flowpaths

Even though there is an obvious need to define flowpath controls on contaminant transport through a watershed, our present views of hydrology, and consequently of how contaminants move within the watershed context, remain slanted by tradition. Research into watershed hydrology was originally founded on the concept of quantifying watershed-scale response to inputs – when rain occurs, the watershed produces a storm hydrograph; when a well is pumped, the ground water table drops. Because water quality was not then a concern, early hydrologists gave little or no consideration to pathways of water movement when attempting to quantify these responses.

Our concern with water quality beginning in the late 1960's required an alternative view; however, this view was not immediately embraced. Quantifying contaminant transport requires knowledge of pathways of water movement, not simply watershed response. We must specify where within the watershed contaminants are introduced, what CSAs they move through and how fast, and finally where they exit the watershed – all controlled by *transport*.

Thus, we must understand and define the transport processes and their controls on flowpath geometry linking all points of the watershed land surface to the watershed outlet, through both the surface and subsurface systems.

A hydrologic basis for nutrient and pathogen transport

The transport and CSA aspects of strategies for nutrient and pathogen management can be developed within the context of dominant characteristics of northeastern upland watershed hydrology – variable-source-area watershed response as it controls surface runoff generation and its related contaminant transport, and bedrock layering and fracturing as it controls ground water flow and related contaminant transport within shallow aquifers. These strategies must also address effects of the riparian zone, and describe relevant interactions between all flow components at the watershed scale.

At the most basic level, the intersection of surface runoff and ground water recharge source-areas with areas of fertilizer or animal waste application over the landscape is what creates the initial CSAs controlling loss of P, N, and pathogens from a watershed. The situation may become more complicated though, when pathways of flow (transport) from these CSAs are considered; the contaminants may undergo transformations downgradient that alter their concentration and/or mass by dilution, reaction, or sink types of processes, adding an additional control on contaminant export.

Surface runoff and phosphorus/pathogen transport

Surface runoff, the primary vehicle for P and pathogen transport, is the direct result of rainfall impacting the land surface. Generation of surface runoff was traditionally thought to be a soil-controlled phenomenon occurring over the entire watershed, and techniques developed to predict storm runoff (e.g., the curve number) were based on this assumption. This process is generally referred to as infiltration excess runoff, and occurs because the infiltration capacity of the soil is less than that of rainfall intensity. More recently however, partial-area hydrology, which then evolved to variable-source-area (VSA) hydrology (Ward, 1984), has become accepted as a descriptor of humid-climate watershed response to precipitation. In this case, runoff is termed saturation excess runoff, and is generated because there is limited storage capacity in the soil. Available porosity is already filled either because of high near-stream water tables or typically wet areas on the landscape (e.g., geologic contacts, areas of convergence, or fragipan soils)

The basic premise of VSA hydrology is that there is a contributing subwatershed within the topographically defined watershed which varies in time – it expands and contracts rapidly during a storm as a function of precipitation, soils, topography, and ground water level and moisture status. VSA runoff is dominated by saturation overland flow and also rapidly responding subsurface flow. The remainder of the watershed provides little or no runoff, only infiltration and ground water recharge. Since surface runoff is the mechanism for P and pathogen transport, the intersections of the VSA with soil/fertility/management combinations that produce excess P or pathogens at the land surface control their loss from the watershed. For loss to occur, there must be coincident available P (or pathogens) and surface runoff to move it to the watershed outlet.

Soil P levels can be high over much of the watershed, but these high-P areas are "activated" to become a source of P to the watershed outlet only by occurrence of surface runoff. The runoff generation portion of a CSA is not easily controlled, but soil P levels in areas conducive to surface runoff formation can be monitored and controlled. Thus, management to limit P loss from a watershed should be focused on areas where runoff is likely to occur that are coincident with high P levels in the soil. This concept is currently being incorporated into the Phosphorus Index (see Gburek et al., 2000), a topic that will be addressed in subsequent papers from this conference.

Ground water and N transport

Ground water flow systems have traditionally been analyzed in terms of deeper water supply aquifers and homogeneous media. Flowpaths developed for these analyses extend deep into the subsurface and are regular in form, while associated travel times are considered to be at scales of years, decades, and even centuries. However, concerns with nonpoint source pollution from agriculture have required us to focus on the characteristics of the shallow surficial aquifers, those most easily and more likely affected by overlying land use. In the Northeast, this leads directly to the need to consider the effects of bedrock layering and fracturing on patterns of flow and contaminant transport, as well as the shallow aquifer's interactions with the surface water flow system.

At the larger scale, the land surface is typically underlain by relatively conductive layers of soil and rock, which are, in turn, underlain by less permeable strata. The more permeable layers forming the surficial aquifer can range from a few meters to tens of meters thick. Ground water moving within the more conductive surficial layers generally forms an unconfined aquifer, i.e., the water table is the upper boundary of the saturated zone. Further, nearly all ground water regions of the eastern U.S. have some degree of fracture control on flow within the shallow bedrock. Bedrock fracturing, when it occurs, exaggerates the hydrogeologic characteristics of the layered aquifer even more, producing extremes in the properties governing flow and transport – very high hydraulic conductivity and relatively low specific yield. The result of these conditions is that the flowpaths tend to be even more constrained within the shallower zones of the aquifer. The shallow zone of fracturing may be overlain by relatively thick glacial deposits or regolith, or by only a thin soil. In all cases though, ground water within the shallow fracture zones is affected directly by overlying and immediately upgradient land use.

Pionke and Urban (1985) developed a comprehensive nitrogen budget within a 7.2 km2 upland agricultural watershed in east-central Pennsylvania, and showed that NO_3 leaving the root zone in percolate at the watershed scale matched that observed in the stream at the watershed outlet. Gburek and Folmar (1999b) extended this finding to show that the NO_3 balance could be maintained down to a subwatershed scale of approximately 0.5 km^2. Thus, at a variety of watershed scales, NO_3 leaving the root zone should be expected to appear in streamflow over the long term. Schnabel et al. (1993a) and Gburek and Folmar (1999a) showed that streamflow patterns of NO_3 are controlled by land use distribution and the layered and fractured geologic system. In total, this research emphasizes the need to focus on subsurface flowpaths to understand nitrate movement and develop corresponding management strategies.

Because of the layering and fracturing prevalent in upland watersheds of the Northeast, ground water quality within the surficial aquifer is strongly affected by land use positioning within the watershed. Land use over only the most upland watershed positions controls water quality within the deeper regional aquifer, while land use distribution over the remainder of the watershed affects water quality patterns within the shallower fractured layers.

From the total watershed perspective, this discussion indicates the potential, or lack thereof, to manage N loss from a watershed. Basically, the subwatersheds producing higher NO_3 concentrations in a stream are those having higher source-area inputs of N to the ground water. The simple answer to reducing N output from a watershed is reducing or managing the N applied to the land surface to minimize loss through the root zone.

Riparian zone considerations

Elevated levels of NO_3 are expected in ground water recharge from agricultural land use, but the resultant high concentrations of NO_3 in ground water may be attenuated enroute to becoming streamflow if the ground water flows through riparian ecosystems (Lowrance et al., 1984; Correll and Weller, 1989). The riparian zone (RZ) is that portion of the watershed where subsurface flow intersects the land surface, usually in near-stream positions, causing high water tables and soil moisture levels. Under humid-climate conditions, ground water convergence to the stream is the prime cause of these high-moisture, high-water-table, riparian conditions effective for nitrate reduction, but aquifer configuration and its controls on the ground water flow system geometry at the watershed-scale also influence the potential for NO_3 removal (Schnabel et al., 1993b).

Cooper (1990) concluded that catchment hydrology, particularly flowpaths through the RZ, determine the control of near-stream environments on pollutant flux. Whether riparian zone processes can substantially reduce nitrogen discharge into a stream depends on both the areal extent of the RZ and the hydrologic linkages between upland nitrate sources (i.e., farm fields) and the RZ. A variety of linkages with varying degrees of complexity are possible on a watershed, ranging from landscape positions with short, shallow, direct paths between cropped areas and riparian zones, to those with deep flow paths through limited riparian zones. When flow paths are direct and shallow, nearly all agricultural drainage passes laterally through the RZ before discharging to the stream. In contrast, when flow paths are less constricted, more of the ground water discharge can bypass some or all of the RZ, either as deeper flow before converging to the channel, or as loss to a deeper ground water system.

As emphasized throughout this paper in other scenarios, flow path and transport definition is critical – here, subsurface flowpaths describe delivery of recharge from the agricultural land use to and through the RZ. Of prime importance is the origin of those flowpaths passing through the RZ, because only land uses from which these flowpaths originate have the potential to be affected by RZ processes. High recharge conditions typical of springtime, results in the most flow passing through the RZ. Lower recharge conditions, more typical of summer and early fall, result in a generalized lowering of the water table over the watershed, and consequently, less flow (and fewer pathways) passing through the active RZ. In evaluating potential RZ effects under these conditions though, the hydrologic status of the watershed must also be considered. The high-recharge configuration represents springtime conditions when denitrification processes are less effective because of cooler temperatures, and residence times within the RZ are also shorter because of higher flowrates. Conversely, when potential for denitrification is greater (summer temperatures and longer residence times), more of the ground water bypasses the RZ. Thus, the effectiveness of local biochemical RZ processes must be integrated with watershed-scale hydrologic conditions to fully evaluate the potential for a landscape to attenuate NO_3 inputs to the ground water.

Implications for nutrient and pathogen management

The concepts presented provide the basis for describing transport and related source-area on nutrient and pathogen loss from the watershed – for loss to occur, there must be available nutrients and/or pathogens as well as water movement to carry them to the watershed outlet. As an example, soil-P levels may be high over much of the watershed, but these high-P areas become a source of P to the stream only by concurrent occurrence of surface runoff.

At the watershed scale, relatively few processes and parameters form the critical source-areas controlling nutrient or pathogen loss – even fewer may be efficiently or readily manageable. In the case of P management, runoff generation itself is not easily controlled at the watershed scale in humid-climate regions, but P levels in areas contributing to P loss via storm runoff can be measured, monitored, and controlled. Erosion may be more manageable, particularly by manipulating cover directly or indirectly through a conservation practice, such as modified tillage. Phosphorus loss may be managed by controlling P fertility level in the primary surface runoff zones, and sediment-related P by controlling erosion and/or P fertility level in the primary erosion zones. Levels of soil P further from the stream may be of less concern since there is less chance of runoff occurring from these areas.

Selection of remediation methods for controlling P export from a watershed should not cause or aggravate other water quality problems though. Phosphorus control strategies based on reducing surface runoff losses by increasing infiltration rate or disposing of excess P in non-runoff zones may well increase NO_3 recharge to ground water, especially where applications of manure or organic materials are the primary source of the excess P. Under these conditions, remedial approaches must be developed and selected to optimize control of P loss relative to achieving N control objectives.

In those watershed positions where P or pathogen loss is not of concern because of minimal surface runoff, management of N from fertilizers and animal wastes becomes critical. N is much less likely than P to be built up or stored in the soil system. Thus, the simple fact is that N excess within the soil in all nitrogen CSAs (the recharge zones) must be carefully controlled to reduce N inputs to the ground water and subsequent export from the watershed. The CSAs for N involve most of the watershed area because they are the inverse of the VSA for runoff. An alternative to management of N in the soil within the nitrogen CSAs is restrictions on types of land use and management – that approach would be less preferable.

Summary

Nutrient and pathogen management strategies must be developed considering the watershed, as well as the farm and field scales. Hydrologic processes that dominate at the watershed scale determine which farms or fields have the potential to contribute most of the exported nutrients or pathogens. Other watershed elements, such as the riparian zone, exist between the individual fields and the watershed outlet that alter the timing, amount, and concentration of the nutrients exported – again though, these are primarily a function of the structure of the watershed-scale flow system. Here, we have delineated P and N CSAs from the hydrologic perspective – considering transport processes – that are not predictable using a field- or farm-scale approach. Only after the hydrologic (runoff and recharge) source-areas are established at the watershed scale can we consider the smaller scale soil, chemistry, and nutrient use information within the farms and fields to define the nutrient and pathogen CSAs.

We remain far removed from development of a single nutrient and pathogen management strategy that can account for all contaminants, as well as all hydrologic implications. Modeling tools and field data are simply not available to integrate all aspects of the hydrologic cycle from the flow perspective alone, much less from that of water quality. However, we can draw conclusions based on what has been shown. Hydrologic implications are that to control P loss, we control P application and build-up primarily in near-stream zones. Levels in the soil at distance from the stream are of minimal concern since there is only a limited chance of runoff occurring to move P from that location to the stream. Thus, the most obvious control from the P point of view is to apply animal wastes on landscape positions at distance from the stream – the further the better.

Concern with nitrate is different though – we must control N in areas of ground water recharge. The ideal animal waste disposal strategy should provide for N in the soil at amounts needed by the plants, both in timing and areal distribution. If excess wastes are applied though, nitrate escaping crop uptake will easily move with any water available to move it. In this case, the management strategy becomes <u>disposal</u> (i.e., excess N), and should be limited to those portions of the landscape which do not affect ground water zones critical to water supply. Lastly, if we can identify those flowpaths that pass through the RZ, we can apply the excess N to watershed areas having the best chance to be affected by RZ processes, thereby realizing maximum reduction of input concentrations.

When all flow components of the hydrologic cycle are intimately connected, as in humid-climate upland watersheds, we must consider interactions between these components when developing effective animal waste management strategies. Management must account for interactions between areas of nutrients and pathogens applied to the landscape, source areas of ground water recharge, patterns of ground water movement and associated contaminant transport, surface runoff source areas and associated transport, and riparian zone controls. To accomplish this, we must continue to do a better job in accurately portraying the hydrologic flow system. External and internal boundaries, controlling geometries, hydraulic properties, major flowpath patterns, and areas of recharge, subsurface discharge, and surface runoff must all be defined and integrated into any nutrient or pathogen management strategy applied to part or all of the watershed.

Finally, the most critical problem of all may be interfacing this objective and technical approach to management with the interests of landowners. The purely hydrologic analysis presented implies that different levels of management for different parts of the watershed may be the most efficient approach to minimize contamination of ground water and/or surface runoff. Thus, there may be differing levels of management suggested from field to field and/or farm to farm. A major challenge in implementing a nutrient or pathogen management strategy derived from these hydrologic implications may be to demonstrate to the affected landowners that the results will be of sufficient benefit to override the apparent inequities associated with its application over the watershed.

Acknowledgements

Contribution from the Pasture Systems and Watershed Management Research Laboratory, U.S. Department of Agriculture, Agricultural Research Service, in cooperation with the Pennsylvania Agricultural Experiment Station, the Pennsylvania State University, University Park, Pennsylvania.

References

Betson, R.P. and C.V. Ardis Jr. 1978. Implications for modeling surface-water hydrology. In: Hillslope Hydrology, M.J. Kirkby (ed.). John Wiley & Sons, New York.

Cooper, A.B. 1990. Nitrate depletion in the riparian zone and stream channel of a small headwater catchment. Hydrobiologia 202:13-26.

Correll, D.L. and D.E. Weller. 1989. Factors limiting processes in freshwater wetlands: An agricultural primary stream riparian forest. Smithsonian Environmental Research Center, Edgewater, MD. DOE Symposium Series, R.R. Sharitz and J.W. Gibbons (eds.), USDOE Office of Scientific and Technical Information, Oak Ridge, TN. pp. 9-23.

Dunne, T. 1978. Field studies of hillslope flow processes. In: Hillslope Hydrology, M.J. Kirkby (ed.). John Wiley & Sons, New York.

Gburek, W.J., and G.F. Folmar. 1999a. Patterns of contaminant transport in a layered fractured aquifer. J. Contaminant Hydrol. 37:87-109.

Gburek, W.J., and G.F. Folmar. 1999b. Flow and chemical contributions to streamflow in an upland watershed: a baseflow survey. J. Hydrol. 217:1-18.

Gburek, W.J., J.B. Urban and R.R. Schnabel. 1986. Nitrate contamination of ground water in an upland Pennsylvania watershed. Proc. Agric. Impacts on Ground Water, Nat. Water Well Assn., pp. 352-380.

Gburek, W.J., A.N. Sharpley, L. Heathwaite, and G.J. Folmar. 2000. Phosphorus management at the watershed scale: A modification of the phosphorus index. J. Environ. Qual. 29:130-144.

Lowrance, R., R. Todd, J. Fail Jr., O. Hendrickson, R. Leonard, and L. Asmussen. 1984. Riparian forests as nutrient filters in agricultural watersheds. BioScience 34:374-377.

Pionke, H.B. and J.B. Urban. 1985. Effects of agricultural land use on ground-water quality in a small Pennsylvania watershed. Ground Water 23:68-80.

Schnabel, R.R., J.B. Urban, and W.J. Gburek. 1993a. Hydrologic controls in nitrate, sulfate, and chloride concentrations. J. Environ. Qual. 22:589-596.

Schnabel, R.R., W.J. Gburek, and W.L. Stout. 1993b. Evaluating riparian zone control on nitrogen entry into Northeast streams. Proc. Riparian Ecosystems in the Humid U.S.: Functions, Values, and Management, March 15-18, 1993, Atlanta, GA.

Shuyler, L.R. 1994. Why nutrient management? In Nutrient Management, supplement to J. Soil and Water Cons. 94(2):3-4.

Ward, R.C. 1984. On the response to precipitation of headwater streams in humid areas. J. Hydrol. 74:171-189.

Source Risk Indicators of Nutrient Loss from Agricultural Lands

Peter J. A. Kleinman, Ph.D.
Soil Scientist
USDA-Agricultural Research Service
Pasture Systems and Watershed Management Research Laboratory
University Park, Pennsylvania 16802-3702

Biographies for most speakers are in alphabetical order after the last paper.

Source risk describes the potential availability of agricultural nutrients to transport mechanisms, such as erosion, runoff and leaching. The combination of source and transport risks quantifies the potential for diffuse losses of nutrients from land to water, and is key to site risk assessment tools such as the Phosphorus Index (Lemunyon and Gilbert, 1993), as well as more recently, the Nitrogen Index (Heathwaite et al., 2000).

The two elements of greatest concern to agricultural and environmental managers, phosphorus (P) and nitrogen (N) behave quite differently in the environment. Whereas N is highly mobile and easily transported via multiple pathways, the affinity of P to soil constituents imposes conditions that restrict its potential for transport. In the past, this affinity supported a perspective that equated P loss with particulate P removal by erosion. Although erosion certainly represents the most drastic means of transporting P from land to water, transport of dissolved P in runoff, and even leachate, is now seen as a major concern. Discrepancies in the chemistry of these elements result in very different approaches to source risk assessment for N and P.

This paper examines source factors, i.e., indicators of source risk, controlling nutrient loss from agricultural lands, with particular emphasis on P, which is currently receiving concentrated scrutiny in research and regulatory arenas.

Nitrogen Source Factors

Nitrogen source factors are the key control of N losses from agricultural lands, as N transport is extremely difficult to manage (Beegle, 2000). For instance, research conducted by Sharpley and Smith (1994) revealed that transport-oriented management simply shifted the pathways by which N was removed from soil. Specifically, imposition of conservation tillage practices on wheat production reduced runoff losses of N but increased leaching losses (Figure 1). Such a "zero sum gain" simplifies the assessment of N source risk; N in excess of crop requirements is likely to be lost. This view is illustrated by Figure 2, which shows the coincidence of agronomic and leaching thresholds for N, based upon research on N fertilization of cereal crops in England (Lord and Mitchell, 1997 as reported by Sharpley and Lord, 1997).

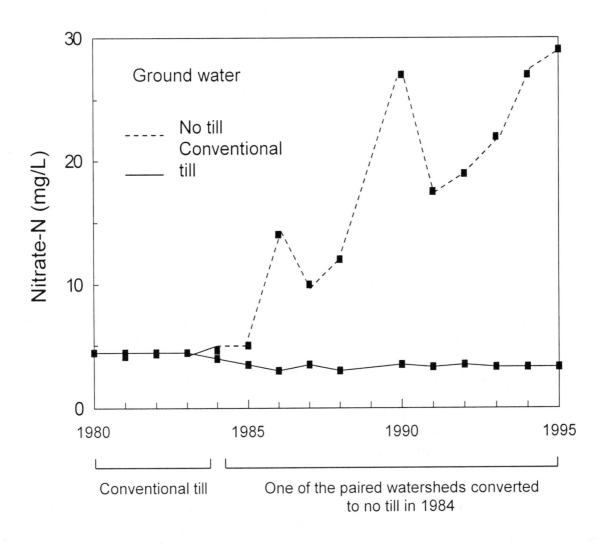

Figure 1.   Mean annual nitrate-N concentration of ground water and dissolved and total P concentration runoff as a function of tillage management of watersheds in Oklahoma, U.S. (data adapted from Sharpley and Smith, 1994).

The risk of nitrate leaching is rooted in cropping system management. Nitrogen source risk assessment must therefore involve consideration of form, rate, timing and method of N application (Sharpley et al., 1998a). For instance, residual soil organic N, which can mineralize when weather conditions favor microbial activity, must be considered in order to assess temporal variation in nitrate leaching from soils. Similarly, crop selection and rotation influence the amount of N in the soil profile, hence source risk, and can also influence transport potential by affecting soil water dynamics. For example, legumes that do not require supplemental N inputs can effectively scavenge residual N from soil that is related to previous crops, thereby reducing source risk (Mathers et al., 1975; Muir et al., 1976).

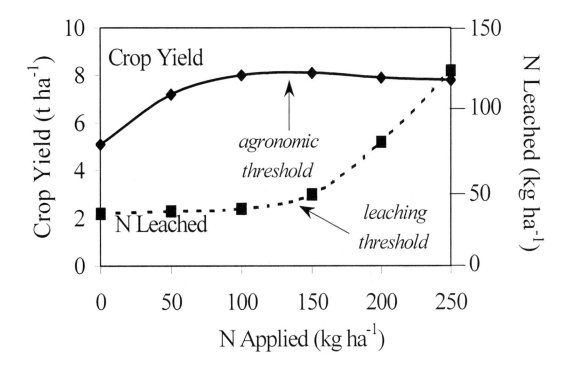

Figure 2. Relationship of N application to crop yield and N leaching, illustrating coincidence of agronomic and environmental thresholds (adapted from Sharpley and Lord, 1997).

The importance of cropping system management to source risk is further illustrated by Figure 3, which overlays hypothetical root development for several crops with predicted N leaching patterns (adapted from Sharpley et al., 1998b). Continuous cropping with corn results in root development that does not overlap with periods of high transport potential (leaching), resulting in increased availability of soil N to sub-surface waters. However, rotational cropping of corn-winter wheat-alfalfa results reduced source risk due to improved efficiency of N retention derived from deeper root distribution at times when leaching risk is higher.

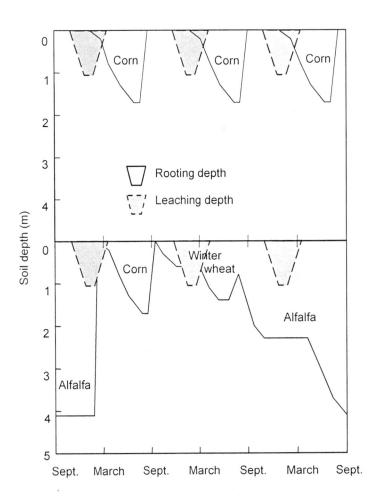

Figure 3. Typical root development of continuous corn and corn-winter wheat-alfalfa rotation in relation to soil drainage over a three year period (adapted from Sharpley et al., 1998b).

Phosphorus Source Factors

As our understanding of P source risk has grown, attention has focused on the paucity of soil-specific data linking soil P with the potential for P loss (Sharpley et al., 1999). Given the absence of such data, a common approach has been to adapt agronomic soil P standards to source risk assessment, following the rationale that soil P in excess of crop requirements is vulnerable to removal by surface runoff or leaching. This rationale applies to N, as described above. Since agronomic standards already exist for soil test P, such an approach requires little investment in new research and can be readily implemented. However, it is clear that we must be careful how we interpret soil test results for source risk assessment.

Interpretations presented on soil test reports (e.g., low, medium, optimum, high) were established based on the expected response of a crop to P. Some would simply extend the levels used for interpretation for crop response and say a soil test that is above the level where a crop response is expected is in excess of crop needs and therefore is potentially polluting (Figure 4). But, it cannot be assumed that there is a direct relationship between the soil test calibration for crop response to P and runoff or leachate enrichment potential (source risk). What is crucial to our estimation of source risk is the interval between the threshold soil P value for crop yield and runoff/leachate P (Figure 4).

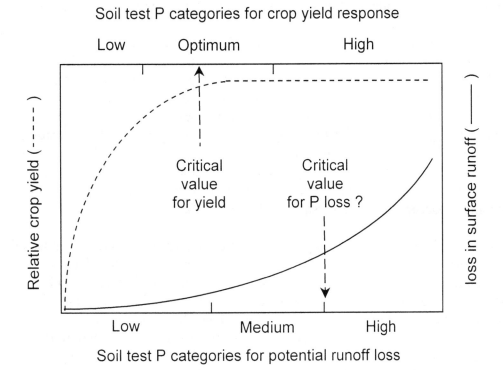

Figure 4.   As soil P increases so does crop yield and the potential for P loss in surface runoff. The interval between the critical soil P value for yield and runoff P will be important for P management.

Environmental soil testing for P

The relationship between soil test P and surface runoff P, shown by the lower curve in Figure 4, is based on studies using traditional soil test methods that estimate the plant availability of soil P. Although these soil test methods show promise in describing the relationship between the level of soil and surface runoff P, they have several drawbacks.

For example, soil test extraction methods were developed to estimate the plant availability of soil P; therefore they may not accurately reflect soil P release to surface or subsurface runoff water (Sharpley et al., 1996).

Soil sampling depths can also be problematic. For routine soil fertility evaluation it is generally recommended that soil samples be collected to plow depth, or the zone of greatest root concentration, which is roughly 6-8 inches deep. When soil testing is used to estimate soil P loss, however, it is the surface inch or two that comes into direct contact with runoff that is important (Sharpley et al., 1996). One exception is the need to consider the amount of subsoil P in soils with high water tables where shallow lateral flow and leaching loss may be a concern. Consequently, different soil sampling procedures may be necessary for a soil test that is used to estimate the potential for P loss in surface runoff. To overcome the limitations of traditional agronomic soil tests and sampling techniques, a variety of approaches are being developed to better represent the amount of P in soil that can be released to runoff water, and, more specifically the amount of P (e.g., algal available P) that is of environmental concern (Sharpley, 1993).

One environmental measure of soil P, developed in the Netherlands by Breeuwsma and Silva (1992) to assess P leaching potential, is soil P saturation (percent saturation = available P / P sorption maximum). The role of soil P saturation as an indicator source risk derives from the observation that soil P saturation is strongly correlated to P desorption, such that P desorption increases at higher degrees of soil P saturation (Sibbeson and Sharpley, 1997). Indeed, many studies have correlated soil P saturation with P in runoff (Pote et al., 1996, 1999; Sharpley, 1995), as well as with P in leachate (Hesketh and Brookes, 2000). In the Netherlands, a threshold soil P saturation of 25% has been established above which the potential for P movement in surface and ground waters becomes unacceptable (Breeuwsma and Silva, 1992).

Unfortunately, source risk measures such as soil P saturation are not offered by most soil testing laboratories. Opportunities exist to relate readily-available soil test data to environmental measures such as soil P saturation. For instance, Kleinman et al. (1999) found a strong correlation between Morgan's extractable P and soil P saturation (Figure 5). Similarly, soil P saturation may be estimated using Mehlich 3 extractable elements, rather than acid, ammonium oxalate extractable elements as is currently standard (Figure 6). As concern over the precise measurement of source risk mounts, soil testing laboratories may need to explore the development of pedotransfer functions, i.e., equations that relate data we *have* to data we *need*, to estimate source risk.

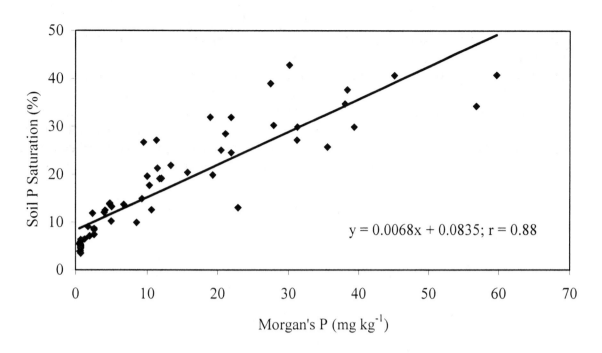

Figure 5. Relationship of Morgan's P (agronomic soil test) to soil P saturation (environmental soil test) (adapted from Kleinman et al., 1999)

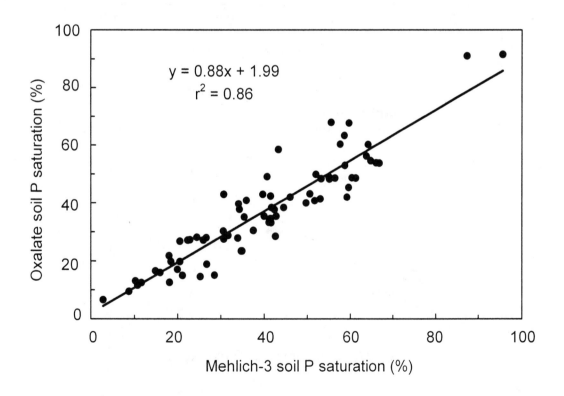

Figure 6. Relationship between soil P sorption saturation calculated from Mehlich-3 and oxalate extractable P, Al, and Fe.

A variety of soil extracts have been evaluated as indicators of source risk by relating extract P to P in surface runoff or sub-surface leachate. Ryden and Syers (1975) stated that to establish the relationship between P additions, fertilizer additions and P in particulate or aqueous phases of runoff, a desorption or "support medium should reflect the cation status as well as the ionic strength of the aqueous phase of the system". With the possible exception of macropore flow, the longer residence time of sub-surface leachate as it flows through the soil implies that a soil extractant of higher ionic strength is required than would be used for surface runoff. Soil extractions with water and 0.01M CaCl$_2$ have shown promise in estimating the dissolved P concentrations of surface runoff and sub-surface leachate, respectively (Figure 7) (McDowell and Condron, 1999;

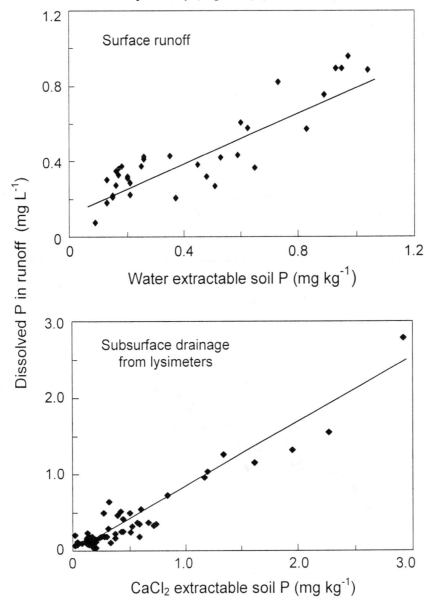

Figure 7. Relationship between the concentration of dissolved P in surface runoff and subsurface drainage from 30 cm deep lysimeters and the water and CaCl$_2$ extractable soil P concentration, respectively of surface soil (0 - 5 cm) from an central PA watershed (see Sharpley et al., 1999).

McDowell and Sharpley, 1999). Using these desorption mediums in conjunction with Fe-oxide strips and gels can determine the quantity of desorbable P in runoff or soils the medium and long-term (Freese et al., 1995; Sharpley, 1993).

Threshold analysis

Threshold analysis of soil P data represents an important step in source risk assessment. One innovative environmental approach uses a split-line model to determine a threshold, termed a "change point," in soil P concentration. The change point separates the relationship between soil phosphorus and dissolved P in runoff into two sections, one with greater P loss per unit increase in STP concentration or percent saturation than the other (Heckrath et al., 1995; Hesketh and Brookes 2000; McDowell and Condron 1999). McDowell and Sharpley (1999) showed that a similar change point occurred between STP concentrations of the 0 – 5 cm soil layer and dissolved P in either surface or sub-surface runoff (Figure 8). The change point identifies a critical soil P level that should not be exceeded, and has potential for use in threshold identification of source risk.

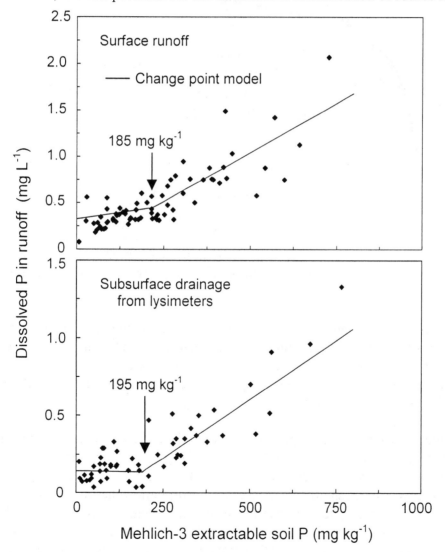

Another approach to threshold identification employs P sorption isotherms that are generated under experimental conditions in the laboratory. The basis for this approach is the assumption that a fundamental property of a soil with a low degree of P saturation is a standard, L- or H-type Q/I curve (see McBride, 1994). Figure 9 illustrates such a curve, divided roughly into two integral stages: an initial, steeply-sloped, "fixation" stage in which the proportion of soluble P additions that are sorbed is high; and, a "saturated" stage of gradual slope in which the proportion of soluble P additions that are sorbed is significantly lower relative to the first stage.

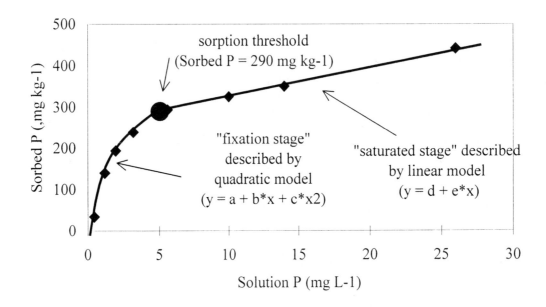

Figure 9. Description of P sorption isotherm by segmented regression. Sorption threshold identifies the intersection of quadratic and linear models

The sorption threshold for an "unsaturated" soil (i.e., a soil with both "fixation" and "saturated") is estimated by determining the point that joins the "fixation" stage with the "saturated" stage of the sorption isotherm. The ordinate value of this join point, the sorbed P value, represents the remaining capacity of a soil to bind soluble P before the P sorption efficiency significantly declines.

Figure 10 illustrates P sorption isotherms for three soils, representing low (Figure 10c), intermediate (Figure 10b) and high soil P saturation (Figure 10a). Notably, as soil P saturation increases, presumably due to increased P loading, the magnitude of the fixation stage (i.e., the sorption threshold value) decreases. Sorption isotherms of soils with high degrees of P saturation (Figure 10a) manifest near-linear relationships (i.e., only the "saturated" portion of the sorption isotherm) and, therefore, possess low P sorption efficiencies. Soluble P added to these "P-saturated" soils is more likely to be removed by runoff or leaching due to the lowered P sorption efficiency.

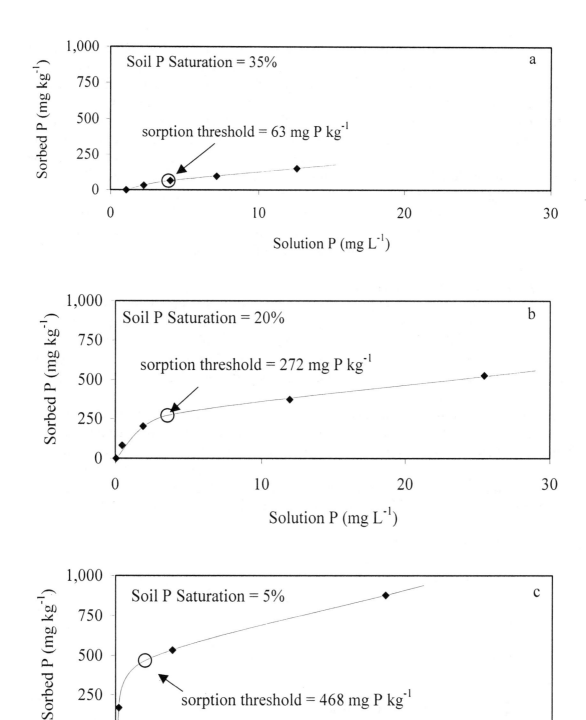

Figure 10. Sorption thresholds of three soils with varying degrees of P saturation.

Soil-Specific Nature of Source Risk

As mentioned earlier, source risk is highly soil specific. A number of studies relating soil P to runoff P have shown the relationships to be dependent on soil properties such as pH, mineralogy and texture. For instance, Sharpley (1995) found the relationship of Mehlich III P and dissolved P in runoff to be related to soil type as differentiated texture (soil P saturation was related to dissolved P in runoff for all soil types by a single regression equation). This study also found calcareous soils to release significantly less P to runoff than non-calcareous soils with similar Mehlich 3 P levels. Indeed, much attention is now focused on the differential release of P from calcareous and non-calcareous soils to runoff, as highlighted by work in Arkansas, Pennsylvania and New York under the National Phosphorus Project (see Sharpley et al., 1999 for a description of the project). At this time, however, further research is required before definitive generalizations can be made concerning calcareous and non-calcareous soils and P source risk.

Role of management in P source risk

Much attention has focused on the relationship between soil P and dissolved P losses. This relationship is certainly fundamental to our understanding of source risk, particularly over the long-term since soil P levels are comparatively recalcitrant. However, the near-term effects of P application, either in manure or mineral fertilizer, generally overwhelm the contribution of soil P to source risk (see Sharpley, this proceedings).

The relative contribution of P additions to source risk is highly time dependent. Over time, added P reacts with the soil, reducing its availability to runoff or leaching waters. Thus, runoff and leaching events that occur near the time of application remove much more added P than events that occur later. Figure 11 shows a typical decay-type curve, relating timing of surface manure application to dissolved P concentration in runoff. Curve parameters vary with factors such as amount and placement of the P addition.

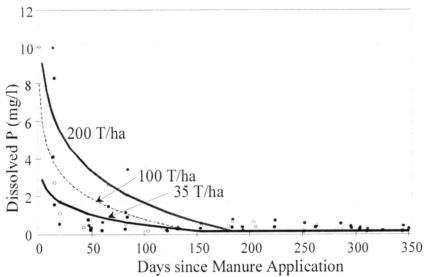

Figure 11. Dissolved phosphorus concentration in runoff vs. days since application and application intensity (data presented in Klausner et al., 1976, and interpreted by Brookes et al., 2000).

Incorporation of added P increases interactions between P and the soil, reducing the availability of that P to transport mechanisms. Surface application of P (e.g., top-dressed manure) greatly increases source risk relative to methods that incorporate added P (e.g., drilling of manure). This applies both to overland transport via surface runoff as well as to sub-surface transport by preferential, or macropore flow: the higher the concentration of P at the soil surface the higher the concentration of P in runoff and leachate. Indeed, deep cultivation can serve to dilute surface P, hence lower source risk (Sharpley, 1999), although the trade off between lowered surface P levels and increased potential for erosion must be assessed.

In addition to application methods, a number of management practices are now being explored to control source risk. The use of phosphorus immobilizing soil and manure amendments (PISMAs) is receiving considerable attention, particularly given the success of alum in simultaneously limiting P solubility and ammonia volatilization in poultry litter (Moore et al., 1999). Stout et al. (1999) examined the utility of low-cost, coal combustion byproducts as PISMAs, with special focus on the combined implications of these PISMAs to agronomic and environmental indicators of soil P. The addition of FBC fly ash and FGD gypsum to soils of near-neutral acidity (pH = 7.2) resulted in significant declines in water-extractable P (environmental indicator) and negligible declines in Mehlich III P (agronomic indicator). The authors conclude that these PISMAs can effectively control source risk (i.e., potential for dissolved P loss) without reducing the availability of P to crops.

Conclusions

Source factors for P and N play an important role in the risk of diffuse nutrient losses from agricultural lands. Because of the differing mobility of P and N in the environment, source factors are much more important to N loss potential than to P loss potential, as P loss is restricted primarily to critical source areas where high P availability and high transport potential overlap. Measurement of source risk requires an understanding of nutrient sources, soil properties and cropping system management. As new management practices, such as PISMAs, are targeted to source risk control, these too must be accounted for in risk assessments. Ultimately, true control of source risk requires balance of nutrient inputs and outputs at the farm gate.

References

Beegle, D. 2000. Integrating phosphorus and nitrogen management at the farm level. p. 159-168. *In* A.N. Sharpley (ed.), Agriculture and Phosphorus Management: The Chesapeake Bay. Lewis Publishers, Boca Raton, FL.

Breeuwsma, A. and S. Silva. 1992. Phosphorus fertilization and environmental effects in The Netherlands and the Po region (Italy). Rep. 57. Agric. Res. Dep. The Winand Staring Centre for Integrated Land, Soil and Water Research. Wageningen, The Netherlands.

Brooks, E.S., M.T. Walter, J. Boll, M.F. Walter, C.A. Scott and T.S. Steenhuis. 2000. Soluble phosphorus transport from manure-applied fields under various spreading strategies. Dept. Agricultural and Biological Engineering, Cornell University, Ithaca, NY.

Freese, D., R. Lookman, R. Merckx, and W.H. van Riemsdijk. 1995. New method for assessment of long-term phosphate desorption from soils. Soil Sci. Soc. Am. J. 59:1295-1300.

Gburek, W.J., A.N. Sharpley and H.B. Pionke. 1996. Identification of critical source areas for phosphorus export from agricultural watersheds. p. 263-282. *In* M.G. Anderson and S. Brookes (eds.), Advances in Hillslope Processes. John Wiley & Sons, New York, NY.

Heathwaite, A.L., T.P Burt and S.T. Trudgill. 1989. Runoff, sediment and solute delivery in agricultural drainage basins – a scale dependent approach. p. 175-191. International Association of Hydrological Sciences Publication 182. International Association of Hydrological Sciences Press, Wallingford, England.

Heathwaite, L., A. Sharpley and W. Gburek. 2000. A conceptual approach for integrating phosphorus and nitrogen management at watershed scales. J. Environ. Qual. 29: 158-166.

Heckrath, G., P.C. Brookes, P.R. Poulton and K.W.T. Goulding. 1995. Phosphorus leaching from soils containing different phosphorus concentrations in the Broadbalk experiment. J. Environ. Qual. 24:904-910.

Hesketh, N. and P.C. Brookes. 2000. Development of an indicator for risk of phosphorus leaching. J. Environ. Qual. 29: 105-110.

Klausner, S.D. P.J. Zwerman, and D.R. Coote. 1976. Design Parameters for the Land Application of Dairy Manure. EPA-600/2-76-187, Project No. S800767. U.S. Environmental Protection Agency, Athens, GA.

Kleinman, P.J.A., R.B Bryant and W.S. Reid. 1999. Development of pedotransfer functions to quantify phosphorus saturation of agricultural soils. J. Environ. Qual. 28: 2026-2030.

Lemunyon, J. L. and R. G. Gilbert. (1993). "The concept and need for a phosphorus assessment tool." J. Prod. Agr. 6: 483-486.

Mathers, A.C., B.A. Stewart and B. Blair. 1975. Nitrate-nitrogen from soil profiles by alfalfa. J. Environ. Qual. 4: 403-405.

McBride, M. 1994. Environmental Chemistry of Soils. Oxford University Press, New York

McDowell, R.W. and L.M. Condron. 1999. Developing a predictor for phosphorus loss from soil. p.153-164. In L.D. Currie (ed.), Best Soil Management Practices for Production (Fertilizer and Lime Research Centre 12th Annual Workshop). Massey University, Palmerston North, New Zealand.

McDowell, R.W. and A.N. Sharpley. 1999. Relating Soil Phosphorus Release to the Potential for Phosphorus Movement in Surface and Subsurface Runoff. p. 336. In Agronomy Abstracts 1999. Am. Soc. Agron., Madison, WI.

Moore, P.A., Jr., T.C. Daniel and D.R. Edwards. 1999. Reducing Phosphorus Runoff and Improving Poultry Production with Alum. Poultry Science 78:692-698.

Muir, J., J.S. Boyce, E.C. Seim, P.N. Mosher, E.J. Deibert and R.A. Olson. 1976. Influence of crop management practices on nutrient movement below the root zone in Nebraska soils. J. Environ. Qual. 5: 255-259.

Pote, D.H., T.C. Daniel, A.N. Sharpley, P.A. Moore, Jr., D.R. Edwards and D.J. Nichols. 1996. Relating extractable soil phosphorus to phosphorus losses in runoff. Soil Sci. Soc. Am. J. 60:855-859.

Pote, D.H., T.C. Daniel, D.J. Nichols, A.N. Sharpley, P.A. Moore, Jr., D.M. Miller and D.R. Edwards. 1999. Relationship between phosphorus levels in three ultisols and phosphorus concentrations in runoff. J. Environ. Qual. 28:170-175.

Ryden, J.C. and J.K. Syers. 1975. Rationalization of ionic strength and cation effects on phosphate sorption by soils. J. Soil Sci. 26:395-406.

Sharpley, A.N. 1999. Soil inversion by plowing decreases surface soil phosphorus content. p. 336. In 1999 Annual Meeting Abstracts. American Society of Agronomy, Crop Science Society of America and Soil Science Society of America, Madison, WI.

Sharpley, A.N. 1995. Dependence of runoff phosphorus on extractable soil phosphorus. J. Environ. Qual. 24:920-926.

Sharpley, A.N. and E. Lord. 1997. The loss of nitrogen and phosphorus in agricultural runoff: processes and management. In van Cleemput, O., S. Haneklaus, G. Hofman, E. Schnug and A. Vermoesen (ed.s), Fertilization for Sustainable Plant Production and Soil Fertility. 11th International World Fertilizer Congress, Ghent, Belgium.

Sharpley, A.N. and S.J. Smith. 1994. Effects of cover crops on surface water quality. p. 41-49. In W.L. Hargrove (ed.), Cover Crops for Clean Water Conference Proceedings. Soil and Water Conservation Society, Ankeny, IA.

Sharpley, A.N., T.C. Daniel, J.T. Sims, and D.H. Pote. 1996. Determining environmentally sound soil phosphorus levels. J. Soil Water Conserv. 51:160-166.

Sharpley, A.N., J.J. Meisinger, A. Breeuwsma, J.T. Sims, T.C. Daniel and J.S. Schepers. 1998a. Impacts of animal manure management on ground and surface water quality. p. 173-242. In J.L. Hatfield and B.A. Stewart (eds.), Animal Waste Utilization: Effective Use of Manure as a Soil Resources. Sleeping Bear Press, Inc., Ann Arbor, MI.

Sharpley, A.N., W.J. Gburek and G.J. Folmar. 1998b. Integrated phosphorus and nitrogen management in animal feeding operations for water quality protection. p. 72-95. In Animal Feeding Operations and Ground Water: Issues, Impacts and Solutions. National Ground Water Association, St. Louis, MO.

Sharpley, A., T. Daniel, B. Wright, P. Kleinman, T. Sobecki, R. Parry and B. Joern. 1999. National research project to identify sources of agricultural phosphorus loss. Better Crops 83:12-15.

Sibbeson, E. and A.N. Sharpley. 1997. Setting and justifying upper critical limits for phosphorus in soils. p. 151-176. In Tunney, H., O.T. Carton, P.C. Brookes and A.E. Johnston (eds.), Phosphorus Loss from Soil to Water. CAB International, Wallingford, England.

Stout, W.L., A.N. Sharpley, W.J. Gburek and H.B. Pionke. 1999. Reducing phosphorus export from croplands with FBC fly ash and FGD gypsum. Fuel 78:175-178.

Session 8

Phosphorus Index

The Phosphorus Index: Assessing Site Vulnerability to Phosphorus Loss

Andrew N. Sharpley
USDA-ARS, Pasture Systems and Watershed Management Research Laboratory
Building 3702, Curtin Road
University Park, Pennsylvania 16802-3702

Biographies for most speakers are in alphabetical order after the last paper.

Introduction

Phosphorus, an essential nutrient for crop and animal production, can accelerate freshwater eutrophication (Carpenter et al., 1998; Sharpley, 2000). Recently, the U.S. Environmental Protection Agency (1996) and Geological Survey (1999) identified eutrophication as the most ubiquitous water quality impairment in the U.S. Eutrophication restricts water use for fisheries, recreation, and industry due to the increased growth of undesirable algae and aquatic weeds and oxygen shortages caused by their death and decomposition. Also, many drinking water supplies throughout the world experience periodic massive surface blooms of *cyanobacteria* and other harmful algal blooms (e.g., *Pfiesteria piscicidia)*, which contribute to summer fish kills, unpalatability of drinking water, and formation of carcinogens during water chlorination (Burkholder and Glasgow, 1997; Kotak et al., 1993; Palmstrom et al., 1988).

Although concern over eutrophication is not new, there has been a profound shift in our understanding of, and focus on, sources of P in U.S. water bodies. Since the late 1960s, the relative contributions of P to U.S. water bodies from point sources and non-point sources has changed dramatically. On one hand, great strides have been made in the control of point source discharges of P, such as the reduction of P in sewage treatment plant effluent. These improvements have been due, in part, to the ease in identifying point sources. On the other hand, less attention has been directed to controlling non-point sources of P, due mainly to the difficulty in their identification and control (Sharpley and Rekolainen, 1997). Thus, control of non-point sources of P is a major hurdle to protecting fresh water bodies from eutrophication (Sharpley and Tunney, 2000; Sharpley et al., 1999a).

While a variety of non-point sources, ranging from suburban lawns to construction sites to golf courses, contribute P to U.S. water bodies, agriculture, particularly intensive livestock agriculture, is receiving more and more attention (Lander et al., 1998; Sharpley, 2000). This may be attributed to the evolution of agricultural systems from net sinks of P (i.e., deficits of P limit crop production) to net sources of P (i.e., P inputs in feed and fertilizer can exceed outputs in farm produce). Before World War II, for example, farming communities tended to be self-sufficient in that they produced enough feed locally to meet livestock requirements and could recycle the manure nutrients effectively to meet crop needs. As a result, sustainable nutrient cycles tended to exist in relatively localized areas. After World War II, farming systems became more specialized with crop and livestock operations in different regions of the country. Today, less than a third of the grain is produced on farms where it is grown (U.S. Department of Agriculture, 1989). This has resulted in a major one-way transfer of P from grain-producing areas to animal-producing areas (Lanyon, 2000; Sharpley et al., 1998b; Sims, 1997).

Most P entering intensive livestock operations is applied to soil. Animal manure can be a valuable resource for improving soil structure and increasing vegetative cover, thereby reducing surface runoff and erosion potential. However, in many areas of intensive confined livestock production, manures are normally applied at rates designed to meet crop N requirements and to avoid groundwater quality problems created by leaching of excess N. This often results in a build up of soil test P above amounts sufficient for optimal crop yields, which can increase the potential for P loss in runoff as well as in leachate (Haygarth et al., 1998; Sharpley et al., 1996; Heckrath et al., 1995).

Assessing the Risk for Phosphorus Loss

Environmental concern has forced many states to consider developing recommendations for land application of P and watershed management based on the potential for P loss in agricultural runoff (Sharpley et al., 1996; U.S. Department of Agriculture and Environmental Protection Agency, 1999). Currently, these recommendations center on the identification of a threshold soil test P level above which the enrichment of P in surface runoff is considered unacceptable. Agronomic soil testing may not be appropriate or results may need to be interpreted differently for environmental purposes (Sims and Sharpley, 1998). Soil test report interpretations (i.e., low, medium, optimum, high) were based on the expected response of a crop to P; therefore, it cannot be assumed a direct relationship exists between the soil test calibration for crop response to P and for runoff P enrichment potential.

However, threshold soil P levels are too limited to be the sole criterion to guide P application and management. For example, adjacent fields having similar soil test P levels, but differing susceptibilities to surface runoff and erosion due to contrasting topography and management, should not face similar P management recommendations (Sharpley and Tunney, 2000). To be most effective, risk assessment must consider "critical source-areas," which are specific identifiable areas within a watershed that are most vulnerable to P loss in surface runoff (Heathwaite and Johnes, 1996; Gburek and Sharpley, 1998). Critical source areas are dependent on the coincidence of transport (runoff, erosion, leaching, and channel processes) and source or site management factors (functions of soil, crop, and management). Transport factors are what translate potential P sources into actual loss from a field or watershed. Site management factors relate to fields or watershed areas that have a high potential to contribute to P export. These are typically well defined and reflect land use patterns related to soil P status, fertilizer and manure P inputs, and tillage.

Generally, most of the P exported from agricultural watersheds comes from only a small part of the landscape during a few relatively large storms, where hydrologically active areas of a watershed contributing surface runoff to streamflow are coincident with areas of high soil P (Gburek and Sharpley, 1998; Pionke et al., 1997). Even in regions where subsurface flow pathways dominate, areas contributing P to drainage waters appear to be localized to soils with high soil P saturation and hydrologic connectivity to the drainage network (Schoumans and Breeuwsma, 1997). Therefore, threshold soil P levels alone have little meaning vis a vis P loss potential unless they are used in conjunction with an estimate of potential surface runoff, erosion, and leaching. To overcome these limitations an alternative approach using critical source area technology was developed (the P index). The P index attempts to account for all factors controlling P loss. This paper describes the development of the P index, the rationale behind the factors used, and some preliminary testing of the index.

Development of the Phosphorus Index

The Natural Resource Conservation Service (NRCS), in cooperation with research scientists, developed the P Index as a screening tool for use by field staff, watershed planners, and farmers to rank the vulnerability of fields as sources of P loss in surface runoff (Lemunyon and Gilbert, 1993). Since its inception, two major changes have been introduced. First, source and transport factors were related in a multiplicative rather than additive fashion, in order to these better represent actual site vulnerability to P loss. For example, if surface runoff does not occur at a particular site, its vulnerability should be low regardless of the soil P content. In the original P index, a site could be ranked as very highly vulnerable based on site management factors alone, even though no surface runoff or erosion occurred. On the other hand, a site with a high potential for runoff, erosion, or leaching but with low soil P, is not at risk for P loss, unless P as fertilizer or manure is applied. Second, an additional transport factor reflecting distance from the stream was incorporated into the P index. The contributing distance categories in the revised P index are based on a hydrologic analysis of the probability (or risk) of occurrence of a rainfall event of a given magnitude which will result in surface runoff to the stream (Gburek et al., 2000).

The P index accounts for and ranks transport and site management factors controlling P loss in surface runoff and sites where the risk of P movement is expected to be higher than that of others (Tables 1 and 2). Site vulnerability to P loss in surface runoff is assessed by selecting rating values for individual transport (Table 1) and site management factors (Table 2) from the P index. A P index value, representing cumulative site vulnerability to P loss, is obtained by multiplying summed transport and site management factors (Table 3).

Table 1.   The transport factors of the P index.

Characteristics	Phosphorus loss rating					Field value
Soil Erosion (tons/acre)	2 X (tons soil loss/acre/year)					
Soil Runoff Class	Very Low 0	Low 2	Medium 4	High 8	Very High 16	
Subsurface Drainage	Very Low 0	Low 2	Medium 4	High 8	Very High 16	
Leaching Potential	Low 0		Medium 2	High 4		
Distance From Edge of Field to Surface Water (feet)	> 30 feet permanent vegetated buffer 0	10 - 30 feet vegetated buffer AND >30 feet no P application zone 2	10 - 30 feet vegetated buffer 4	< 10 feet AND >30 feet no P application zone 8	< 10 feet from water 16	

Total Site Value:_____

258

Table 2.   Site management factors of the P index.

Site Characteristics	Phosphorus Loss: Site Management Factors				
	Very Low	Low	Medium	High	Very High
Soil Test P	Soil test P (ppm)				
Loss Rating Value	Soil Test P * 0.05				
Fertilizer P Rate	Fertilizer Rate (lbs P$_2$O$_5$/ acre)				
P Fertilizer Application Method and Timing	Placed with planter or injected more than 2" deep	Incorporated <1 week after application	Incorporated >1 week or not incorporated >1 following application in May - October	Incorporated >1 week or not incorporated following application in Nov - April	Surface applied on frozen or snow covered soil
	0.2	**0.4**	**0.6**	**0.8**	**1.0**
Loss Rating Value	Fertilizer P Application Rate * Loss Rating for Fertilizer P Application Method and Timing				
Manure P Rate	Manure application (lbs P$_2$O$_5$/ acre)				
P Fertilizer Application Method and Timing	Placed with planter or injected more than 2" deep	Incorporated <1 week after application	Incorporated >1 week or not incorporated >1 following application in May - October	Incorporated >1 week or not incorporated following application in Nov - April	Surface applied on frozen or snow covered soil
	0.2	**0.4**	**0.6**	**0.8**	**1.0**
Loss Rating Value	Fertilizer P Application Rate * Loss Rating for Fertilizer P Application Method and Timing				

Total Management Value:_____

Table 3.   Worksheet and generalized interpretation of the P Index.

To solve for P loss rating - add all numbers on Part A and all numbers on Part B. Write these numbers on the worksheet. Multiply Part A x Part B. This is your final P loss rating.

Part A Value:_____

Part B Value:_____

Multiply A X B = _____ = _____ P Index Rating

P Index	Generalized interpretation of the P index
< 15	**LOW** potential for P loss. If current farming practices are maintained, there is a low probability of adverse impacts on surface waters.
15 - 150	**MEDIUM** potential for P loss. The chance for adverse impacts on surface waters exists, and some remediation should be taken to minimize the probability of P loss.
150 - 300	**HIGH** potential for P loss and adverse impacts on surface waters. Soil and water conservation measures and a P management plan are needed to minimize the probability of P loss.
> 300	**VERY HIGH** potential for P loss and adverse impacts on surface waters. All necessary soil and water conservation measures and a P management plan must be implemented to minimize the P loss.

The P index is intended to serve as a practical screening tool for use by extension agents, watershed planners, and farmers. The index can also be used to help identify agricultural areas or management practices that have the greatest potential to accelerate eutrophication. As such, the P index will identify alternative management options available to land users, providing flexibility in developing remedial strategies. Some general recommendations are given in Table 4; however, P management is very site-specific and requires a well-planned, coordinated effort between farmers, extension agronomists, and soil conservation specialists.

Table 4.   Management options to minimize nonpoint source pollution of surface waters by soil P.

Phosphorus index	Management options to minimize nonpoint source pollution of surface waters by soil P
< 15 (LOW)	*Soil testing*: Have soils tested for P at least every three years to monitor build-up or decline in soil P. *Soil conservation*: Follow good soil conservation practices. Consider effects of changes in tillage practices or land use on potential for increased transport of P from site. *Nutrient management*: Consider effects of any major changes in agricultural practices on P losses before implementing them on the farm. Examples include increasing the number of animal units on a farm or changing to crops with a high demand for fertilizer P.
15 - 150 (MEDIUM)	*Soil testing*: Have soils tested for P at least every three years to monitor build-up or decline in soil P. Conduct a more comprehensive soil testing program in areas that have been identified by the P Index as being most sensitive to P loss by surface runoff, subsurface flow, and erosion. *Soil conservation*: Implement practices to reduce P losses by surface runoff, subsurface flow, and erosion in the most sensitive fields (i.e., reduced tillage, field borders, grassed waterways, and improved irrigation and drainage management). *Nutrient management*: Any changes in agricultural practices may affect P loss; carefully consider the sensitivity of fields to P loss before implementing any activity that will increase soil P. Avoid broadcast applications of P fertilizers and apply manures only to fields with lower P Index values.
150 - 300 (HIGH)	*Soil testing*: A comprehensive soil testing program should be conducted on the entire farm to determine fields that are most suitable for further additions of P. For fields that are excessive in P, estimates of the time required to deplete soil P to optimum levels should be made for use in long range planning. *Soil conservation*: Implement practices to reduce P losses by surface runoff, subsurface flow, and erosion in the most sensitive fields (i.e., reduced tillage, field borders, grassed waterways, and improved irrigation and drainage management). Consider using crops with high P removal capacities in fields with high P Index values. *Nutrient management*: In most situations fertilizer P, other than a small amount used in starter fertilizers, will not be needed. Manure may be in excess on the farm and should only be applied to fields with lower P Index values. A long-term P management plan should be considered.
> 300 (VERY HIGH)	*Soil testing*: For fields that are excessive in P, estimate the time required to deplete soil P to optimum levels for use in long range planning. Consider using new soil testing methods that provide more information on environmental impact of soil P. *Soil conservation*: Implement practices to reduce P losses by surface runoff, subsurface flow, and erosion in the most sensitive fields (i.e., reduced tillage, field borders, grassed waterways, and improved irrigation and drainage management). Consider using crops with high P removal capacities in fields with high P Index values. *Nutrient management*: Fertilizer and manure P should not be applied for at least three years and perhaps longer. A comprehensive, long-term P management plan must be developed and implemented.

It is important to note that, as it is currently constructed, the P index is not a quantitative predictor of P loss in surface runoff or leaching from a watershed. Rather it is a qualitative assessment tool to rank site vulnerability to P loss. Ultimately, the P index is an educational tool that brings interaction between the planner and farmer in assessing environmental management decisions required to improve the farming system on a watershed rather than political basis.

Transport Factors

Transport factors are critical to site assessment as they translate potential P sources into actual loss from a field or watershed. Factors controlling the transport of P within agricultural watersheds are conceptualized in Figure 1. The main controlling factors and those considered in the P index are erosion, surface runoff, leaching, and distance or connectivity of the site to the stream channel. The justification for inclusion of each of these factors is given below.

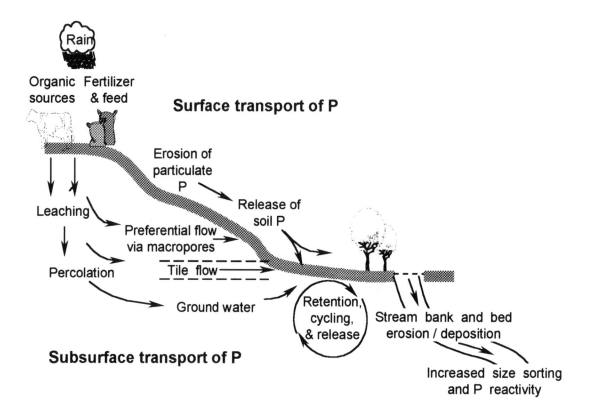

Figure 1. Transport and site management factors influencing the potential for P loss
from agricultural land to surface waters.

Erosion

Erosion preferentially moves the finer-sized soil particles. As a result, the P content and reactivity of eroded material is usually greater than source soil. For example, Sharpley (1985b) found that the enrichment of soil test P (Bray-1 P) and total P content of sediment in runoff from several soils under simulated rainfall ranged from 1.2 to 6.0 and 1.2 to 2.5, respectively. The enrichment of P increased as erosion decreased and the relative movement of fine-particles (<2 μm) of greater P content than coarser ones (> 5 μm) increased.

The effect of erosion on particulate P movement is illustrated by a 15-yr study of runoff from several grassed and cropped watersheds in the Southern Plains (Sharpley et al., 1991; Smith et al., 1991). As erosion from native grass and no-till and conventional-till wheat (*Triticum aestivum* L.) at El Reno, Oklahoma increased, particulate P was a greater portion of P transported in runoff, although amounts transported varied (0.08 to 10 Mg/ha/yr) with management and associated fertilizer P application (Fig. 2). Accompanying the increase in particulate P movement, is a relative decrease in dissolved P movement (Fig. 2).

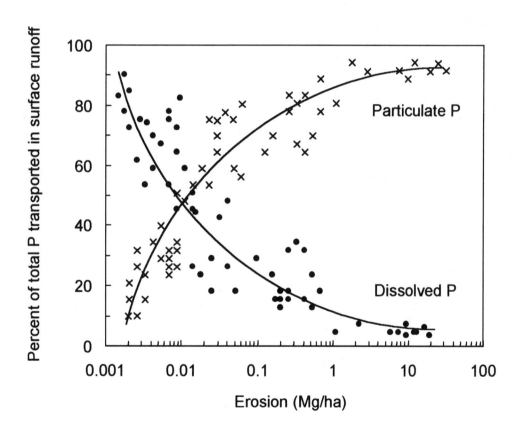

Figure 2. Percentage of total P as dissolved and particulate P as a function of erosion in surface runoff from watersheds at El Reno, OK.

Surface Runoff

The potential for surface runoff from a given site is a critical component of the P index, as this is typically the main pathway by which dissolved P loss occurs. The transport of dissolved P in runoff is initiated by the desorption, dissolution, and extraction of P from soil and plant material (Fig. 1). These processes occur when rainfall interacts with a thin layer of surface soil (1 to 5 cm) before leaving the field as runoff (Sharpley, 1985a). Although the proportion of rainfall and depth of soil involved are difficult to quantify in the field, they will be highly dynamic due to variations in rainfall intensity, soil tilth, and vegetative cover.

Leaching

Generally the P concentration in water percolating through the soil profile is small due to sorption of P by P-deficient subsoils. Exceptions occur in acid organic or peaty soils, where the adsorption affinity and capacity for P is low due to the predominantly negative charged surfaces and the complexing of Al and Fe by organic matter (Duxbury and Peverly, 1978; Miller, 1979; White and Thomas, 1981). Similarly, P is more susceptible to movement through sandy soils with low P sorption capacities; in soils which have become waterlogged, leading to conversion of Fe (III) to FE (II) and the mineralization of organic P; and with preferential flow through macropores and earthworm holes (Bengston et al., 1992; Sharpley and Smith, 1979; Sims et al., 1998).

Because of the variable paths and time of water flow through a soil with subsurface drainage, factors controlling P loss in subsurface waters are more complex than for surface runoff. Subsurface runoff includes tile drainage and natural subsurface flow, where tile drainage is percolating water intercepted by artificial systems, such as mole and tile drains (Fig. 1). In general, the greater contact time between subsoil and natural subsurface flow than tile drainage, results in lower losses of dissolved P in natural subsurface flow than in tile drainage (Sharpley and Rekolainen, 1997; Sims et al., 1998).

The leaching category in the P index represents a modification of the nitrogen leaching classes developed for Kansas soils by Kissel et al. (1982). The transport of P by leaching depends on soil properties (primarily texture and permeability), and the amount of water percolating through the soil profile (rainfall, irrigation). More detail on the development of the leaching classes for P are given by Sharpley et al. (1998a).

Distance or Connectivity to the Stream Channel

In order to translate the potential for P transport in erosion, surface runoff or leaching from a given site to the potential for P loss in stream flow, it is necessary to account for whether water leaving a site actually reaches the stream channel. For instance surface runoff and leaching may occur at various locations in a watershed and not reach the stream channel. Thus, an additional transport characteristic reflecting distance from the stream based on the hydrologic return period concept is now incorporated into the P index . The return period, a commonly accepted hydrologic design criterion, represents the probability (or risk) of occurrence of a rainfall or flood event of a given magnitude. Return period is typically expressed in terms of years, and is most simply understood to imply the particular event occurring once within that return period on the average. Thus, a flood event having a 10-yr return period will occur, on the average over the long term, once every 10 years, but not necessarily once within every 10-yr period.

Because of their high frequency of occurrence, shorter return periods and small storms represent a higher risk of surface runoff contributing P to the stream. These storms contribute surface runoff from more limited watershed areas designated with higher loss ratings, and consequently must be well managed to minimize P loss to the stream. Larger storms associated with longer return periods occur much less often, and therefore, pose a lower risk of P loss to the stream from the larger watershed areas affected.

Ongoing research by ARS evaluates how this return period approach can be applied to a wide geographic area, with readily available information. We are also pursuing characterization of a site's contribution to stream flow in terms of connectivity to the channel. For a simple assessment, a site can be categorized as not connected to the stream channel or connected to the channel by direct runoff, drainage ditch, or similar topographic feature. Intermediate categories of ephemeral or temporary connectivity may be appropriate.

Site Management Factors

The P index includes the following site management factors controlling P loss; soil test P concentration and the rate, type (fertilizer or manure), and method of P applied (Fig. 1). These factors reflect day-to-day farm operations, while the transport factors discussed earlier tend to represent inherent soil and topographic properties. As such, soil management factors are critical to the P index in determining if a site is a high or low source of P.

The following review of site management factors is based upon data that were obtained using a portable rainfall simulator and a combination of either field plots (1 m wide and 2 m long) or packed boxes of soil (15 cm wide and 1 m long). In all cases, the experimental protocol developed for the National P project was used (Sharpley et al., 1999b). The rainfall simulator was based on design of Miller (1987). Each simulator has one TeeJet^{TM} ½HH-SS50WSQ nozzle placed in the center of the simulator and 305 cm (10 ft) above the soil surface. The nozzle and associated water plumbing, pressure gauge, and electrical wiring is mounted on an aluminum frame, which in turn is fitted with plastic tarps to provide a windscreen.

Local tap water was used as the water source for the simulator and had a dissolved inorganic P concentration of 0.01 mg L^{-1}, nitrate-N of 3.1 mg L^{-1}, and pH of 5.7. Water pressure at the nozzle was regulated to 28 kPa (4.1 psi) to establish a water flow rate of 126 mL sec^{-1} at each nozzle. Shelton et al. (1985) found this pressure to give the best coefficient of uniformity and produce drops with size, velocity, and impact angles approximating natural rainfall. The simulator used in the present study produced a rainfall distribution with a uniformity coefficient of 85%. A rainfall intensity of 6.5 cm hr^{-1} for 30 min was used. This intensity for 30 min has an approximate 5-yr return frequency in south-central Pennsylvania. In all cases reported in this paper, dissolved P concentrations of surface runoff or subsurface flow represents the value determined on the total runoff volume; an event concentration.

Soil Phosphorus

The loss of dissolved P in surface runoff is dependent on the P content of surface soil (Fig. 3). These data were obtained from several locations within a 40 ha watershed (FD-36) in south-central Pennsylvania (Northumberland Co.). Locations were selected to give a wide range in soil test P concentration as Mehlich-3 P (15 to 500 mg/kg). A change point in the relationship between soil and surface runoff P was observed (220 and 175 mg/kg; Fig.3). The change point was determined as the interception of significantly different regression slopes ($p<0.5$). The potential for soil P release above this point is greater than below it.

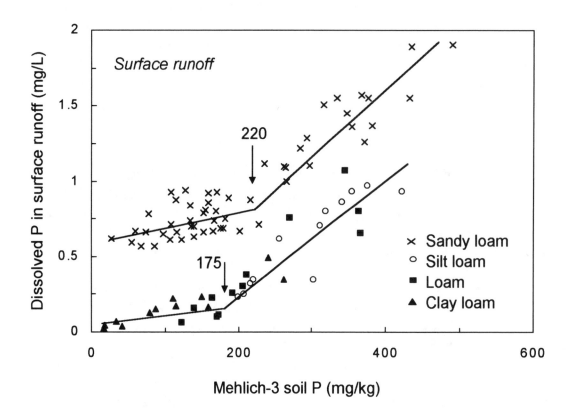

Figure 3. Relationship between the concentration of dissolved P in surface runoff and Mehlich-3 extractable soil P concentration of surface soil (0 - 5 cm) from a central PA watershed.

In a review of earlier studies, Sharpley et al. (1996) found that the relationship between soil P and surface runoff P varies with soil type and management. Relationship slopes were flatter for grass (4.1 to 7.0, mean 6.0) than for cultivated land (8.3 to 12.5, mean 10.5), but slopes were too variable to allow use of a single or average relationship for P management. Also, the variation in soil P change point or threshold among the Pennsylvania soils in Figure 3, shows that the ability of soils to release P to runoff is a function of soil type. Clearly, several soil and site management factors will determine not only the relationship between soil and surface runoff P but the amount of P lost.

The concentration of P in subsurface flow is also related to surface soil P (McDowell and Sharpley, 1999; Fig. 4). Thirty-cm deep lysimeters were taken from the FD-36 watershed and subjected to simulated rainfall (6.5 cm/hr for 30 min). The concentration of dissolved P in drainage from the lysimeter increased (0.07 to 2.02 mg/L) as the Mehlich-3 P concentration of surface soil increased (15 to 775 mg/kg; Fig. 4). This data manifest a change point that was similar to the change point identified for surface runoff. The dependence of leachate P on surface soil P is evidence of a the importance of P transport in preferential flow pathways such as macropores, earthworm holes, and old root channels.

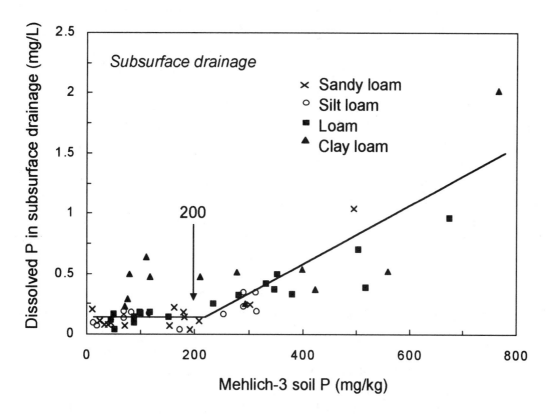

Figure 4. Relationship between the concentration of dissolved P in subsurface drainage from 30 cm deep lysimeters and the Mehlich-3 extractable soil P concentration of surface soil (0 - 5 cm) from a central PA watershed (adapted from McDowell and Sharpley, 1999).

Other studies have found a similar relationship between surface soil P and P loss in subsurface flow. For example, Heckrath et al. (1995) found that soil test P (Olsen P) >60 mg/kg in the plow layer of a silt loam, caused the dissolved P concentration in tile drainage water to increase dramatically (0.15 to 2.75 mg/L). They postulated that this level, which is well above that needed by major crops for optimum yield (about 20 mg/kg; Ministry of Agriculture, Food and Fisheries, 1994), is a critical point above which the potential for P movement in land drains greatly increases. Similar studies suggest that this change point can vary threefold as a function of site hydrology, relative drainage volumes, and soil P release (desorption) characteristics (Sharpley and Syers, 1979).

Application of Phosphorus as Fertilizer or Manure

Increased P loss in runoff has been measured after the application of fertilizer P and manure (Table 5). For example, the dissolved P concentration of surface runoff (6.5 cm/hr rainfall for 30 min), 14 days after applying 0, 50, or 100 kg P/ha as dairy manure to a Berks silt loam with a Mehlich-3 P content of 75 mg/kg, was 0.25, 1.35, and 2.42 mg/L, respectively (Fig. 5).

The loss of P is influenced by the rate, time, and method of application; form of fertilizer or manure, amount and time of rainfall after application; and vegetative cover (Table 5). The portion of applied P transported in runoff was greater from conventional- than conservation-tilled watersheds. Elsewhere however, McDowell and McGregor (1984) found fertilizer P application to no-till corn reduced P transport, probably due to an increased vegetative cover afforded by fertilization. As expected, the loss of applied P in subsurface tile drainage is appreciably lower than in surface runoff (Table 5). Although it is difficult to distinguish between losses of fertilizer, manure, or native soil P, without the use of expensive and hazardous radio tracers, total losses of applied P in runoff are generally less than 10% of that applied, unless rainfall immediately follows application or where runoff has occurred on steeply sloping, poorly drained, and/or frozen soils. The high proportion of manurial P in runoff may result from high manure application and generally less flexibility in application timing than for fertilizer (Table 5). This inflexibility results from the continuous production of manure throughout the year and a frequent lack of manure storage facilities.

Although we have shown soil P is important in determining P loss in surface runoff, applying P to soil can override soil P in determining P loss. For example, the dissolved P concentration of surface runoff increased with an increase in the Mehlich-3 soil P concentration in the surface 5 cm of a Berks silt loam (Fig. 5). When dairy manure was broadcast on these grassed soils, the dissolved P concentration of surface runoff 14 days later, was greater than with no manure (Fig. 5). However, soil P had little effect on the dissolved P concentration of surface runoff at the 50 kg P/ha/yr manure rate (r^2 of 0.71; $p>0.155$) and no effect at the 100 kg P/ha/yr rate (r^2 of 0.06; $p>0.751$), compared to when no manure was applied (r^2 of 0.99; $p<0.005$).

Table 5. Effect of fertilizer and manure application on P loss in surface runoff and fertilizer application on P loss in tile drainage.

Land use	P added	Phosphorus loss		Percent applied [a]	Reference and location
		Dissolved	Total		
Surface runoff	--------------- kg ha^{-1} yr^{-1}-------------				
Fertilizer					
Grass	0	0.02	0.22		McColl *et al.*, 1977;
	75	0.04	0.33	0.1	New Zealand
No-till corn	0	0.70	2.00		McDowell and McGregor, 1984;
	30	0.80	1.80		Mississippi
Conventional corn	0	0.10	13.89		
	30	0.20	17.70	12.7	
Wheat	0	0.20	1.60		Nicolaichuk and Read, 1978;
	54	1.20	4.10	4.6	Saskatchewan, Canada
Grass	0	0.50	1.17		Sharpley and Syers, 1976;
	50	2.80	5.54	8.7	New Zealand
Grass	0	0.17	0.23		Uhlen, 1988;
	24	0.25	0.31	1.2	Norway
	48	0.42	0.49	1.0	
Dairy Manure [b]					
Alfalfa	0	0.10	0.10		Young and Mutchler, 1976;
-spring	21	1.90	3.70	17.1	Minnesota
-autumn	55	4.80	7.40	13.3	
Corn	0	0.20	0.10		
-spring	21	0.20	0.60	2.4	
-autumn	55	1.00	1.60	4.7	
Poultry Manure					
Grass	0	0.00	0.10		Edwards and Daniel, 1992;
	76	1.10	2.10	2.6	Arkansas
Grass	0	0.10	0.40		Westerman *et al.*, 1983;
	95	1.40	12.4	12.6	North Carolina
Swine Manure					
Fescue	0	0.10	0.10		Edwards and Daniel, 1993a;
	19	1.50	1.50	7.4	Arkansas
	38	4.80	3.30	8.4	
Aritifical Drainage					
Corn	0	0.13	0.42		Culley *et al.*, 1983;
	30	0.20	0.62	0.7	Ontario, Canada
Oats	0	0.10	0.29		
	30	0.20	0.50	0.7	
Potatoes + Wheat + Barley					Catt *et al.*, 1997; Woburn, England
Minimal till	102	0.26	8.97	8.8	
Conventional till	102	0.35	14.38	14.1	
Alfalfa	0	0.12	0.32		
	30	0.20	0.51	0.6	
Grass - 0-30 cm	32	0.12	0.38	1.1	Heathwaite *et al.*, 1997;
- 30-80 cm	32	0.76	1.77	5.5	Devon, U.K.
Grass	0	0.08	0.17		Sharpley and Syers, 1979;
	50	0.44	0.81	1.3	New Zealand

[a] Percent P applied lost in runoff.

[b] Manure applied in either spring or autumn.

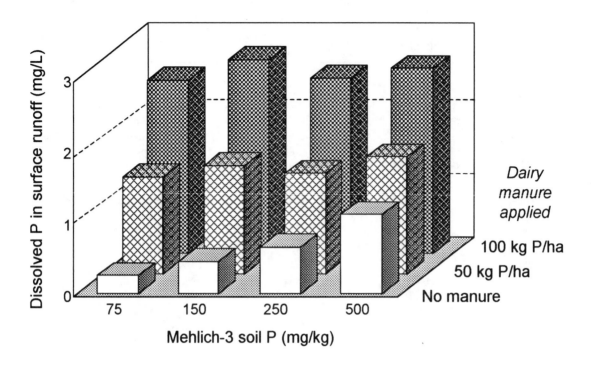

Figure 5. The concentration of P in surface runoff from a grassed Berks silt loam, as a function of Mehlich-3 soil P concentration and amount of dairy manure applied two weeks before the rainfall.

Phosphorus application method and timing relative to rainfall influences the loss of P in runoff. For example, several studies have shown a decrease in P loss with an increase in the length of time between P application and surface runoff (Edwards and Daniel, 1993b; Sharpley, 1997; Westerman et al., 1983). This decrease can be attributed to the reaction of added P with soil and dilution of applied P by infiltrating water from rainfall that did not cause surface runoff. The dissolved P concentration of surface runoff from a Berks silt loam (6.5 cm/hr rainfall for 30 min) decreased from 2.75 to 0.40 mg/L when rainfall occurred 35 days rather than 2 days after a surface broadcast application of 100 kg P/ha as dairy manure (Fig. 6).

Incorporation of manure into the soil profile either by tillage or subsurface placement, reduces the potential for P loss in runoff (Fig. 6). For example, the dissolved P concentration of surface runoff 2 days after 100 kg P/ha dairy manure was surface applied to a Berks silt loam was 2.75 mg/L. When the same amount of manure was incorporated by plowing to a depth of 10 cm, surface runoff dissolved P was 1.70 mg/L and when placed 5 cm below the soil surface, dissolved P in surface runoff was only 0.15 mg/L (Fig. 6).

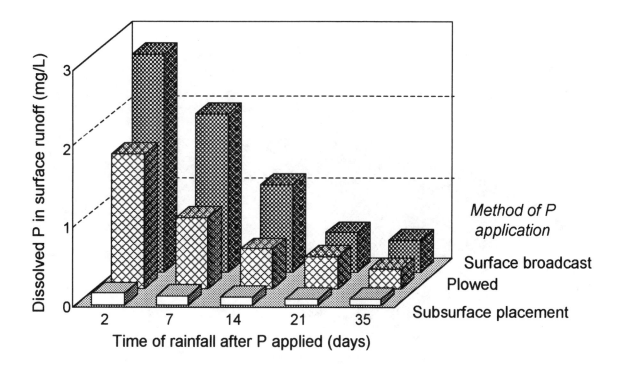

Figure 6.   The effect of application method and timing of rainfall after application of dairy manure (100 kg P/ha) on the concentration of P in surface runoff from a grassed Berks silt loam.

In an earlier field study, Mueller et al. (1984) found that incorporation of dairy manure by chisel plowing reduced total P loss in runoff from corn 20-fold compared to no-till areas receiving surface applications. However, the concentration of P in runoff did not decrease as dramatically as the mass of P lost. This was due to an increase in infiltration rate with manure incorporation and consequent decrease in runoff volume. In fact, runoff volume from no-till corn was greater than from conventional-till corn (Mueller et al., 1984). Thus, P loss in runoff is decreased by a dilution of P at the soil surface and reduction in runoff with incorporation of manure.

Testing the P Index

Although there is a great deal of research documenting the justification of the transport and source factors included in the P index, there has been little site evaluation of the index ratings. The original and modified versions of the P index have been used to assess the potential for P loss in several regions including the Delmarva Peninsula (Leytem et al., 1999; Sims, 1996), Oklahoma (Sharpley, 1995), Texas (McFarland et al., 1998), Vermont (Jokela et al., 1997), and Canada (Bolinder et al., 1998). However, few comparisons of P index ratings and measured P loss have been made. In Nebraska, Eghball and Gilley (1999) found r values between total P loss from simulated rainfall-runoff plots and P index ratings as high as 0.84, when erosion factor weighting was increased from 1.5 to 7.5.

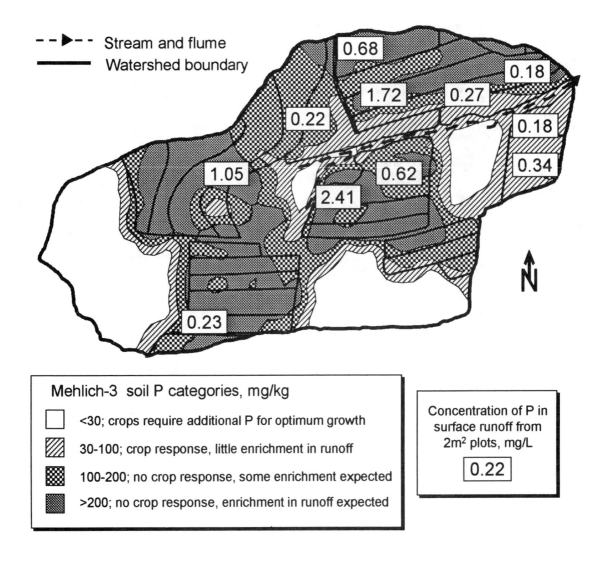

Figure 7. Distribution of Mehlich-3 soil P (0-5 cm soil depth) and concentration dissolved P in surface runoff from 2 m² plots within the FD-36 watershed, Northumberland Co., PA.

Using the portable rainfall simulator, we measured the dissolved P concentration in surface runoff from the 2 m² plots within FD-36 watershed in an attempt to evaluate the P index over the watershed. Surface runoff from a total of 48 plot locations was measured. A selection of dissolved P concentrations of surface runoff within FD-36 is given in Figure 7, along with surface soil (0 to 5 cm depth) Mehlich-3 P to demonstrate the large variation in concentration among plot locations. At some sites, rainfall was applied and surface runoff collected approximately 2 weeks after manure application. At other sites, no manure had been applied for at least 9 months. Thus, the range in dissolved P concentration is a function of soil P concentration and manure application (Fig. 7).

The P index was calculated for each plot location with FD-36. Using soil survey, land management, and topographic information, erosion was calculated using RUSLE and runoff using the curve number approach as in Table 1 (Sharpley et al., 1998a). Site management factors of the P index were calculated from Mehlich-3 P concentration of surface soil (0-5 cm depth) and P application rate, method, and timing as in Table 2. The P index rating for each plot location was calculated as the product of transport and site management factors as described in Table 3.

The P index rating for each plot location was closely related to the concentration of dissolved P in surface runoff ($r^2 = 0.78$; Fig. 8). This evaluation of the P index did not account for site position within the watershed relative to the stream channel or plot connectivity to the channel. Thus, the rating values of Figure 8 cannot be compared to the management categories given in Tables 3 and 4. Even so, the close relationship between index ratings and surface runoff P indicates the P index can accurately account for and describe a site's potential for P loss if surface runoff were to occur (Fig. 8).

Future Development of the P Index

The P index has been adopted by NRCS in their revised nutrient management planning guidelines to address P management issues. There are still, however, several areas where the index needs to be further evaluated and refined. Some factors included in the original P index may not be appropriate for certain regions and should be deleted, while some important factors influencing the pathways of P loss are not adequately represented. For example, subsurface flow may be the main pathway of P loss in some areas (notably the Delmarva Peninsula and Florida) and surface runoff in others (e.g., western Maryland, upper Chesapeake Bay watershed).

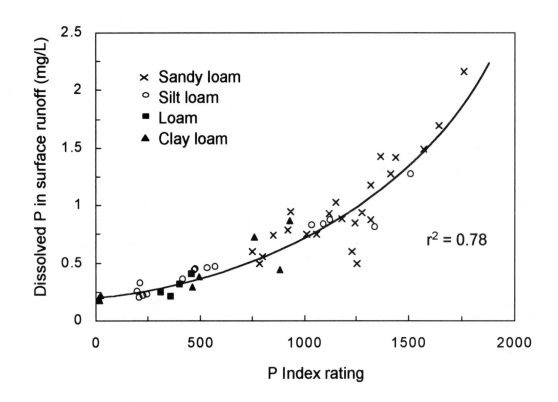

Figure 8.   Relationship between the concentration dissolved P in surface
runoff and P index rating for 2 m² plots within the FD-36
watershed, Northumberland Co., PA.

Transport Factors

1.   Use of the curve number approach to identify the potential for surface runoff to occur
should be evaluated. This approach may not adequately address the spatial and temporal
variability in runoff generation on a watershed scale.
1.   We know even less about the role of subsurface flow in P transport within a watershed
and to a stream channel. This is mainly due to the inherent difficulties in field measurement
and quantification of subsurface flow. This has led to a major information gap in terms of the
relative importance of subsurface flow and surface runoff as pathways of P transport from a
watershed.
2.   Site impact on P transport in terms of its position in a watershed relative to the stream
channel is addressed somewhat arbitrarily compared to other factors. Because it is critical
that transport processes be accurately represented.

3.	Phosphorus transport within a watershed is a dynamic process, with distinct areas that act as sources and sinks or depositional zones for P. For example, eroded P may be redeposited prior to reaching the channel. Although these translocation processes will be important in determining P export from a watershed, the P index framework does not address their highly dynamic nature or occurrence.

Site Management Factors

1.	The soil test P value used in the index will vary with the extraction method used, of which there are several (e.g., Mehlich 1 and 3, Bray 1, Morgan, and Olsen). Currently, local state methods are used to quantify this factor, which may lead to localized differences in index ratings depending on how the soil was analyzed. It would not be practical or appropriate to recommend the use of one soil test method where ever the index is applied. Thus, the potential for differences in index ratings due to soil test methodology should be addressed. As the P index is an environmental risk assessment tool, there may be justification in the future to use an environmentally based method (e.g., water, resin, or iron oxide strip), rather than one based on crop response.

2.	Soil sampling depth can vary with tillage, with shallower depths often recommended for no till (0 to 5 cm) than conventional till (0 to 15 or 20 cm). If stratification of high soil P occurs at the surface due to broadcast applications of P without incorporation, a 15-cm deep soil sample will underestimate the actual concentration that could be released to surface runoff. Thus, future development of the index should account for any variation on soil sampling depths among management practices.

3.	Soil P availability and release to runoff is influenced by several chemical properties that are not necessarily represented by soil test P, which vary among major soil types within the U.S. In Vermont, extractable soil Al influences residual P availability and has been included in their version of the P index (Jokela, 1999). Inclusion of other soil properties may improve the accuracy of using soil test P to describe P release to runoff.

4.	Several amendments (e.g., lime, flyash, gypsum) have been used to reduce the solubility of soil P and possibly loss in surface runoff (O'Reilly and Sims, 1995; Stout et al., 1998). Amendments are also used to reduce the solubility of P in solid and liquid manures (e.g., alum, Fe chelates) (Moore et al., 2000; Shreve et al., 1995). Consideration of them may be necessary where soil amendments have been used to more accurately reflect P loss potential.

5.	The original index of Lemunyon and Gilbert (1993)assigned weights to each site factor to account for their relative importance in P loss. Weights were 1.0 for soil test P, 0.75 for fertilizer rate, 0.5 for fertilizer application method, and 1.0 for both manure rate and application method. The magnitude of these weightings has become a contentious issue with little research available to support any differences between soil P, fertilizer and manure factors. Recent adaptations f the P index in Maryland and Vermont have assumed similar transport potentials for fertilizer and manure (Coale, 1999; Leytem et al., 1999; Jokela, 1999). Within the next several years, we must also be able to answer the question of where and when soil P has a greater effect than applied P in P loss and *visa-versa*. In Pennsylvania's P index, soil and applied P weightings are used so that the magnitude of no one site management factor dominates the total management value (see Table 2). The weightings used in any index, however, must be verified with field research.

6.   Where biosolids are applied to agricultural land, weightings for the manure management factor may be modified to reflect a reduction in P solubility during waste-water treatment. Maryland's P index for example, provides several weightings dependent on the type of biosolid or manure treatment considered (Coale, 1999). Although this theoretically sound, research information is needed on the relative susceptibilities of these different materials to release of to runoff, to validate and refine these weighting factors.

· 7.   Current versions of the P index do not directly consider the impact of any Best Management Practice (BMPs) on the potential P loss from a watershed. Some BMPs may be indirectly accounted for by transport factors, via reduced erosion or runoff potential. However, others such as buffers strips, feed amendments (phytase), innovative crop rotations, or crop hybrid P accumulators, may not be adequately addressed by the index.

8.   The P index was developed to consider one application of P annually. In many cases, however, split applications of fertilizer and manure are made and daily spreading of manure is common, particularly on dairy farms. On the other hand, manure may be applied only once in a 3-year crop rotation for example. Information is needed in the effect of P application on P loss potential for use of the index in these farming systems. Further, the index should be able to address the effect of multi-year rotations in the lang-term vulnerability for P loss.

In terms of the overall P index, watershed-scale validation is needed. Are the areas identified to be at greatest risk for P loss, actually sources of most of the P exported? In the same vein, will remediation of high risk areas identified by the index, decrease P export in stream flow from a watershed? Conversely, can low vulnerability areas receive more liberal P management without increasing P export?

Finally, and perhaps most critical to implementation of the P index in terms of recommending beneficial options for P management, will be development of the overall risk assessment classifications (see Tables 3 and 4). These classifications and interpretations must be developed with careful consideration of local management options, industry infrastructures, and State and Federal policy programs. With further development and testing, the P index will be a valuable tool to identify critical areas of P export, so that alternative management options can be identified. Limited resources and assistance can then be targeted to areas where they will have the most benefit.

References

Bengston, L., P. Seuna, A. Lepisto, and R.K. Saxena. 1992. Particle movement of meltwater in a subdrained agricultural basin. J. Hydrol. 135:383-398.

Bolinder, M.A., R.R. Simard, S. Beauchemin, and K.B. MacDonald. 1998. Indicator of risk of water contamination: methodology for the phosphorus component. p. 11-21, Agri-environmental Indicator Project Report No. 24, Agriculture and Agri-food Canada.

Burkholder, J.A., and H.B. Glasgow, Jr. 1997. *Pfiesteria piscicidia* and other Pfiesteria-dinoflagellates behaviors, impacts, and environmental controls. Limnol. Oceanogr. 42:1052-1075.

Carpenter, S.R., N.F. Caraco, D.L. Correll, R.W. Howarth, A.N. Sharpley, and V.H. Smith. 1998. Nonpoint pollution of surface waters with phosphorus and nitrogen. Ecol. Applic. 8:559-568.

Catt J.A., A.E. Johnston, and J.N. Quinton. 1997. Phosphate losses in the Woburn erosion reference experiment. p. 374-377. In: H. Tunney, O.T. Carton, P.C. Brookes, and A.E. Johnston (eds), Phosphorus loss from soil to water, CAB International, Wallingford, England.

Coale, F.J. 1999. The Maryland phosphorus site index: A technical user's guide. College of Agriculture and Natural Sciences, University of Maryland, College Park, MD. 14 pp.

Culley J.L.B., E.F. Bolton, and V. Bernyk. 1983. Suspended solids and phosphorus loads from a clay soil: I. Plot studies. J. Environ. Qual. 12:493-498.

Duxbury, J.M., and J.H. Peverly. 1978. Nitrogen and phosphorus losses from organic soils. J. Environ. Qual. 7:566-570.

Edwards, D.R., and T.C. Daniel. 1992. Potential runoff quality effects of poultry manure slurry applied to fescue plots. Trans ASAE 35:1827-1832.

Edwards, D.R., and T.C. Daniel. 1993a. Runoff quality impacts of swine manure applied to fescue plots. Trans Am. Soc. Agric. Eng. 36:81-80.

Edwards, D.R., and T.C. Daniel. 1993b. Drying interval effects on runoff from fescue plots receiving swine manure. Trans. Am. Soc. Agric. Eng. 36:1673-1678.

Eghball, B., and J.E. Gilley. 1999. Phosphorus risk assessment index and comparison with results of runoff studies. p. 30. In: Agronomy Abstracts 1999. Am. Soc. Agron., Madison, WI.

Gburek, W.J., and A.N. Sharpley. 1998. Hydrologic controls on phosphorus loss from upland agricultural watersheds. J. Environ. Qual. 27:267-277.

Gburek, W.J., A.N. Sharpley, A.L. Heathwaite, and G.J. Folmar. 2000. Phosphorus management at the watershed scale: A modification of the phosphorus index. J. Environ. Qual. 29:In press.

Haygarth, P.M., L. Hepworth, and S.C. Jarvis. 1998. Form of phosphorus transfer and hydrological pathways from soil under grazed grassland. Eur. J. Soil Sci. 49:65-72.

Heathwaite, A.L., and P.J. Johnes. 1996. The contribution of nitrogen species and phosphorus fractions to stream water quality in agricultural catchments. Hydrol. Proc. 10:971-983.

Heathwaite, A.L., P. Griffiths, P.M. Haygarth, S.C. Jarvis, and R.J. Parkinson. 1997. Phosphorus loss from grassland soils: implications of land management for the quality of receiving waters. p. 177-186. In: Freshwater Contamination, Proc. Rabat Symp., April-May 1997, IHAS Publ. No 243.

Heckrath, G., P.C. Brookes, P.R. Poulton, and K.W.T. Goulding. 1995. Phosphorus leaching from soils containing different phosphorus concentrations in the Broadbalk experiment. J. Environ. Qual. 24:904-910.

Jokela, W., F. Magdoff, and R. Durieux. 1997. Soil testing for improved phosphorus management. Vermont Cooperative Extension Service, Burlington, VT.

Jokela, W. 1999. The phosphorus index: A tool for management of agricultural phosphorus in Vermont. Vermont Cooperative Extension Service, Burlington, VT. 7 pp.

Kissel, D.E., O.W. Bidwell, and J.F. Kientz. 1982. Leaching classes of Kansas soils. Kansas Experimental Station Bulletin 6641, 10pp.

Kotak, B.G., S.L. Kenefick, D.L. Fritz, C.G. Rousseaux, E.E. Prepas, and S.E. Hrudey. 1993. Occurrence and toxicological evaluation of cyanobacterial toxins in Alberta lakes and farm dugouts. Water Res. 27:495-506.

Lander, C.H., D. Moffitt, and K. Alt. 1998. Nutrients Available from Livestock Manure Relative to Crop Growth Requirements. Resource Assessment and Strategic Planning Working Paper 98-1. U.S. Department of Agriculture, Natural Resources Conservation Service, Washington, DC. (http://www.nhq.nrcs.usda.gov/land/pubs /nlweb.html).

Lanyon, L.E. 2000. Nutrient management: Regional issues affecting the Chesapeake Bay. In: A.N. Sharpley (ed.), Agriculture and Phosphorus Management: The Chesapeake Bay. CRC Press, Boca Raton, FL.

Lemunyon, J.L., and R.G. Gilbert. 1993. The concept and need for a phosphorus assessment tool. J. Prod. Agric. 6:483-496.

Leytem, A.B., J.T. Sims, F.J. Coale, A.N. Sharpley, and W.J. Gburek. 1999. Implementing a phosphorus site index for the Delmarva Peninsula: Challenges and research needs. p. 336. In: Agronomy Abstracts 1999. Am. Soc. Agron., Madison, WI.

McColl, R.H.S., E. White, and A.R. Gibson. 1977. Phosphorus and nitrate runoff in hill pasture and forest catchments, Taita, New Zealand. N.Z. J. Mar. Freshwater Res. 11: 729-744.

McDowell, L.L., and K.C. McGregor. 1984. Plant nutrient losses in runoff from conservation tillage corn. Soil Tillage Res. 4:79-91.

McDowell, R.W., and A.N. Sharpley. 1999. Relating soil phosphorus release to the potential for phosphorus movement in surface and subsurface runoff. p. 336. In: Agronomy Abstracts 1999. Am. Soc. Agron., Madison, WI.

McFarland, A., L. Hauck, J. White, W. Donham, J. Lemunyon, and S. Jones. 1998. Nutrient management using a phosphorus risk index for manure application fields. p. 241-244. Proceedings of Manure Management in Harmony with the Environment and Society, Soil and Water Conservation Society, Feb. 10-12, 1998, Ames, IA.

Miller, M.H. 1979. Contribution of nitrogen and phosphorus to subsurface drainage water from intensively cropped mineral and organic soils in Ontario. J. Environ. Qual. 8:42-48.

Miller, W.P. 1987. A solenoid-operated, variable intensity rainfall simulator. Soil Sci. Soc. Am. J. 51:832-834.

Ministry of Agriculture, Food and Fisheries. 1994. Fertilizer recommendations for agricultural and horticultural crops. Ministry of Agriculture, Fisheries, and Food, Reference Book 209. HMSO, London, England. 48 p.

Moore, P.A., Jr., T.C. Daniel, and D.R. Edwards. 2000. Reducing phosphorus runoff and inhibiting ammonia loss from poultry manure with aluminum sulfate. J. Environ. Qual. 29:In press.

Mueller, D.H., R.C. Wendt, and T.C. Daniel. 1984. Phosphorus losses as affected by tillage and manure application. Soil Sci. Soc. Am. J. 48:901-905.

Nicholaichuk, W., and D.W.L. Read. 1978. Nutrient runoff from fertilized and unfertilized fields in western Canada. J. Environ. Qual. 7:542-544.

O'Reilly, S.E., and J.T. Sims. 1995. Phosphorus adsorption and desorption in a sandy soil amended with high rates of coal fly ash. Commun. Soil Sci. Plant Anal. 26:2983-2993.

Palmstrom, N.S., R.E. Carlson, and G.D. Cooke. 1988. Potential links between eutrophication and formation of carcinogens in drinking water. Lake and Reservoir Managt. 4:1-15.

Pionke, H.B., W.J. Gburek, A.N. Sharpley, and J.A. Zollweg. 1997. Hydrologic and chemical controls on phosphorus losses from catchments. p. 225-242. In: H. Tunney, O. Carton, and P. Brookes (eds.), Phosphorus Loss to Water from Agriculture. CAB International, Cambridge, England.

Schoumans, O.F., and A. Breeuwsma. 1997. The relation between accumulation and leaching of phosphorus: Laboratory, field and modelling results. p. 361-363. In: H. Tunney, O.T. Carton, P.C. Brookes, and A.E. Johnston (eds.), Phosphorus loss from soil to water. CAB International Press, Cambridge, England.

Sharpley, A.N. 1985a. Depth of surface soil-runoff interaction as affected by rainfall, soil slope, and management. Soil Sci. Soc. Am. J. 49:1010-1015.

Sharpley, A.N. 1985b. The selective erosion of plant nutrients in runoff. Soil Sci. Soc. Am. J. 49:1527-1534.

Sharpley, A.N. 1995. Identifying sites vulnerable to phosphorus loss in agricultural runoff. J. Environ. Qual. 24:947-951.

Sharpley, A.N. 1997. Rainfall frequency and nitrogen and phosphorus in runoff from soil amended with poultry litter. J. Environ. Qual. 26:1127-1132.

Sharpley, A.N. Editor. 2000. Agriculture and Phosphorus Management: The Chesapeake Bay. CRC Press, Boca Raton, FL.

Sharpley, A.N., and S. Rekolainen. 1997. Phosphorus in agriculture and its environmental implications. p. 1-54. In: H. Tunney, O.T. Carton, P.C. Brookes, and A.E. Johnston (eds.), Phosphorus loss from soil to water. CAB International Press, Cambridge, England.

Sharpley, A.N., and J.K. Syers. 1976. Phosphorus transport in surface runoff as influenced by fertilizer and grazing cattle. N.Z. J. Sci. 19:277-282.

Sharpley, A.N., and J.K. Syers. 1979. Loss of nitrogen and phosphorus in tile drainage as influenced by urea application and grazing animals. N.Z. J. Agric. Res. 22:127-131.

Sharpley, A.N., and H. Tunney. 2000. Phosphorus research strategies to meet agricultural and environmental challenges of the 21st century. J. Environ. Qual. 29:In press.

Sharpley, A.N., W.J. Gburek, and G. Folmar. 1998a. Integrated phosphorus and nitrogen management in animal feeding operations for water quality protection. p. 72-95. In: R.W. Masters and D. Goldman (eds.), Animal Feeding Operations and Ground Water: Issues, Impacts, and Solutions. National Ground Water Association, Westerville, OH.

Sharpley, A.N., W.J. Gburek, and A.L. Heathwaite. 1998b. Agricultural phosphorus and water quality: Sources, transport and management. J. Agricultural and Food Chemistry, Finland 7:297-314.

Sharpley, A.N., T.C. Daniel, J.T. Sims, and D.H. Pote. 1996. Determining environmentally sound soil phosphorus levels. J. Soil Water Conserv. 51:160-166.

Sharpley, A.N., S.J. Smith, J.R. Williams, O.R. Jones, and G.A. Coleman. 1991. Water quality impacts associated with sorghum culture in the Southern Plains. J. Environ. Qual. 20:239-244.

Sharpley, A.N., T.C. Daniel, J.T. Sims, J. Lemunyon, R.A. Steven, and R. Parry. 1999a. Agricultural phosphorus and eutrophication. U.S. Department of Agriculture - Agricultural Research Service, ARS-149 U.S. Govt. Printing Office, Washington, D.C. 34pp.

Sharpley, A., T. Daniel, B. Wright, P. Kleinman, T. Sobecki, R. Parry, and B. Joern. 1999b. National research project to identify sources of agricultural phosphorus loss. Better Crops 83:12-15.

Shelton, C.H., R.D. von Bernuth, and S.P. Rajbhandari. 1985. A continuous-application rainfall simulator. Trans. Am. Soc. Agric. Eng. 28:1115-1119.

Shreve, B.R., P.A. Moore, Jr., T.C. Daniel, D.R. Edwards, and D.M. Miller. 1995. Reduction of phosphorus in runoff from field-applied poultry litter using chemical amendments. J. Environ. Qual. 24:106-111.

Sims, J.T. 1996. The Phosphorus Index: a phosphorus management strategy for Delaware's agricultural soils. Fact Sheet ST-05, Delaware Cooperative Extension Service, Newark. DE.

Sims, J.T. 1997. Agricultural and environmental issues in the management of poultry wastes: Recent innovations and long-term challenges. p. 72-90. In: J. Rechcigl and H.C. MacKinnon (eds.), Uses of by-products and wastes in agriculture. Am. Chem. Soc., Washington, D.C.

Sims, J.T., and A.N. Sharpley. 1998. Managing agricultural phosphorus for water quality protection: future challenges. p. 41-43. In: J.T. Sims (ed), Soil Testing for Phosphorus: environmental uses and implications. Southern Cooperative Series Bulletin No. 389, SERA-IEG 17, USDA-CSREES.

Sims, J.T., R.R. Simard, and B.C. Joern. 1998. Phosphorus losses on agricultural drainage: Historical perspectives and current research. J. Environ. Qual. 27:277-293.

Smith, S.J., A.N. Sharpley, J.W. Naney, W.A. Berg, and O.R. Jones. 1991. Water quality impacts associated with wheat culture in the Southern Plains. J. Environ. Qual. 20:244-249.

Stout, W.L., A.N. Sharpley, and H.B. Pionke. 1998. Reducing soil phosphorus solubility with coal combustion by-products. J. Environ. Qual. 27:111-118.

Uhlen, G. 1988. Surface runoff losses of phosphorus and other nutrient elements from fertilized grassland. Norwegian J. Agric. Sci. 3:47-55.

U.S. Department of Agriculture. 1989. Fact book of agriculture. Misc. Publ. No. 1063, Office of Public Affairs, Washington, DC.

U.S. Department of Agriculture and U.S. Environmental Protection Agency. 1999. Unified national strategy for Animal Feeding Operations. March 9, 1999. (http://www.epa.gov/owm/finafost.htm).

U.S. Environmental Protection Agency. 1996. Environmental indicators of water quality in the United States. EPA 841-R-96-002. U.S. EPA, Office of Water (4503F), U.S. Govt. Printing Office, Washington, DC.

U.S. Geological Survey. 1999. The quality of our nation's waters: Nutrients and pesticides. U.S. Geological Survey Circular 1225, 82pp. USGS Information Services, Denver, CO. http://www.usgs.gov.

Westerman, P.W., T.L. Donnely, and M.R. Overcash. 1983. Erosion of soil and poultry manure - a laboratory study. Trans. Am. Soc. Agric. Eng. 26:1070-1078, 1084.

White, R.E., and G.W. Thomas. 1981. Hydrolysis of aluminum on weakly acidic organic exchangers: implications for phosphorus adsorption. Fert. Res. 2:159-167.

Young, R.A., and C.K. Mutchler. 1976. Pollution potential of manure spread on frozen ground. J. Environ. Qual. 5:174-179.

●●●●●●●●●●●●●●●●●●●●●●●●●●●●●●●●

Adapting the Phosphorus Site Index to the Delmarva Peninsula: Delaware's Experience

April B. Leytem, Ph.D., Postdoctoral Research Associate
Department of Plant and Soil Sciences, University of Delaware

J. Thomas Sims, Ph.D., Professor of Soil and Environmental Chemistry
Department of Plant and Soil Sciences, University of Delaware

Frank J. Coale, Ph.D., Associate Professor of Soil Fertility
Department of Natural Resources Sciences and Landscape Architecture
University of Maryland

Biographies for most speakers are in alphabetical order after the last paper.

●●

Introduction

Agriculture on the Delmarva Peninsula

Agriculture on the Delmarva Peninsula is heavily dependent on animal based agriculture, which is dominated by poultry production. Over 600,000,000 broiler chickens are produced annually on the peninsula, primarily in four counties in Delaware and Maryland which have, in total, ~175,000 ha of cropland (Sims, 1997a). The poultry industry provides a stable market for commercial growers of corn, soybeans, small grains and sorghum. This "poultry-grain agriculture" has the potential to be a completely self-contained system, in which the animal manures generated provide virtually all of the fertilizer nutrients needed to produce the grain consumed by the poultry. Unfortunately, a comprehensive, workable, manure management program has not been developed on the peninsula.

For example, the poultry industry in Delaware has nearly tripled in size since the mid-1950's and now produces about 260,000 million broiler and roaster chickens each year, primarily in Sussex County which is one of the largest poultry producing counties in the U.S. (DDA, 1997). During this time, the number of farms and hectares of cropland available to efficiently use the by-products created by the poultry industry has decreased. Intensification of animal agriculture in this manner generates excess nutrients, which are not exported from the region as outputs in crops and animal products. Compounding the situation is the continual use of commercial fertilizers when excess nutrients are available as litters/manures.

A simplified statewide mass balance for nitrogen (N) and phosphorus (P) for Delaware indicates that there is a yearly N surplus of 83 kg/ha/yr and a P surplus of 30 kg/ha/yr (Sims, 1998a). These excess nutrients, when improperly managed, can significantly contribute to water quality problems.

The topography, soil, and hydrology of the Delmarva Peninsula

The Atlantic Coastal Plain of the U.S. extends westward from the Atlantic Ocean to the Piedmont (an elevated, rolling plain separating the mountains from the ocean). The Coastal Plain ranges from a narrow strip of land in New England to a broad belt covering much of North and South Carolina, Georgia, and Florida. Many rivers cross the plain flowing into the ocean or into important coastal estuaries such as the Chesapeake Bay and Delaware Bays. The Delmarva (Delaware-Maryland-Virginia) peninsula, located in the mid-Atlantic region of the Coastal Plain, is bordered on the west and east by the Chesapeake and Delaware Bays; it is also the site of a national estuary (Delaware's Inland Bays). The Delmarva Peninsula is dominated by flat topography, having hydrogeomorphic features that range from a mix of well-drained and poorly drained uplands in the northern region of the peninsula to poorly drained and fine-grained lowlands in the southern region of the peninsula. Rainfall is plentiful, with an annual precipitation rate of approximately 115 cm/yr. In some poorly drained areas on the peninsula, such as parts of Delaware's Inland Bays watershed, farming is only possible due to an extensive network of open drainage ditches which were constructed decades ago to lower the water table and remove excess surface water from fields. These drainage systems are direct conduits for surface runoff and subsurface discharges of nutrients from agricultural land to nearby surface water systems.

Why is P a concern for the Delmarva Peninsula?

One of the more pressing concerns today is the eutrophication of many of Delmarva's surface waters, particularly in the Chesapeake Bay and Delaware's Inland Bays. Nutrients entering streams, rivers, ponds, lakes, and estuaries via runoff and subsurface groundwater flow are known causative factors in eutrophication. Eutrophication restricts water use for fisheries, recreation, industry, and drinking, due to the increased growth of undesirable algae, aquatic weeds and oxygen shortages caused by the death and decomposition of these organic materials.

For example, previous nonpoint source pollution studies in southern Delaware have identified the long-term accumulation of soil P to exceedingly high levels as a potential concern for eutrophication in lakes, ponds, and bays (Gartley and Sims, 1994; Mozaffari and Sims, 1994). Soil test results in Delaware found that in Sussex County (site of the Inland Bays watershed) approximately 84% of soils tested from commercial cropland were rated as "high" (32%) or "excessive" (52%) in P (Sims and Leytem, 1999). This accumulation of soil P is a result of long term application of animal manures to agricultural lands. Management of animal manures based on meeting crop N needs, which has been the practice in the past, results in long-term increases in soil P. Sims (1997b) estimated that a typical poultry grain farm in Delaware produced annual surpluses of P in the range of 90 to 120 kg P/ha/yr.

Efforts to reduce nutrient enrichment of ground and surface waters have become a high priority for state and federal agencies and a matter of considerable importance to all nutrient users and generators on the peninsula. In the spring of 1998 the state of Maryland passed legislation requiring that N and P based management plans be developed by farmers and for large-scale non-agricultural users (e.g. commercial lawn care companies fertilizing an aggregate of >1.25 ha). In January of 1999, Virginia passed a poultry waste management bill requiring the development and implementation of nutrient management plans for "any person owning or operating a confined poultry feeding operation" (Sims, 2000). In Delaware, as a result of a lawsuit filed by environmental action groups, the U.S. Environmental Protection Agency (USEPA) recently negotiated a Total Maximum Daily Load (TMDL) agreement with Delaware's Department of Natural Resources and Environmental Control (DNREC). The agreement mandates that the state establish TMDLs for nutrients (N, P), sediments, and pathogens for all impacted water bodies and calls for pollution control strategies to make these waters "fishable and swimmable" by 2007. Also in Delaware, an *Agricultural Industry Advisory Committee on Nutrient Management* (AIACNM) was appointed by Governor Carper to address the issue of agricultural nonpoint source pollution. This committee issued a series of recommendations that led to the passage in 1999 of House Substitute Bill 1 for House Bill 250 which established a *Delaware Nutrient Management Commission* (DNMC) to develop and implement a *State Nutrient Management Program*. All three states have identified the need for P based nutrient management planning in those areas having significant potential for P transport to surface waters. Each state has also identified the need for a *Phosphorus Site Index* to identify areas having higher risks for P transport from fields to waterways, based on the properties and management of all P sources (fertilizers, manures, biosolids), soil properties, hydrology, and soil and crop management practices.

What is a Phosphorus Site Index?

In the early 1990's the U.S. Department of Agriculture (USDA) began to develop assessment tools for areas with water quality problems. While some models, such as the Universal Soil Loss Equation (USLE) for erosion and GLEAMS (Groundwater Loading Effects of Agricultural Management Systems) for ground water pollution, were already being used to screen watersheds for potential agricultural impacts on water quality, there was no model considered suitable for the field-scale assessment of the potential movement of P from soil to water. A group of scientists from universities and governmental agencies met in 1990 to discuss the potential movement of P from soil to water and later formed a national work group (PICT: Phosphorus Index Core Team) to more formally address this problem. Members of PICT soon realized that despite the many scientists conducting independent research on soil P, there was a lack of integrated research that could be used to develop the field-scale assessment tool for P needed by USDA. Consequently, the first priority of PICT was a simple, field-based, planning tool that could integrate, through a multi-parameter matrix, the soil properties, hydrology, and agricultural management practices within a defined geographic area, and thus to assess, in a relative way, the risk of P movement from soil to water. The initial goals of the PICT team were:

- *To develop an easily used field rating system (the Phosphorus Site Index) for Cooperative Extension, NRCS technical staff, crop consultants, farmers or others that rates soils according to the relative potential for P loss to surface waters.*

- *To relate the P Site Index to the sensitivity of receiving waters to eutrophication.* This is a vital task because soil P is primarily an environmental concern if a transport process

284

exists that can carry particulate or soluble P to surface waters where eutrophication is limited by P.

- *To facilitate adaptation of the P Site Index to site specific situations.* The variability in soils, crops, climates and surface waters makes it essential that each state or region modify the parameters and interpretation given in the original *P Site Index* to best fit local conditions.

- *To develop agricultural management practices that will minimize the buildup of soil P to excessive levels and the transport of P from soils to sensitive water bodies.*

The *P Site Index* is designed to provide a systematic assessment of the risks of P loss from soils, but does not attempt to estimate the actual quantity of P lost in runoff. Knowledge of this risk not only allows us to design best management practices (BMPs) that can reduce agricultural P losses to surface waters, but to more effectively prioritize the locations where their implementation will have the greatest water quality benefits. When assessing the risk of P loss from soil to water, it is important that we not focus on strictly one measure of P, such as agronomic soil test P value. Rather, a much broader, multi-disciplinary approach is needed, one that recognizes that P loss will vary among watersheds and soils, due to the rate and type of soil amendments used, and due to the wide diversity in soils, crop management practices, topography, and hydrology (Sims, 1998b; Sims et al., 1998). <u>At a minimum</u>, any risk assessment process for soil P should include the following:

- Characteristics of the P source (fertilizer, manure, biosolids) that influence its solubility and thus the potential for movement or retention of P once the source has been applied to a soil.

- The concentration and bioavailability of P in soils susceptible to loss by erosion.

- The potential for soluble P release from soils into surface runoff or subsurface drainage.

- The effect of other factors, such as hydrology, topography, soil, crop, and P source management practices, on the potential for P movement from soil to water.

- Any "channel processes" occurring in streams, field ditches, etc. that mitigate or enhance P transport into surface waters.

- The sensitivity of surface waters to inputs of P and the proximity of these waters to agricultural soils.

In summary, when resources are limited, it is critical to target them at areas where the interaction of P source, P management, and P transport processes result in the most serious risk of losses of P to surface and shallow ground waters. <u>This is the fundamental goal of the P Site Index.</u>

How did the Phosphorus Site Index evolve?

The Original P Site Index:
The first *P Site Index* developed by the PICT team was published by Lemunyon and Gilbert (1993) and included the following parameters known to influence P availability, retention, management, movement, and uptake (see Table 1):

I.   Soil erosion (1.5)

II.   Irrigation erosion (1.5)

III.   Soil runoff class (0.5)

IV.   Soil test P (1.0)

V.   P fertilizer application rate (0.75)

VI.   P fertilizer application method (0.5)

VII.   Organic P source application rate (1.0)

VIII.   Organic P source application method (1.0)

Each site characteristic was assigned a weighting factor (shown in parentheses above) based on the reasoning that some site characteristics will be more important than others in controlling the potential for P movement from a site. Each site characteristic was also assigned a relative loss rating of low (=1), medium (=2), high (=4) or very high (=8) that is used to make a site-specific ranking of the severity of conditions found at individual locations. To make an assessment using *the P Site Index*, the weighting factor for each of the eight site characteristics is first multiplied by the site-specific relative loss rating. Then, the resulting values for all eight characteristics (weighting factor x loss rating) are summed to determine the *P Site Index* for an individual site. Comparison of the final *P Site Index* value with the site vulnerability chart (Table 2) is then done to categorize the risk of P loss as low, medium, high or very high. Interpretations and recommendations for soil and nutrient management can then be developed in accordance with the level of risk.

Table 1. Original Phosphorus Site Index. (Lemunyon and Gilbert, 1993)

SITE CHARACTERISTIC (Weighting Factor)	PHOSPHORUS LOSS RATING (Value)				
	NONE (0)	LOW (1)	MEDIUM (2)	HIGH (4)	VERY HIGH (8)
Soil Erosion (1.5)	N/A	< 5 tons/acre	5-10 tons/acre	10-15 tons/acre	> 15 tons/acre
Irrigation Erosion (1.5)	N/A	Infrequent irrigation on well-drained soils	Moderate irrigation on soils with slopes < 5%	Frequent irrigation on soils with slopes of 2-5%	Frequent irrigation on soils with slopes > 5%
Soil Runoff Class (0.5)	N/A	Very Low or Low	Medium	High	Very High
Soil Test P (1.0)	N/A	Low	Medium	Optimum	Excessive
P Fertilizer Application Rate (lb P_2O_5/acre) (0.75)	None Applied	< 31	31-90	91-150	> 150
P Fertilizer Application Method (0.5)	None Applied	Placed with planter deeper than 2 inches	Incorporate immediately before crop	Incorporate > 3 months before crop or surface applied < 3 months before crop	Surface applied to pasture or applied > 3 months before crop
Organic P Source Application Rate (lb P_2O_5/acre) (1.0)	None Applied	< 31	31-90	91-150	> 150
Organic P Source Application Method (1.0)	None	Injected deeper than 2 inches	Incorporate immediately before crop	Incorporate > 3 months before crop or surface applied < 3 months before crop	Surface applied to pasture or applied > 3 months before crop

Table 2. Site vulnerability ratings and the interpretations obtained from the original Phosphorus Site Index (Lemunyon and Gilbert, 1993).

Total of Weighted Rating Values	Site Vulnerability
< 8	LOW
8 - 14	MEDIUM
15 - 32	HIGH
> 32	VERY HIGH

The Phosphorus Site Index: An Ongoing National Effort by SERA-IEG17

After the development of the original *Phosphorus Site Index,* interest grew within PICT to expand the scope of research and extension activities related to P management for water quality protection. In 1992 PICT organized a symposium at the national meetings of the American Society of Agronomy (published in the *Journal of Production Agriculture*, 1993) highlighting the *Phosphorus Site Index* and the need to expand our knowledge on the role of agricultural P in eutrophication. The original PICT soon grew to over 50 scientists from the U.S. and other countries. The efforts of PICT were formalized in 1993 by establishing a USDA research and information group (SERA-IEG 17). A major goal of the group has been to bring together a greater diversity of disciplines to discuss the research and management needs related to agricultural P and water quality. SERA-IEG 17 has expanded rapidly since 1993 and now has over 100 members with expertise in disciplines ranging from soil science and corn genetics to hydrology and limnology. It has become a valuable informational resource for agencies (USEPA, USDA) and state universities that are addressing the need for best management practices (BMPs) to prevent nonpoint source pollution of surface waters by agricultural P. In 1996 SERA-IEG 17 co-sponsored a symposium entitled *Agricultural Phosphorus and Eutrophication* at the national meetings of the American Society of Agronomy and the Soil Science Society of America. Topics included hydrologic controls on P loss from uplands, P losses in agricultural drainage, watershed modeling of P transport, and plant genetic approaches to P management for agriculture. The symposium was published in the *Journal of Environmental Quality* in 1998.

SERA-IEG 17 has adopted a broad, long-term perspective on the issue of minimizing P losses from agriculture for water quality protection. It has also identified the following specific objectives, now being addressed by separate task forces:

a) To develop an interdisciplinary approach to identify P sensitive watersheds and water bodies, expanding and improving upon the *Phosphorus Site Index.*

b) To develop BMPs to reduce agricultural P losses to surface and ground waters by erosion and runoff (surface and subsurface).

c) To develop an animal manure application strategy based on both P and N.

d) To develop upper, environmentally based, critical limits for soil test P and new soil testing methods that can more accurately identify soils where P loss will be of environmental concern.

Developing a Phosphorus Site Index for the Delmarva Peninsula

It has always been recognized, and strongly recommended, by PICT and SERA-IEG 17, that the *P Site Index* must be modified as needed to reflect local or regional conditions. In 1997 a regional effort was initiated by scientists from Delaware, Maryland, Pennsylvania and Virginia to develop a *P Site Index* that would more accurately predict the potential for P loss from agricultural sites in the Chesapeake Bay watershed. In particular, Delaware and Maryland have worked together to develop a *P Site Index* for the Delmarva Peninsula (Table 3). One of the most significant changes that has resulted from this cooperative effort has been the separation of the *P Site Index* into two components:

(i) <u>Part A: Site and Transport Factors</u>: soil erosion, soil surface runoff, subsurface drainage, leaching potential, distance from edge of water, and priority of receiving water.

(ii) <u>Part B: P Source and Management Factors</u>: soil test P, P fertilizer application rate and application method, and organic P source application rate and application method.

Separating the *P Site Index* into two parts makes it possible to separate the risk assessment for a site into (i) site and transport factors that affect P loss, and (ii) P source and management practices that affect P loss. Instead of adding these together, as in the original *P Site Index*, the sums of each Part are multiplied together to prevent overemphasis of one set of factors. For example, a field with a very high P source potential (i.e., a high soil test P value) but with a low or moderate transport potential would not likely receive a high *P Site Index* rating because there is low probability that P would be transported from the field to ground or surface waters.

Table 3. The Phosphorus Site Index adopted for use in Maryland and proposed for use in Delaware.

Part A: Phosphorus loss potential due to site and transport characteristics

Site Characteristics	Phosphorus Loss Rating					Field Value
Soil Erosion (tons/acre)	2 x (tons soil loss/acre/year)					
Soil Surface Runoff Class	Very Low 0	Low 2	Medium 4	High 8	Very High 16	
Subsurface Drainage	Very Low 0	Low 2	Medium 4	High 8	Very High 16	
Leaching Potential	Low 0		Medium 2		High 4	
Distance from Edge of Field to Surface Water	>25' vegetated buffer AND >25' no P application zone 0	10-25' vegetated buffer AND >25' no P application zone 2	10-25' vegetated buffer AND 10-25' no P application zone 4	<10' from water AND >25' no P application zone 8	<10' from water 16	
Priority of Receiving Water	Very Low 0	Low 1	Medium 2	High 4	Very High 8	

Part B: Phosphorus loss potential due to P source and management practices.

Site Characteristics	Phosphorus Loss Management					Field Value
	None	Low	Medium	High	Very High	
Soil Test P Fertility Index Value	0.05 x FIV					
P Fertilizer Application Rate	0.10 x (lb P_2O_5/ acre)					
P Fertilizer Application Method	None applied 0	Injected/ banded below surface at least 2" 2	Incorporated within 5 days of application 4	Surface applied March through November OR incorporated in >5 days 8	Surface applied December through February 16	
Organic P Application Rate	PAC x (lb P_2O_5/acre)					
Organic P Application Method	None applied 0	Injected/ banded below surface at least 2" 2	Incorporated within 5 days of application 4	Surface applied March through November or incorporated in >5 days 8	Surface applied December through February 16	

Modifications of the P Site Index for Delmarva

Soil Erosion. In many areas, particularly when dealing with fine textured soils on sloping landscapes, erosion dominates the transport of P from soils to surface waters. Up to 90% of P transported from cropland is attached to sediment (Sharpley and Beegle, 1999). In areas of the inner coastal plain (northern Delaware and northeastern Maryland), there is a larger potential for P transport by erosion due to silty soil textures and areas with steeper slopes. In contrast, the lower coastal plain region (southern Delaware and southeastern Maryland) has a flatter topography with little to no slope and predominately sandy soils. The infiltration rates on many of these sandy soils are high. Due to high infiltration rates and flat topography, there is little runoff produced and the potential for erosion of soil particles from cropland is limited. Therefore, the impact of soil erosion on P transport can be less in much of the lower coastal plain region. There are also areas of the lower coastal plain that have poorly drained soils with the potential for soil erosion. When water tables rise to the soil surface (during rainfall events and seasonally high water tables), overland flow of water to drainage areas occur, which can transport particulate matter. In the *P Site Index*, erosion is calculated using the Revised Universal Soil Loss Equation (RUSLE) and is multiplied by a weighting factor of 2 to obtain the field value for soil erosion.

Surface Runoff. In areas of the inner coastal plain, the potential for surface runoff is also a concern. Areas with silt loam soils having significant slope generate a great deal of surface runoff that carries both particulate and dissolved P to surface waters. In contrast to this, the well-drained sandy soils of the lower coastal plain will tend to have a minimal amount of surface runoff, contributing little to this mode of P transport. However, in the poorly drained areas, surface runoff can be a significant transport mechanism. In poorly drained areas, storm events and seasonally high water tables (during the winter) can cause water table levels to reach the surface, producing overland flow to drainage ditches and potentially transporting large amounts of P to nearby surface waters. When soil test P levels are high or there are high rates of P applied as fertilizer or manure (particularly surface applications), the potential for transport of significant amounts of P can become a concern.

Leaching and Subsurface Drainage. Subsurface drainage and leaching potential components were also added to the *P Site Index*. While P leaching is typically considered to be small, there is potential for significant movement of P through the soil profile when soil P levels increase to very high or excessive values due to long term overfertilization or manuring (Sims et al., 1998). Mozaffari and Sims (1994) measured soil test P (Mehlich 1) values with depth in cultivated and wooded soils on farms in a coastal plain watershed dominated by intensive poultry production, and reported that P leaching to depths of ~60 to 75 cm was commonly observed in agricultural fields. Whether this leached P will reach surface waters depends on the depth to which it has leached and the hydrology of the site in question. Soils that are poorly drained with high water tables have a higher possibility of P loss than soils that are well drained with deep water tables. It is common in poorly drained soils to have water tables rise to the soil surface during the winter and spring months, when precipitation is greater than evaporation. During periods when water tables are high, there is the potential for release of P into subsurface waters and transport of P to nearby streams and drainage ditches via subsurface flow.

Distance from Edge of Field to Surface Water. Another factor that affects the risk of P transport from soils to surface waters is the distance between the source (i.e., the field) and the receiving waters. In some areas, the nearest waterbody may be a mile or more from the field being evaluated. In these cases, even high levels of soil P may have low risk for nonpoint source pollution in the near term since the potential for transport to the waterbody is low. In addition, many studies have shown that vegetated filter strips can remove P, especially particulate P, from water running off agricultural fields (Mikkelsen and Gilliam, 1995). Therefore, fields having grassed filter strips or riparian buffers may be less of a threat to water quality than fields with no buffer present. To accurately reflect regional landscapes and field management, a category was added to the *P Site Index* to take into account the distance from field edge (or edge of the P application zone) to nearby waterbodies and whether or not these areas are vegetated. A waterbody is defined as any permanent conduit that can transport surface water, including permanent streams and drainage ditches. Realizing that in many areas drainage ditches can be located within approximately a hundred meters of each other, distance categories had to be chosen that would be practical to use for these fields without causing the entire field to become unusable. It has been demonstrated that even a 3.1 meter buffer can decrease the dissolved P (DP) and total P in runoff by 68 and 70% respectively (Chaubey et al., 1993). In areas where the close proximity of drainage ditches makes establishing a vegetated buffer physically or financially impractical, utilizing simple management practices, such as application setbacks, may be useful. By not applying P fertilizer or organic P sources in a setback area (no P application zone) P loss potential can be reduced simply by keeping manures or fertilizers from being spread adjacent to or directly into ditches and streams

Soil Test P. The use of soil test P values to determine whether a soil can become a source of P has also been taken into consideration. Research on many Delaware soils has demonstrated that there is a positive relationship between soil test P and soluble P (Pautler and Sims, 2000). As soil test P increases there is an increase in soluble P. Thus, as soils are fertilized to levels exceeding soil test P values considered optimum for plant growth, the potential for P to be released into the soil solution and be transported by surface runoff, leaching, subsurface movement, and even groundwater increases. Soil test P results must be reported in University of Delaware or University of Maryland fertility index values (FIV) to be used in the *P Site Index*. Soil test P values reported in other units must first be converted to an FIV in order to obtain an accurate field value.

P Source Application Rate. The addition of fertilizer P or organic P sources to a field will usually increase the amount of P available for transport to surface and ground waters. The potential for P loss when fertilizers, manures, or other P sources are applied is influenced by the rate, timing, and method of application and by the form of the P source (e.g., organic vs. inorganic). These factors also interact with others, such as the timing and duration of subsequent rainfall, snowmelt, or irrigation and the type of soil cover present (vegetation, crop residues, etc.) (Sharpley et al., 1993). Past research has established a relationship between the rate of P added as either fertilizer or organic P sources and the amount of P transported in runoff (Romkens and Nelson, 1974; Baker and Laflen, 1982; Westerman et al., 1983). Since all sources of P may not be equally as susceptible to dissolution and transport in runoff or equally bio-available to aquatic species, the concept of phosphorus availability coefficients (PACs) was introduced into the *P Site Index*. A PAC assigns a relative weighting based on the solubility and bio-availablilty of the P in the organic material being utilized. For example, broiler litter that has been amended with alum, which is known to form stable P complexes that are less soluble and susceptible to loss in runoff, would have a smaller PAC than unamended broiler litter. Coefficients will also be used to differentiate between types of manures (poultry vs. cattle vs. swine), whether or not the manure has been composted, and for biosolids. By assigning PACs, farmers can obtain a lower *P Site Index* value by using management strategies that stabilize P in organic materials and thereby decrease the potential for transport of P.

P Source Application Method and Timing. Directly related to the amount of fertilizer and organic P applied to a field is the method and timing of the application. Baker and Laflen (1982) determined that the DP concentrations in runoff from areas receiving broadcast fertilizer P averaged 100 times more than from areas where comparable rates were applied five centimeters below the soil surface. Mueller et al. (1984) showed that incorporation of dairy manure reduced total P losses in runoff five-fold compared with areas receiving broadcast applications. Surface application of fertilizers and manures decreases the potential interaction of P with the soil, and therefore increases the potential for P loss in runoff. When fertilizers and manures are incorporated or banded, there is greater interaction between P sources and the soil, which decreases the likelihood of P loss. It is particularly important that fertilizers and manures are not surface applied during times when there is no plant growth, when the soil is frozen, during or shortly before periods of intense storms, or during times of the year when fields are generally flooded due to snowmelt or recharge periods. The major portion of annual P loss in runoff generally results from one or two intense storms (Sharpley et al. 1994). If P applications are made during periods of the year when intense storms are likely, then the percentage of applied P lost would be higher than if applications are made when runoff probabilities are lower (Edwards et al. 1992). Also the time between application of P and the first runoff event is important. Westerman and Overcash (1980) applied both swine and poultry manures to plots and simulated rainfall at intervals ranging from one to three days following manure application. Total P concentrations in the runoff were reduced by 90% when the first runoff event was delayed by three days. When managing manures and fertilizers to decrease the potential for P transport off site, they should either be applied below the surface, or incorporated into the soil within a short period of time. In addition, fertilizers and manures should be applied shortly before the growing season when available P can best be utilized by the plant.

Evaluation of the P Site Index in Delaware and Maryland

Maryland

The first field evaluations of the *P Site Index* were performed in Maryland. The *P Site Index* was used to evaluate the potential P loss for 182 fields in 13 counties with varying site characteristics and crop/nutrient management practices. Preliminary interpretive categories were then developed to illustrate how the *P Site Index* could be used to group soils according to the relative risk of P loss from soil to water (Table 4; Figure 1). Associated with these interpretive categories were recommendations for the level of P-based nutrient management planning that would be required. It is important to recognize that the interpretive categories shown in Table 4 have been delineated based on the best professional judgment of those involved in the development of the *P Site Index*, not actual field-scale measures of P loss in erosion and/or runoff. Further research will be needed to verify their accuracy in identifying the actual risk of P loss. However, at this stage of development of the *P Site Index*, delineating these categories demonstrates how the *P Site Index* can be used in watershed or regional planning. For instance, categories could be established based on the short-term need to direct the limited funding available for water quality protection to a certain percentage of land in each watershed or in a state. This would increase the likelihood that expenditure of these funds would result in water quality improvement. Consider the situation in Maryland as an example. There are approximately 810,000 ha of agricultural land in Maryland (MDA, 1997). The average cost of a P based nutrient management plan in this region is approximately $12/ha, thus at least $10 million would be required to develop P-based plans for every farm in the state. By identifying areas that are most critical in terms of P losses using the preliminary categories in Table 4, funding can be channeled to the highest priority areas first. Based on the data in Figure 1, about 25% of fields evaluated fall into the medium, high, and very high categories, where some form of P-based plan would be needed. By identifying these high priority areas, the first $2.5 million available should be spent in these areas to see the greatest improvements in water quality.

Table 4. Preliminary interpretation of P Site Index Values obtained using the Maryland/Delaware P Site Index

P Site Index Value	Generalized Interpretation of P Site Index Value
< 600	**LOW** potential for P movement from this site given current management practices and site characteristics. There is a low probability of an adverse impact to surface waters from P losses from this site. Nitrogen-based nutrient management planning is satisfactory for this site. Soil P levels and P loss potential may increase in the future due to N-based nutrient management.
600 - 1200	**MEDIUM** potential for P movement from this site given current management practices and site characteristics. Practices should be implemented to reduce P losses by surface runoff, subsurface flow, and erosion. Phosphorus applications should be limited to the amount expected to be removed from the field by harvest or soil test based P application recommendations, whichever is greater.
1200 - 1800	**HIGH** potential for P movement from this site given current management practices and site characteristics. Phosphorus-based nutrient management planning should be used for this site. Phosphorus applications should be limited to soil test based P application recommendations. All practical management practices for reducing P losses by surface runoff, subsurface flow, or erosion should be implemented.
> 1800	**VERY HIGH** potential for P movement from this site given current management practices and site characteristics. No P should be applied to this site. Active remediation techniques should be implemented in an effort to reduce the P loss potential.

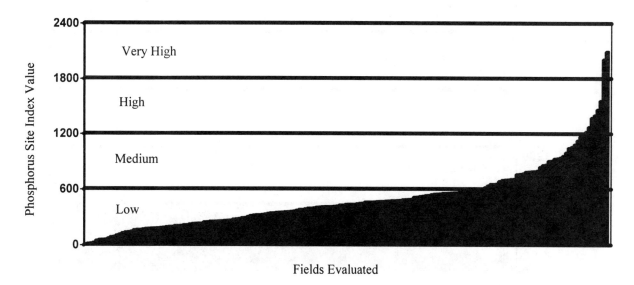

Figure 1. Phosphorus Site Index Values for 182 fields evaluated throughout Maryland in 1999. (Low = 0-600; Medium = 600-1200; High = 1200-1800; Very High = >1800)

Delaware

Three farms in Delaware were used in the fall of 1999 to evaluate the *P Site Index*. Evaluations were performed on whole farms from different regions of the state. The overall goal was to evaluate different types of farming operations to better understand the implications of farm type and management on the practical utility and economic implications of the *P Site Index*.

Farm A: This farm is located in Sussex County, Delaware, in the lower coastal plain. It is a ~2400 ha poultry-grain farm with predominately loamy sand soils, that range from well to poorly drained. One area, composed of approximately 10 fields requires artificial drainage (tax ditches) to enable cultivation. This farm has a large enough land base that poultry manure is generally spread on any given field only once during a three-year crop rotation at a rate of ~7 Mg/ha. The manure is usually incorporated within three days of application. Starter fertilizer containing P is used for fields that are planted in corn. Soil test data showed that fields ranged from optimum to excessive in P. Only one field fell into the medium category of the *P Site Index* (Figure 2). This field was located on a poorly drained soil with drainage ditches that had no buffers. The other field evaluated in this area was close to the medium category, having a *P Site Index* value of 580, but fell in the low category due to a lower soil test P value.

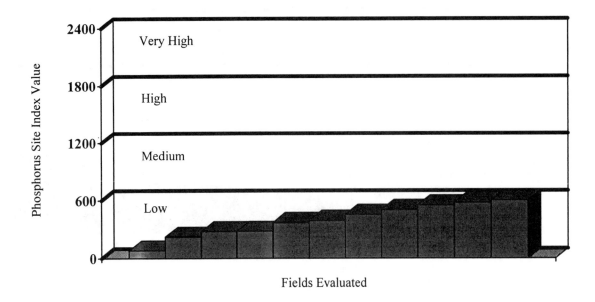

Figure 2. Distribution of P Site Index Values from Farm A (Sussex County, Delaware).
(Low = 0-600; Medium = 600-1200; High = 1200-1800; Very High = >1800)

The recommendations for this farm would be to follow a N-based nutrient management plan on all fields in the low category. The one field falling into the medium category would have to limit P applications to the amount expected to be removed from the field in harvest or soil test based P application recommendations, whichever is greater. However, by simply placing fertilizers and manures 25 feet away from the drainage ditches, this field would then fall into the low category of the *P Site Index*, in which case a N-based plan could also be used. It would also be recommended that the farmer review the need for soil conservation measures that reduce the risk of P loss in certain areas on the farm (e.g. consider widening buffer strips in some areas near to surface waters) and monitor soil test P values to prevent the buildup of soil P to higher levels (some fields are already at excessive levels). And finally, the farmer should consider the effects of any major changes in agricultural practices on P losses before implementing them on the farm.

Farm B: This farm is located in New Castle County, Delaware, in the inner coastal plain, but has attributes similar to the Piedmont region. The soils consist of a mix of well-drained sandy loams and silt loams and have slopes reaching up to 12% in some areas. This farm is ~1200 ha in size and mainly produces small grains. Starter fertilizers are used, both banded and broadcast, on all fields. Soil test P data ranges from low to excessive. Of the fields evaluated, 83% fell into the low *P Site Index* category and the remaining 17% fell into the medium category (Figure 3). One of the fields in the medium category was rated as such due to the application of an excessive amount of liquid poultry manure for disposal purposes. The other two fields in the medium category were areas with fairly steep slopes, where both fertilizer and manure were applied, and were located adjacent to a freshwater impoundment.

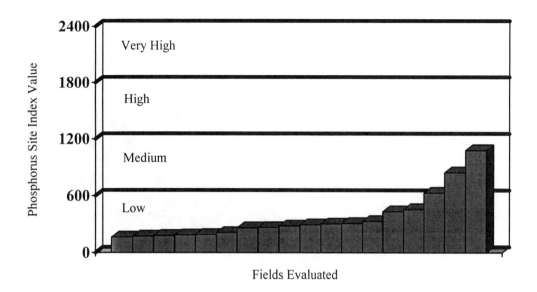

Figure 3. Distribution of P Site Index Values from Farm B (New Castle County, Delaware). (Low = 0-600; Medium = 600-1200; High = 1200-1800; Very High = >1800)

The recommendations for this farm would be to follow N-based nutrient management plans for those areas falling into the low category. Due to the potential for erosion and surface runoff on many of these sites, soil conservation measures (such as conservation tillage and the use of buffer strips) must also be used on these fields to reduce the risk of P losses. Those areas falling into the medium category will have to follow a P-based nutrient management plan, limiting P applications to that removed in harvest or soil test based P application recommendations, whichever is greater. In all cases, it is important to monitor soil test P levels to prevent the buildup of P to excessive levels (only two fields at present have excessive soil test P values). Any changes in agricultural practices will need to be evaluated for potential impacts on P losses before implementation, particularly in areas having steeper slopes and fine textured soils.

Farm C is also located in New Castle County, Delaware, and has mainly well-drained silt loam soils with some slopes of up to 8%. This is a small farm, approximately 243 ha, which produces poultry and small grains. Application of poultry litter is very heavy on fields adjacent to the poultry houses, and is mainly for disposal purposes, not crop production. The fields that are not receiving poultry litter have starter fertilizer added, which is banded. Soil test P values range from optimum to excessive. Those fields that receive no poultry litter fell in to the low category, while those receiving litter were in the high and very high categories (Figure 4). Poultry litter is surface applied at rates exceeding ~13 Mg/ha, throughout the year (includes surface applications during winter months). In the fields rated as very high, there is an adjacent stream, with a small feeder stream, having little to no buffer.

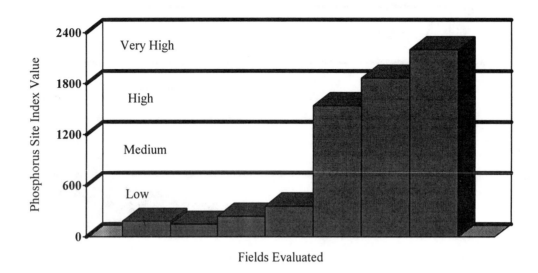

Figure 4. Distribution of P Site Index values from Farm C (New Castle County, Delaware). (Low = 0-600; Medium = 600-1200; High = 1200-1800; Very High = >1800)

The recommendations for this farm would be to follow nitrogen-based nutrient management plans for those areas falling into the low category. The fields falling into the high and very high categories will have to follow P-based nutrient management planning. The field rated as high could still receive P applications if soil test results indicated a need for P. The fields rated as very high should avoid any application of P from any source (fertilizers or poultry litters). Poultry litter should be applied to fields with lower *P Site Index* values, and at lower application rates. Poultry litter should not be surface applied in winter months when there is greater potential for P loss due to runoff and erosion. The establishment of buffers along the stream will also help reduce potential P losses from this site. The farmer should continually monitor changes in soil test P with time to determine when N- and P-based management practices can be followed.

Conclusions

The *P Site Index* is a valuable planning tool that can be used to identify areas of greatest concern for P loss. It is also a very flexible tool for use in decision-making processes. *Phosphorus Site Index* values rank agricultural land into **relative risk** categories. The value ranges in each category can be determined depending on the goal of those involved at that time, and can be changed in response to new research or new priorities for nutrient management. This can be useful when making decisions regarding spending of monies on water quality improvement projects as well as cost-share programs. Greater water quality benefits can be obtained from allocating resources to areas having the greatest potential for P losses.

In August of 1999, Maryland began use of this version of the *P Site Index* (Version 1.0; Tables 3 and 4) in their statewide nutrient management training program. Maryland has also referred to the use of a *P Site Index* in their proposed nutrient management regulations. These proposed regulations state "If the soil sample analysis results show a phosphorus fertility index value (FIV) of 150 or greater, a phosphorus site index or other phosphorus risk assessment method acceptable to the Department, as provided in the <u>Maryland Nutrient Management Manual</u>, Section II-B shall be used to determine the potential risk of phosphorus loss due to site characteristics". An FIV of 150 (which equals approximately 75 mg/kg Mehlich 1 P or three times the average crop removal rate) was chosen as a cutoff point above which the *P Site Index* should be used to evaluate the potential for P loss. The University of Delaware and USDA- Natural Resource Conservation Service (USDA-NRCS) are currently assessing the suitability of Version 1.0 for use in Delaware. One of the main concerns surrounding the current version of the *P Site Index* is how well relative risk categories reflect actual risks of P losses in the field. In response to this, cooperative efforts to develop and validate a final, reliable *P Site Index* for Delaware and other Mid-Atlantic states are planned for the next two years. During this time, laboratory, field and rainfall simulator studies will be used to validate portions of the *P Site Index* and to determine what modifications are necessary to obtain more reliable risk assessments

References

Baker, J.L. and J.M. Laflen. 1982. Effect of crop residue and fertilizer management on soluble nutrient runoff losses. Trans. ASAE. 25:344-348.

Chaubey, I., D.R. Edwards, T.C. Daniel, and D.J. Nichols. 1993. Effectiveness of vegetative filter strips in controlling losses of surface-applied poultry litter constituents, Paper No. 932011, ASAE, St. Joseph, MI.

Delaware Department of Agriculture. 1997. Poultry highlights [online]. Available at http://www.nass.usda.gov/de/p2097.htm (verified 21 Jan. 2000).

Edwards, D.R., T.C. Daniel, and O. Marbun. 1992. Determination of best timing for poultry waste disposal: A modeling approach. Water Res. Bul. 28:487-494.

Gartley, K.L. and J.T. Sims. 1994. Phosphorus soil testing: Environmental uses and implications. Commun. Soil Sci. Plant Anal. 25:1565-1582.

Lemunyon, J.L., and R.G. Gilbert. 1993. Concept and need for a phosphorus assessment tool. J. Prod. Agric. 6:483-486.

Maryland Department of Agriculture. 1997. Land use [online]. Available at http://www.nass.usda.gov/census/census97/highlights/md/graphics.htm (verified 21 Jan. 2000).

Mikkelsen, R.L. and J.W. Gilliam. 1995. Animal waste management and edge of field losses. p. 57-68. *In* K. Steele (ed). Animal Waste and the Land-Water Interface. Lewis Publishers. Boca Raton.

Mozaffari, P.M. and J.T. Sims. 1994. Phosphorus availability and sorption in an Atlantic Coastal Plain watershed dominated by intensive, animal-based agriculture. Soil Sci. 157:97-107.

Muller, D.H., R.C. Wendt, and T.C. Daniel. 1984. Phosphorus losses as affected by tillage and manure application. Soil Sci. Soc. Am. J.48:901-905.

Pautler, M.C. and J.T. Sims. 2000. Relationships between soil test phosphorus, soluble phosphorus, and phosphorus saturation in soils of the Mid-Atlantic region of the U.S. J. Environ. Qual. *In press.*

Romkens, J.M. and D.W. Nelson. 1974. Phosphorus relationships in runoff from fertilized soils. J. Environ. Qual. 3:10-13.

Sharpley, A.N. and D. Beegle. 1999. Managing phosphorus for agriculture and the environment. College of Agricultural Sciences Cooperative Extension. Pennsylvania State Universiy, University Park, PA.

Sharpley, A.N., S.C. Chapra, R. Wedepohl, J.T. Sims, T.C. Daniel, and K.R. Reddy. 1994. Managing agricultural phosphorus for the protection of surface waters: Issues and options. J. Environ. Qual. 23:437-451.

Sharpley, A.N., T.C. Daniel and D.R. Edwards. 1993. Phosphorus movement in the landscape. J. Prod. Agric. 6:492-500.

Sims, J.T. 1997a. Agricultural phosphorus and water quality: Environmental challenges in the U.S. Atlantic Coastal Plain. Proc. of the 2nd Int. Conf. on Diffuse Pollution from Agric., Edinburgh, Scotland, April 6th – 9th.

Sims, J.T. 1997b. Agricultural and environmental issues in the management of poultry wastes: Recent innovations and long-term challenges. P. 72-90. *In* J. Rechcigl (ed.) Uses of By-Products and Wastes in Agriculture. Am. Chem. Soc., Washington, D.C.

Sims, J.T. 1998a. Assessing the impact of agricultural drainage on ground and surface water quality in Delaware. Final Project Report. Prepared March 30, 1998.

Sims, J.T. 1998b. The role of soil testing in environmental risk assessment for phosphorus in the Chesapeake Bay watershed. *In* A.N. Sharpley (ed.) Proc. Conf. Agric. Phosphorus Chesapeake Bay Watershed, University Park, PA, April 4-6, 1998.

Sims, J.T. and A.B. Leytem. 1999. The phosphorus site index: A phosphorus management strategy for Delaware's agricultural soils. Dep. of Plant and Soil Sci. ST-05. Univ. Delaware, Newark, DE.

Sims, J.T., R.R. Simard, and B.C. Joern. 1998. Phosphorus losses in agricultural drainage: Historical perspectives and current research. J. Environ. Qual. 27:277-293.

Westerman, P.W., T.L. Donnely, and M.R. Overcash. 1983. Erosion of soil and poultry manure-a laboratory study. Trans. ASAE. 26:1070-1078, 1084.

Westerman, P.W. and M.R. Overcash. 1980. Short-term attenuation of runoff pollution potential for land-applied swine and poultry manure. p. 289-292. *In* R.J. Smith et al. (eds.) Livestock waste-A renewable resource. Proc. 4th Int. Symp. on Livestock Wastes, Amarillo TX, 15-17 Apr. ASAE, St. Joseph, MI.

●●●●●●●●●●●●●●●●●●●●●●●●●●●●●●

A Phosphorus Index for Vermont: Adapting the Index to Landscapes, Soil Chemistry, and Management Practices in Vermont

William E. Jokela
Extension Soils Specialist
Plant and Soil Science Department
University of Vermont

Biographies for most speakers are in alphabetical order after the last paper.

●●●

Introduction

The Phosphorus Index is a tool developed to assess the potential for phosphorus runoff from individual fields based on soil and field characteristics and management practices. The relative P Index ratings can be used to prioritize fields as to the need for conservation and nutrient and management practices, especially those with a high risk of serious P runoff. The Phosphorus Index does not attempt to predict actual quantity of P lost in runoff, but P Index rankings have been shown to be closely related to total P loss (Sharpley, 1995).

The Phosphorus Index for Vermont is based on the original Phosphorus Index developed several years ago by a USDA-university working group (Lemunyon and Gilbert,1993; NRCS, 1994). It also incorporates a number of ideas from modified P Index versions from other states (McFarland et al., 1998; Klausner et al.,1997; Gburek et al, 2000; and Coale and Layton, 1999), as well as some from our own work in Vermont (Magdoff et al., 1999; Jokela et al., 1998b). The version described in this paper is only slightly revised from one presented and distributed at the annual meeting of the SERA-17 Workgroup: Minimizing P Losses from Agriculture, Quebec City, July, 1999 (Jokela, 1999). This version of the Vermont P Index is a work in progress and will continue to evolve as we gain more research data and practical experience. It will undergo further review and field testing over the next months, most likely followed by modification and refinement.

Phosphorus Source and Transport Potentials

Ten site characteristics are included in the Vermont version of the P Index (Table 1). They are grouped into P Source and P Transport categories, which are multiplied to get a P Index (as proposed by Gburek et al, 2000). Thus, the P Index combines an estimate of phosphorus available for loss via runoff and erosion (P Source Potential) with that of transport mechanisms that can move phosphorus from the field in runoff (P Transport Potential) (Table 2). Each site characteristic in the P Index is weighted to account for the relative importance in contributing to P runoff potential. The relative weighting is reflected in the different values for the maximum, or very high category (Table 2). (Most earlier P Index versions include a separate weighting coefficient for each site characteristic that is used in the calculation.) Based on Phosphorus Index, fields are given a classification (Low, Medium, High, Very High), each with associated interpretations and recommendations (Table 3).

Three sources of phosphorus are considered in determining the ***P Source Potential*** – soil test P (STP), fertilizer P (FP), and manure P (MP) (Table 2). Soil test phosphorus is a measure of plant-available P and has been shown to be well correlated with the concentration of soluble P in runoff (Pote et al., 1996). Applied fertilizer and manure represent readily available sources of P that are susceptible to runoff; but the values for fertilizer and manure P application rate need to be adjusted depending on how the applied P is managed – timing and method, or placement. Phosphorus applied during the non-growing season period (October-April) and left on the surface has maximum availability for loss in runoff and receives full value (1.0). If improved P management methods are used, P Source Potential values are reduced to account for the estimated reduction in availability for runoff. Fields with incorporated manure or fertilizer are further modified by a factor for Reactive Aluminum (Lee and Bartlett, 1977), which is a good indicator of how much the addition of incorporated P (as manure or fertilizer) will increase the soil test P level.

The P Source Potential, then, is calculated as follows (Al factor to be used only for non-surface-applied fertilizer or manure):

P Source Potential = STP + (FP Rate x FP Method x Al) + (MP Rate x MP Method x Al)

The ***P Transport Potential*** is calculated by adding values for three major types of field transport – soil erosion (E), runoff (R), and flooding (F) – and modifying the result by a factor for the distance of vegetated buffer between field and adjacent drainage path or waterway (Table 2). The values for each field transport mechanism are added because the total amount of P transported from the field would be expected to be the sum from all three methods. However, we express the field transport potential as a fraction by dividing the sum by the sum of maximum or Very High values (15 + 10 + 5 = 30), thus expressing it as a proportion of the full transport potential. The resultant value (for edge-of-field P transport) is multiplied by a coefficient for Buffer Width (BW) since the existence of a vegetated buffer would be expected to reduce the amount of P entering the waterway by a some percentage, which is a function of the width of the buffer.

Table 1. Description of Individual P Index Site Characteristics.

Site Characteristic	Description
P Source	
Soil Test P	Available P in Modified Morgan (or Morgan) extractant, expressed as ppm.
Fertilizer P Rate	Rate of fertilizer P, expressed as lb P_2O_5/acre.
Fertilizer P Application Method/Timing	Sub-surface is 2 inches or more below surface. Incorporation means with tillage to 3-inch or greater depth within 3 days of application.
Manure P Rate	Rate of P applied as manure, compost, or other organic material, expressed as lb P_2O_5/acre.
Manure P Application Method/Timing	Same as fertilizer P.
Reactive Aluminum	Aluminum in Modified Morgan extractant, expressed as ppm.
P Transport	
Soil Erosion	Average annual erosion rate in tons/acre as estimated by NRCS Revised Universal Soil Loss Equation (RUSLE).
Soil Runoff Class	Runoff Class as determined from % slope and either Runoff Curve Number (NRCS, 1985) or saturated soil conductivity (NRCS, 1993)
Flooding Frequency	Designation as defined in NRCS soil survey database for each soil mapping unit.
Buffer width	Distance, in feet, of grass or other vegetated buffer from field edge to waterway or path of seasonal concentrated flow.

$$P\ Transport\ Potential = [(E+R+F)/30]\ x\ BW \quad (Value\ from\ 0.1\ to\ 1.0)$$

The P Transport Potential is considered a measure of the relative effectiveness of this site for transporting P that is potentially available from various sources. It has a value less than one unless all factors are at their maximum, in which the P Transport Potential would be 1.0.

The final Phosphorus Index is calculated by multiplying the P Source Potential by the P Transport Potential.

$$P\ Index = P\ Transport\ Potential\ x\ P\ Source\ Potential$$

Table 2. The Phosphorus Index for Vermont: site characteristics and P runoff potential ratings.

Site Characteristic	Phosphorus Source Potential Rating			
	Low/None	Medium	High	Very High
Soil Test P, ppm, Mod. Morgan's	(0-8 ppm)	(8.1-20)	(20-40)	(>40)
	0.25 x Soil Test P			
Fertilizer P Rate lb P_2O_5/acre	**0.2 x lb P_2O_5/acre**			
Fertilizer P Application Method/Timing	Inject/sub-surface band **0.4**	Broadcast and incorporate **0.6**	Surf-applied May-Sept. **0.8**	Surf-applied Oct.-April **1.0**
Manure P Rate lb P_2O_5/acre	**0.2 x lb P_2O_5/acre**			
Manure P Application Method/Timing	Inject or sub-surface band **0.4**	Broadcast and incorporate **0.6**	Surf-applied May-Sept **0.8**	Surf-applied Oct-April **1.0**
Reactive Aluminum, Mod. Morgan's, ppm	>80 ppm **0.7**	41-80 **0.75**	21-40 **0.8**	<20 **1.0**

> **P Source Potential =**
> **STP + (FP Rate x FP Method x Al) + (M Rate x M Method x Al)**
> Note: Use Al factor only for non-surface-applied fertilizer or manure.

Site Characteristic	Phosphorus Transport Potential Rating			
	Low/None	Medium	High	Very High
Soil Erosion (E) tons/acre/yr	(0-4)	(4-8)	(8-12)	(>12)
	1.25 x tons soil loss/acre/yr			
Soil Runoff Class (R)	Low/ V. Low **4**	Medium **6**	High **8**	Very High **10**
Flooding Frequency (F)	Rare/None **0**	Occasional **3**	---	Frequent **5**
Buffer Width (BW), ft.	>100 **0.7**	41-100 **0.8**	16-40 **0.9**	<15 **1.0**

> **P Transport Potential = [(E+R+F)/30] x BW**   (Value from 0.1 to 1.0)

> **P Index = P Transport Potential x P Source Potential**

Table 3. Interpretation for the Phosphorus Index. (from NRCS, 1994)

Phosphorus Index	Site Interpretations and Recommendations
<6	**LOW** potential for P movement from site. If farming practices are maintained at the current level there is a low probability of an adverse impact to surface waters from P loss.
6-12	**MEDIUM** potential for P movement from site. Chance for an adverse impact to surface water exists. Some remedial action should be taken to lessen probability of P loss.
12-25	**HIGH** potential for P movement from site and for an adverse impact on surface waters to occur unless remedial action is taken. Soil and water conservation as well as P management practices are necessary to reduce the risk of P movement and water quality degradation.
>25	**VERY HIGH** potential for P movement from site and for an adverse impact on surface waters. Remedial action is required to reduce the risk of P movement. Soil and water conservation practices, plus a P management plan must be put in place to reduce potential for water quality degradation.

P Index Interpretation and Management Recommendations

Depending on the P Index, a field is rated as having low, medium, high, or very high potential for P movement from the site (Table 3). Some general management practices are recommended for each category, but a more site-specific nutrient management program should be developed for each field. If the P Index is adapted for use in other areas, especially if source or transport factors are added or changed, the limits for risk categories would need to be modified as well.

Features of the Vermont Phosphorus Index

The version of the P Index has been built on the design and features of earlier versions, but there are some features that are unique or have been added to account for the soil chemistry, landscape, and management practices common to Vermont.

Modified Morgan Soil Test P Extractant

The University of Vermont has used the Modified Morgan extractant (1.25 M NH$_4$Acetate, pH 4.8) for nutrient recommendations since the 1960's when it was introduced (McIntosh, 1969; Jokela et al., 1998a). Either the Morgan (NaAcetate) or the Modified Morgan extractants, which are equivalent for phosphorus, are used in New York and most of New England (ME, MA, CT, RI, and VT). Modified Morgan P is typically less than one-tenth the value obtained with Melich 3 extractant although there is not a strong linear correlation.

Modified Morgan P has shown a very strong correlation with water soluble and $CaCl_2$ solution P (Figure 1. Magdoff et al, 1999; Jokela et al, 1998b), both of which have been used as indicators of soil P available for release into runoff (Pote et al, 1996).

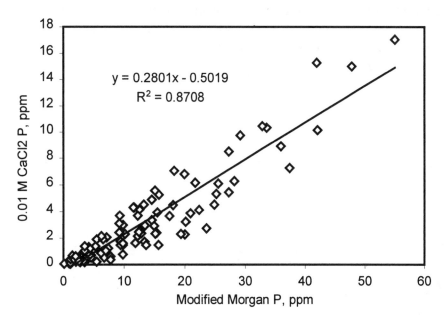

Figure 1. P extracted by 0.01 M $CaCl_2$ solution as a function of Modified Morgan P in 54 soils. (Magdoff et al., 1999; Jokela et al., 1998b).

Use of P Application Management Factors

While P Application Method/Timing is included in the P source group, we see it as a modifier of the manure and fertilizer P rate factors, rather than as a separate additive term. Phosphorus management is not an independent factor, but one that interacts with P application rate to determine the availability of the applied P for transport in runoff. More specifically, application of manure or fertilizer P on the surface in the most vulnerable time period is assumed to have full, or maximum, susceptibility for runoff and receives a coefficient of 1; whereas application with improved methods or timing is assigned a value less than 1 to reflect a reduced availability for runoff (Table 2). Each P management coefficient is multiplied times P rate values for manure and fertilizer P to create adjusted P application ratings.

Reactive Aluminum

When manure or fertilizer is incorporated into the soil, the effect of the applied P on runoff P concentration is primarily a function of increased soil P enrichment as indicated by an increase in P soil test. Different soils, however, vary greatly in the P test increase that results from a given P addition. Aluminum extracted by Modified Morgan solution, termed

"Reactive Aluminum" (Lee and Bartlett, 1977), is a good indicator of how much the addition of P as manure or fertilizer will increase soil test P and is part of the Vermont soil testing system (Jokela et al., 1998a). Depending on the reactive Al test level, a field is assigned a coefficient ranging from 0.7 to 1.0, which is multiplied times the P Rate-Method value (Table 2). Soils testing lowest in reactive Al (<20 ppm) show the greatest increase in soil test P (Fig. 2; Magdoff et al., 1999; Jokela et al., 1998b) and receive a value of 1.0. Those testing higher in Al receive lower coefficients to reflect the lower expected soil test P increase. If manure is not incorporated, the presence of manure on the surface will tend to outweigh the effect on increased soil test P, so no adjustment for Al is made.

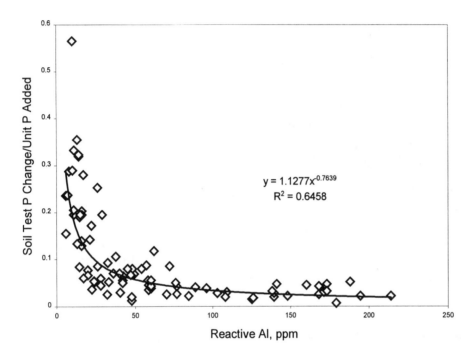

Fig. 2. Change in soil test P (Mod. Morgans) with P addition as a function of reactive Al. (Magdoff et al., 1999; Jokela et al., 1998b)

Flooding Frequency

Spring flooding is common on some Vermont fields adjacent to streams and rivers. Particulate and soluble P removed from these fields during flooding events can contribute significantly to P loading of a stream. Therefore, we have added a "flooding frequency" factor (none/rare, occasional, or frequent), as defined for soil mapping units in the NRCS soil survey database, to account for this transport mechanism (Tables 1 and 2). This term was suggested by Klausner et al. (1997) as one of five factors in a Hydrologic Sensitivity factor.

Vegetated Field Buffer

A vegetated buffer strip at the edge of a cropped field can retain a portion of the phosphorus in runoff, especially that in the particulate form (Schmitt et al., 1999; Uusi-Kämppä et al., 2000). Preliminary results from a field study in Addison, VT, show significant reductions in runoff P and sediment concentrations from the establishment of grass-legume buffer strips at the field edge (Jokela et al, 1999). The P Index buffer width is defined as the distance of grass or other close-seeded vegetation, woody species, or a combination from field-edge to waterway or path of seasonal concentrated flow. No manure or P fertilizer is to be applied to the buffer area. The value for the buffer width factor ranges from 0.7 to 1.0, depending on width (Table 2). This coefficient is multiplied times the preliminary P Transport Potential to reflect the portion of P in runoff that is retained in the buffer strip.

Additional Features Under Development or Consideration

Adjustment for Other Best Management Practices

The role of a field-edge buffer strip in retaining runoff P that leaves the field is accounted for by multiplying the preliminary transport potential by a coefficient that is a function of its effectiveness (Table 2). This same approach could be used to adjust source or transport ratings in the P Index for the beneficial effects of other BMPs not already accounted for in the site characteristic ratings (excluding such practices as those incorporated in the RUSLE estimate of erosion).

Leaching, or Subsurface Flow, of Phosphorus

Our P Index assumes that the dominant loss mechanism for loading of P into surface waters is surface runoff. Limited observational data in Vermont have suggested only minimal amounts of P lost via leaching. However, subsurface P losses, primarily via preferential flow, have been shown to be quite important in other areas parts of the region, as well as in other countries (Sims et al., 1998; Simard et al., 2000), and have been incorporated into some P Indices. Subsurface P losses to tile lines have also been reported in eastern New York (H.M van Es, L.D. Geohring, personal communication, 1999). We are in the process of assessing the likely significance and prevalence of subsurface P losses from Vermont soils and will add a component to the P Index if the evidence warrants.

Bioavailability of Manure Phosphorus

An analysis for total P is recommended to determine the application rate of manure P. However, different fractions of manure P vary in their availability to algae and other aquatic plants and, therefore, in their effect on water quality. This is especially true when changes are made in concentration of P in livestock rations or when additives such as alum are mixed with manure (Valk et al., 2000; Moore et al., 2000; Powell et al., 1999). This suggests that another manure analysis for soluble or bioavailable P would be useful. The results could be used to adjust the manure P application rating in the P Index to better reflect the expected effect on surface waters.

Distance or Connection to Surface Water

Most parameters in the P Index contribute to an estimate of P runoff loss at the edge of the field, whereas the ultimate objective is to assess the impact of P runoff from the field on water bodies some distance away. Consequently, a parameter that indicates the probability of phosphorus that leaves the field actually reaching a surface water body could be an important addition. This would likely involve some combination of distance, as included in some P Indices, and directness of connection between field and water (A. Sharpley, personal communication, 2000). It could be expressed as a coefficient to adjust transport potential, in the same way as we have used the vegetated buffer factor (Table 2). In fact, a logical approach might be to combine the two in a single parameter to modify the preliminary transport rating to account for the nature of the pathway between field and water.

Return Period or Contributing Distance

Gburek et al. (2000) proposed use of the concept of return period, or contributing distance, a hydrologic term representing the probability of runoff as a function of distance from a stream. This would be a valuable addition to the P Index, but we were unable to determine how to apply it to a range of fields in different watersheds without monitoring data to support the relationship between distance-to-stream and runoff probability. Consequently, we have not included it in our P Index at this time but will consider it in the future if there is a system for implementation.

Priority or Phosphorus Sensitivity of Watershed

We have not included a parameter for watershed priority or P sensitivity of surface waters. Rather than seeing this characteristic as one of many that influence the P Index value, we expect watershed priority to be either a starting point to determine need for the P Index or a factor in interpretation of the index. For example, in a high priority, high P sensitive watershed use of a P Index may be required on all fields as part of nutrient management planning. But if the watershed is not considered P sensitive or is low priority the P Index may not be required or may be required only on selected fields or areas. Alternatively, the categories for interpretation of the P Index might be shifted in a high priority watershed to require more intensive P management.

Vermont P Index Applied to Different Field Scenarios

The Vermont P Index was run on a set of eight hypothetical fields to assess what range of P Index ratings would result from fields with a wide range of source and transport site characteristics (Table 4). The phosphorus application rates remained constant for all scenarios at 50 P_2O_5 /acre as fertilizer and 100 lb P_2O_5/ acre as manure, but combinations of other site characteristics and management practices were changed. The first field scenario (A) had high soil test P, low reactive aluminum test, surface P applications, high levels of erosion and runoff, and a very narrow vegetative buffer. Scenario A received a "Very High"

Table 4. Site characteristics for each of eight hypothetical fields. Shading indicates which site characteristics were changed from the previous scenario.

Site Characteristic	A	B	C	D	E	F	G	H
Soil Test P, ppm	35	35	35	35	35	35	7	7
Fert P Rate, lb/acre	50	50	50	50	50	50	50	50
Fert Method	Surf Nov	Inc Band	Surf Nov	Surf Nov	Surf Nov	Inc Band	Inc Band	Inc Band
Manure P Rate, lb/acre	100	100	100	100	100	100	100	100
Manure P Method	Surf Nov	Inc Bdcst	Surf Nov	Surf Nov	Surf Nov	Inc Bdcst	Inc Bdcst	Inc Bdcst
React Al, ppm	15	15	15	15	15	15	15	55
Erosion, T/a	12	12	3	3	3	3	3	3
Runoff	VH	VH	Med	Med	Med	Med	Med	Med
Flood Freq	None	None	None	None	Freq	None	None	None
Buffer Width, ft	10	10	10	50	50	50	50	50
P Index	32.3	20.6	12.6	10.1	15.2	5.1	3.7	2.8
Interpretation	VH	H	H	M	H	L	L	L

P Index rating of 32. Improved manure and fertilizer P management (incorporation) brought it down into the "High" range (B), while practices to reduce erosion and runoff (C) reduced it even further though still in the High category. Increasing buffer width to 40 ft lowered the PI further, into the "Medium" category (D). A field with the same situation as D but with frequent flooding raised the PI into "High" (E). Implementation of both improved P management and erosion and runoff controls (F) dropped PI to the "Low" category. The same scenario with a lower soil test P (G) and a higher reactive aluminum test (H) lowered the P Index even further.

This exercise showed that over a range of site characteristics representative of what might be found on farm fields the P Index varied from Low to Very High. It also showed that substantial changes in the P Index can be made by improved management practices such as erosion and runoff control practices, improved manure management, and establishment of vegetative field buffers. A true evaluation of the P Index would involve running P Indices on actual farm fields or groups of fields within a drainage area and comparing results with measured P runoff loads.

Application of a P Index to Nutrient Management Planning

General Nutrient Management Planning

A P Index can play an important role in nutrient management planning by calling attention to those fields that require additional conservation practices or more careful nutrient management to avoid P runoff problems. It can also indicate on which fields current practices can be maintained without the likelihood of water quality problems. A sound approach is as follows: a) develop a whole-farm nutrient management plan, which includes application rates and methods of manure and fertilizer P for all fields; b) calculate a P Index for each field based on planned management; c) modify management practices on fields with P Index ratings higher than acceptable and rerun P Index

Because of the time requirement to develop P Indices for all fields on a farm, especially the in-field measurements needed to estimate soil erosion with RUSLE, an alternative would be to limit P Index calculations to those fields that are suspected of having serious P runoff problems. (See later discussion.)

P Index as a Guide to Manure Application

A more specific use of the P Index is as a guide for determining acceptable application rates of manure on cropland. Recent guidelines by NRCS state that manure application rates are to be based on one of three options – P Index, P threshold, or soil test P (NRCS, 1999). If the P Index is used, the guidance provides that for Low or Medium Risk (P Index rating) manure rate can be N-based, for High Risk is to be P-based (e.g. crop removal), and for Very High Risk P-based (e.g. no application). Maryland has followed a similar approach in recent nutrient management legislation (Coale and Layton, 1999).

Screening Tool Options: Soil Test P or P Transport Potential?

Developing a P Index requires a site visit to measure slope and other field characteristics needed for estimating soil erosion and other parameters. This may make it impractical to run a P Index on all fields where nutrient management planning is being implemented on a large scale, whether by legislation, cost-share requirements, or other reasons. Consequently, it is desirable to have a screening tool to prioritize those fields most likely to have P runoff problems and, therefore, having the greatest need for a P Index.

The most commonly used screening tool is soil test P, which is being used to determine the need for a P Index in Maryland (Coale and Layton, 1999) and used directly to restrict manure application in several other states (CO, KS, ME, MS, OK, TX ; Lory and Scharf, 1999). Use of a P soil test has the advantage of being an easily quantifiable measure that is already in common usage; and fields with higher P soil tests tend to have runoff with higher concentrations of soluble P (though not necessarily higher quantities of P).

However, the main consideration should be to avoid applying manure on fields where application would result in the greatest *increase* in P runoff. This is primarily a function of runoff and erosion potential (or the P Transport part of the P Index) and is not primarily a function of soil test P. Manure application increases the P source available for transport in runoff, so the greatest impact of the manure P addition will be on fields that have the highest potential to transport that P in runoff. Consequently, where a full P Index can not be run on all fields, we are proposing use of a quick P runoff estimate by means of a Phosphorus Runoff Screening Matrix (PRSM). It includes parameters that can be determined directly from soil survey database (runoff class, HEL classification, and flooding frequency) along with soil test P and, consequently, does not require on-site field measurements (except for a soil test). If a field ranks High with the PRSM then the full P Index must be determined to obtain a more complete estimate of P runoff potential.

Summary

A Phosphorus Index has been adapted for use in Vermont, accounting for the soil chemistry, landscapes, and management practices of the state. Features include use of the Modified Morgan P and Reactive Aluminum tests and incorporation of factors for flooding frequency and vegetative field buffers. The P Index is currently undergoing review and field testing and further improvements are under consideration. The Vermont P Index was run on eight hypothetical fields with a large range of field site characteristics, resulting in a large range of P Index values – from Low to Very High. The P Index can serve an important function in nutrient management planning and prioritizing manure applications.

References

Coale, F, and S. Layton. 1999. Phosphorus site Index for Maryland. Report to Northeast Phosphorus Index Work Group. Univ. of Maryland., College Park, MD.

Gburek, W.J., A.N. Sharpley, L. Heathwaite, and G.J. Folmar. 2000. Phosphorus management at the watershed scale: a modification of the phosphorus index. J. Environ. Qual. 29:130-144.

Jokela, W.E. 1999. The phosphorus index: a tool for management of agricultural phosphorus in Vermont. Report to SERA-17: Minimizing P Losses from Agriculture. Quebec City, July, 1999. Web: http://pss.uvm.edu/vtcrops/NutrientMgt.html#TOP

Jokela, B., F. Magdoff, R. Bartlett, S. Bosworth, and D. Ross. 1998a. Nutrient recommendations for field crops in Vermont. Br. 1390. Univ. of Vermont Extension, Burlington, VT. Web: http://ctr.uvm.edu/pubs/nutrientrec/

Jokela, W.E., F. R. Magdoff, and R. P. Durieux. 1998b. Improved phosphorus recommendations using modified Morgan phosphorus and aluminum soil tests. Comm. Soil. Sci. Plant Anal. 29:1739-1749.

Jokela, W.E., J.W. Hughes, D. Tobi, and D.W. Meals. 1999. Managed vegetative riparian buffers to control P runoff losses from corn fields. Agronomy Abstracts. Amer. Soc. of Agron., Madison, WI.

Klausner, S., D. Flaherty, and S. Pacenka. 1997. Working paper: Field phosphorus index tools for the NYC watershed agricultural program. Cornell University. Ithaca, NY.

Lee, Y. S. and R. J. Bartlett. 1977. Assessing phosphorus fertilizer need based on intensity-capacity relationships. Soil Sci. Soc. Amer J. 41:710-712.

Lemunyon, J.L., and R.G. Gilbert. 1993. The concept and need for a phosphorus assessment tool. J.Prod. Agric. 6:483-486.

Lory, J.A., and P.C. Scharf. 1999. Threshold P survey. *On* Web page for SERA-17, Minimizing P losses from agriculture: http://ces.soil.ncsu.edu/sera17/publications/P_Threshhold/Threshold_P_Survey_3_1_99.htm

Magdoff, F.R. C. Hryshko, W.E. Jokela, R.P. Durieux, and Y. Bu. 1999. Comparison of phosphorus soil test extractants for plant availability and environmental assessment. Soil Sci. Soc. Am. J. 63:999-1006

McFarland, A., L. Hauck, J. White, W. Donham, J. Lemunyon, and S. Jones. 1998. Manure management in harmony with the environment and society. SWCS, Ames, IA.

McIntosh, J. L. 1969. Bray and Morgan soil test extractions modified for testing acid soils from different parent materials. Agron. J. 61:259-265

Moore. P.A., Jr., T.C. Daniel, and D.R. Edwards. 2000. Reducing phosphorus runoff and inhibiting ammonia loss from poultry manure with aluminum sulfate. J. Environ. Qual. 29:37-49.

Natural Resource Conservation Service (NRCS or SCS). 1985. Runoff curve number method. National Engineer. Handbook No. 4.

Natural Resource Conservation Service (NRCS). 1993. Index surface runoff classes. Soil survey manual. Agric. Handbook No. 18.

Natural Resource Conservation Service (NRCS). 1994. The Phosphorus Index: A Phosphorus Assessment Tool. Technical Note. Series No. 1901. Web: http://www.nhq.nrcs.usda.gov/BCS/nutri/phosphor.html

Natural Resource Conservation Service. NRCS. 1999. Nutrient Management. 190-GM, Issue 9, 3/99; Part 402. Web: http://www.nhq.nrcs.usda.gov/BCS/nutri/gm-190.html

Pote, D.H., T.C. Daniel, A.N. Sharpley, P.A. Moore, D.R. Edwards, and D.J. Nichols. 1996. Relating extractable soil phosphorus to phosphorus losses in runoff. Soil Sci. Soc. Amer. J. 60:855-859.

Powell, J.M., Z. Wu, and L.D. Satter. 1999. Dairy diet effects on phosphorus cycles of cropland. P. 62 *In* Agronomy Abstracts. Amer. Soc. of Agron., Madison, WI.

Schmitt, T.J., M.G. Dosskey, and K.D. Hoagland. 1999. Filter strip performance and processes for different vegetation, widths, and contaminants. J. Environ. Qual. 28:1479-1489.

Sharpley, A.N. 1995. Indentifying sites vulnerable to phosphorus loss in agricultural runoff. J. Environ. Qual. 24:947-951.

Simard, R.R., S. Beauchemin, and P.M. Haygarth. 2000. Potential for preferential pathways of phosphorus transport. J. Environ. Qual. 29:97-105.

Sims, J.T., R.R. Simard, and B.C. Joern. 1998. Phosphorus loss in agricultural drainage: historical perspective and current research. J. Environ. Qual. 27:277-293.

Uusi-Kämppä, J., B. Braskerud, H. Jansson, N. Syverson, and R. Uusitalo. 2000. Buffer zones and constructed wetlands as filters for agricultural phosphorus. J. Environ. Qual. 29:151-158.

Valk, H., J.A. Metcalf, and P.J.A. Withers. 2000. Prospects for minimizing phosphorus excretion in ruminants by dietary manipulation. J. Environ. Qual. 29:28-36.

316

Session 9

Land Application

Organic Nitrogen Decay Rates

Gregory K. Evanylo, Ph.D.
Associate Professor
Department of Crop and Soil Environmental Sciences
Virginia Polytechnic Institute and State University

Biographies for most speakers are in alphabetical order after the last paper.

Introduction

Animal manures can have both beneficial and detrimental effects when applied to agricultural land. Properly applied manure can furnish the nitrogen (N) required to meet a crop's nutritional need. Conversely, manure can load soil with higher amounts of N than can be used by plants when applied at excessive rates or at improper times. Soil N not utilized by plants can be transported to groundwater and surface water, where it may impact the health of aquatic systems, humans, and livestock.

Animal manure is composed of both inorganic (mineral) and organic forms of N. Inorganic forms of N are readily available for plant uptake, but organic forms of N must be converted (mineralized) to inorganic forms prior to utilization by plants. The rate and amount of mineralization vary as a result of the heterogeneous nature of manures and environmental effects. Reliable estimates of manure N mineralization rates will reduce the risk of applying insufficient or excessive amounts of plant available N.

The purpose of this paper is to summarize the factors that influence organic N mineralization. Topics include characteristics of manures, factors that influence N mineralization, approaches for estimating N mineralization, and considerations for developing an approach for predicting manure N mineralization rates.

Characteristics of Manures

Animal manure is a mixture of metabolic products such as urea and uric acid, living and dead organisms, and partially decomposed residues from feed (Wilkinson, 1979). Variations in N content and availability occur because of dietary and metabolic differences between animal types. About 75% of the N fed to domestic farm animals is excreted in the manure.

Examples of nutrient content of freshly excreted manures are provided in Table 1. Dairy, beef, and swine excrete similar concentrations of N in fresh manure. Poultry are less efficient than other livestock in utilizing dietary protein and excrete higher concentrations of N in their manure than dairy, beef, and swine.

Table 1. Manure production and nutrient content. (Source: Barker, 1980)

Animal	Animal fresh manure		N in fresh Manure
	Weight	Production	
	lbs	lbs/yr	lbs/ton
Dairy	640	37,400	12
Beef	440	24,200	12
Swine	45	3,080	12
Caged layer	2	100	28
Dry litter:			
broiler	1	15	52
turkey	7	48	36

Nearly all of the N initially excreted is organically complexed (Bouldin et al., 1984). About half of these compounds include urea and uric acid, which are rapidly hydrolyzed and/or decomposed into ammoniacal (ammonia-containing) compounds. Furthermore, some of the more complex organic compounds are converted to ammoniacal N when environmental conditions are favorable for microbial activity. The net effect is that the N in manure is divided approximately equally between insoluble organic and ammoniacal forms within several days of excretion.

The processing, handling, and storage of animal manures influence N transformations. Manure moisture content, temperature, and bedding materials will determine the fractions of total N in organic and ammoniacal forms. The greatest impact on N content of manures is loss of ammonia (NH_3) through volatilization during handling and storage. On average, about 50% of the total N in manure is lost by NH_3 volatilization (Bouldin et al., 1984). The importance of understanding the extent of NH_3 volatilization affects the prediction of N mineralization rates only when plant available N is calculated from an analysis of N that combines ammonium, ammonia, and organic N (i.e., total Kjeldahl N).

The variability in the nutrient content of animal manure is too great to recommend the use of an average value for the calculation of accurate nutrient loading rates. In the absence of a manure nutrient analysis, suggested manure application rates must rely on general and, probably, inaccurate estimates of the forms and amounts of N. Table 2 illustrates the wide range in nutrient composition of manures sampled from numerous farms in Virginia following the effects of considerable collection and storage losses of N. The minimum analysis for determination of available N should include: percent dry matter, ammonium N (NH_4-N), and total Kjeldahl N (TKN). Organic N can be calculated as the difference between total Kjeldahl N and ammonium N (i.e., TKN - NH_4-N). Nitrate N is normally so low in manure that its concentration is not determined.

Table 2. Mean, minimum and maximum amounts of N in manure from various animal types and handling systems tested by the Virginia Tech Water Quality Laboratory, January 1989 to November 1992. (Source: E.R. Collins, personal communication)

Manure Type	TKN			NH_4-N		
(No. of samples)	Mean	Min	Max	Mean	Min	Max
Liquid:						
Dairy (434)[1]	22.6	1.0	52.5	9.6	0.0	44.6
Swine (109)[1]	10.0	0.6	58.5	5.3	0.3	25.8
Poultry (14)[1]	51.1	4.5	89.1	33.0	0.6	66.3
Semi-solid:						
Dairy (46)	10.5	3.4	23.6	3.2	0.1	7.8
Beef (18)	12.8	7.8	23.9	2.6	0.2	25.5
Dry:						
Broiler Litter (254)	62.6	5.7	99.5	11.8	0.2	25.8
Layer/breeder (54)	36.5	9.1	110.6	9.0	0.2	29.6

[1] Values presented in lbs/1000 gals. All other values in lbs/ton.

Nitrogen Transformations in Manure-Amended Soils

Heterotrophic soil organisms that utilize nitrogenous organic substances as energy sources convert organic N to NH_4^+ or NH_3 in the process known as mineralization (Jansson and Person, 1982). The supply of plant-available N in the soil is determined not only by the process of mineralization but also by the process of immobilization, whereby, as soil microbes multiply, they assimilate inorganic N compounds and transform them into organic forms of N that cannot be utilized by plants (immobilization). Mineralization and immobilization work in opposite directions by building up and breaking down organic matter. The supply of N is determined by the additive effects of the two processes and is expressed as either net mineralization or net immobilization. Thus, the change in inorganic N upon the addition of manure to soil is a function of the competition between mineralization and immobilization processes, and any increases in soil inorganic N is a result of net mineralization.

Nitrogen transformations in soil follow the same principles regardless of the source of the N (i.e., inorganic fertilizers, biosolids, manures), but the rate or extent of mineralization, immobilization, and other soil biological processes (i.e., volatilization, nitrification and denitrification) will be functions of the composition (e.g., organic compounds, trace elements, soluble salts) of the N source applied and the environmental conditions (i.e., soil properties, microclimate effects) under which the soil organisms operate.

Estimating Plant Available N

Calculations of plant available N (PAN) from manure should include the portion of the inorganic N that is not lost by ammonia volatilization and a portion of the stable organic N. The amount of PAN in manure can be estimated by:

$$PAN = [NO_3\text{-}N] + A \times [NH_4\text{-}N] + B \times [Organic\ N],$$

where $[NO_3\text{-}N]$, $[NH_4\text{-}N]$, and $[Organic\ N]$ equal the concentrations of nitrate, ammonium, and organic forms of N, A is the fraction of $NH_4\text{-}N$ that does not volatilize and B is the fraction of organic N expected to mineralize.

Mineralization of organic N occurs in two phases. The first phase includes the less resistant organic N, which mineralizes during the first year of application. The second phase includes the more resistant organic N, which mineralizes very slowly in future years. Repeated annual manure applications to the same field result in an accumulation of slow-release N that will reduce the future amounts of manure required to supply the same amount of N.

Factors that Affect Mineralization

Mineralization of organic N from manure applied to the soil is influenced by direct (i.e., waste composition) and indirect (e.g., soil properties, microclimate) factors. This section provides research results that demonstrate the effects of these factors on N mineralization.

Direct factors

Manure composition: Manure N concentration, C:N ratio, and stability of C and N compounds are the main factors governing the amount of organic N that will mineralize.

The N availability will increase with the concentration of N in a manure, assuming that all other factors (e.g., decomposability) are equal.

- Chae and Tabatabai (1985) found that mineralization rates of cow manure in five Iowa soils varied between 13% and 51%, with a mean of 35%. The mineralization rate increased in proportion to the N content of the manure.

- Gordillo and Cabrera (1997a) compared the kinetics of N mineralization of 15 broiler litter samples in a 112-day incubation study. Total mineralizable N ranged from

46.5% to 86.8% of the organic N and could be predicted (R^2=0.91) from uric acid-N and total N concentrations.

The ratio of carbon (C) to N of the material undergoing mineralization or immobilization has a great effect on these N transformations. Mineralization yields more N than required by the microbial biomass for protein when the C:N ratio of the manure is less than about 20:1, and the net result is an addition of inorganic N to the soil pool. When the C:N ratio of the waste is greater than about 25:1, more N than contained in the mineralized organic matter is required by microbes for protein production. Microbes will obtain additionally needed N from the existing soil inorganic N pool, resulting in a net immobilization of soil N. Animal manures generally have C:N ratios of between 5:1 and 20:1 and are net mineralizers of N. The C:N ratio of straw, sawdust, and wood shavings (materials most often used as animal bedding) range from 30:1 to greater than 1,000:1. Highly bedded systems increase the C:N ratio, which favors immobilization over mineralization.

- Reddy et al. (1980) reported mineralization rates for a house floor swine manure with a C:N ratio of 11:1 as 25% to 36% in a ten-week incubation study. Castellanos (1980) estimated first year mineralization of organic N from paved corral swine manure having a C:N ratio of 10:1 as 35% to 45% in a greenhouse study and 33% in an incubation study. By comparison, a drying floor poultry manure with a lower average C:N ratio (6.5:1) than the swine manure mineralized more N (46% to 64%) than the swine waste (Castellanos, 1980). Composting the poultry manure increased the C:N ratio and reduced the N mineralization rate.

- Gordillo and Cabrera (1997a) were able to predict (R^2=0.95) the total mineralizable N in 15 broiler litter samples in a 112-day incubation study with a two-pool, first order model that included the C:N ratio and the uric acid-N concentration in the litter.

- Chescheir et al. (1986) found a very good inverse relationship between available N and the C:N ratio of beef cattle and dairy cattle.

As organic wastes are decomposed by microorganisms, they are broken down to simple compounds and then reconstructed into complex, more stable constituents. The forms of C and N in highly decomposed materials are more resistant to subsequent breakdown and release of N. Lower portions of N in highly processed manure (i.e., well digested, composted) is contained in readily mineralizable forms.

- Willrich et al. (1974) determined that fresh dairy manure generated more first year plant-available inorganic N than did an anaerobically-digested dairy manure from the same source (50% vs. 30%).

- Douglas and Magdoff (1991) evaluated the mineralization potential of four cow and beef manures in various stages of decomposition. Only the anaerobically digested cow manure mineralized more N (20% of added organic N) than the control, indicating a strong effect of waste processing on N availability.

- Eghball et al. (1997) reported that composting can result in an additional 20 to 40% NH_3 volatilization loss beyond the N lost from cattle feedlots, and that the N in compost is in even more stable forms than in typical beef cattle feedlot manure.

- Tyson and Cabrera (1993) demonstrated that composting poultry manure reduced the organic N mineralization rate from 25.4-39.8% in the uncomposted broiler litter to 0.4-5.5% in the composted manure. Composting reduces N mineralization rate by reducing N concentration, increasing C:N ratio, and increasing C and N stability. Composting reduced the N content and increased the C:N ratio of the uncomposted manure. The initial N concentrations of 5.1-6.5% at C:N ratios of 5:1 to 7:1 were increased to 0.9-1.4% (C:N = 10.3:1 to 20.4:1) in the composted manure.

- Hadas and Portnoy (1994) determined the decomposition rate constants for four composted cattle manures in a 32-week incubation study. Three of the composts exhibited similar values, but one compost was considerably more stable. Insoluble constituents of the composted manure must be better characterized to develop universal rate constants.

- Chescheir et al. (1986) found that composted manure did not exhibit an inverse relationship found between available N and the C:N ratio of uncomposted beef and cattle manures. *The stability of the N, not simply the N concentration or the C:N ratio, is important in determining mineralization.*

Animal type: Many of the manure compositional effects are associated with type of animal. While the variability in composition of manure among different species is great, there are more similarities in the composition of manure of the same animals than between different types of animals. A great deal of the similarity is due to the composition of the feed provided to poultry and livestock, which largely affects the composition of the manure.

- Chescheir et al. (1986) found that plant-available N indices, expressed as fractions of the manure organic N available during the first year were higher for beef cattle (15% to 49%) than for dairy cattle (7% to 27%).

- Castellanos (1980) and Reddy et al. (1980) estimated the first year mineralization of organic N from swine manure as somewhat higher (25% to 45%) than dairy and beef.

- The proportion of organic N mineralized from poultry manure has been consistently higher than that for other animal manures. Bitzer and Sims (1988) found that 66% of the organic N is mineralized in 140 days in an incubation study. Castellanos (1980) reported rates of between 46% and 64% in a drying floor poultry waste.

- Real-world variability allows for considerable overlap among the N mineralization rates of animal types. Schepers and Mosier (1991) found first-year mineralization rates of 60-90% for various poultry manures, 30-50% for dairy cattle manures, 20-

75% for beef cattle manures, and 90% for swine manure. Killorn (1993) cited values of 25-50% for various dairy cattle manures and 35-75% for swine and beef cattle wastes.

▸ Castellanos and Pratt (1981) and Reddy et al. (1979) found that about 50% of the N available for mineralization was converted in 18 weeks for beef cattle manure and in 3 to 6 weeks for swine and poultry manures.

Despite the variability of N mineralization rates for any given type of animal, the general order of N mineralization rate for different farm animal manures is: poultry > swine > beef > dairy.

Indirect factors

Environmental factors: Soil temperature and moisture strongly affect mineralization because of their influence on biological activity. The populations and activity of heterotrophic, N-mineralizing suites of microorganisms respond to temperature and moisture optima, which are greatly influenced by soil physical properties.

▸ Soil texture plays an important role in determining the amounts of N mineralized through its effects on local soil environmental conditions. Chescheir et al. (1986) found 40-67% of TKN from various manures and biosolids to be plant available in a Norfolk sand, in contrast to 17-38% in a Cecil sandy loam.

▸ Castellanos (1980) and Castellanos and Pratt (1981) determined that first-year mineralization of beef, sheep, swine and chicken manures were greater in a coarse-textured fine sand than in a fine-textured silty clay.

▸ Gordillo and Cabrera (1997b) compared the kinetics of N mineralization of the same broiler litter in nine soil of varying characteristics. Multiple regression analysis of the cumulative net N mineralized from the litter, which ranged from 36.4 to 78.4% of the total organic N, identified the ratio of sand:water content at field capacity as an important factor in predicting short and long term N mineralization.

▸ Han and Wolf (1992) determined that temperature and soil type were important in determining the kinetics of N release during a 70-day incubation of a swine lagoon effluent in two ultisols. Net mineralization rates ranged from 50% in a loam soil at 20°C to 68% in a silt loam at 35°C.

In all of the above studies, *increased soil aeration and/or higher temperatures increased mineralization*, although greater mineralization rates would be expected in fine-textured soils than coarse-textured soils during periods of drought owing to creation of more favorable moisture conditions for microbial activity (i.e., decay rates of manure N in coarse-textured soils are more sensitive to weather).

Other soil factors: Soil pH, soluble salt content, concentrations of toxic chemicals and heavy metals, and the effects of fertilizers, herbicides, and pesticides may reduce mineralization by inhibiting microbial activity. Gordillo and Cabrera (1997b) identified soil pH as an important factor in predicting short and long term N mineralization. Wetting and drying or freezing and thawing may reduce mineralization by diminishing microbial viability or may enhance mineralization by increasing the availability of potentially mineralizable N to microbes. The total amount of nitrogen mineralized should not be influenced by the C:N ratio of the soil because the C:N ratio of native soil organic matter is approximately 10:1 (Brady, 1974) and probably exerts little control over the total amount of decomposition of applied waste.

Decay Series Concept

The concept of a "decay" series was developed by Pratt et al. (1973, 1976) to describe the amount of inorganic N that becomes available from manure with time. This concept recognizes that plant available inorganic N is the sum of readily available inorganic N and organic N that is gradually mineralized over several years. Pratt developed a series of decreasing fractions to represent the portions of total N in organic wastes that become available to plants over time. For example, a decay series of 0.5, 0.2, 0.1, and 0.05 indicates that 50% of the initial total N mineralizes during the first year, 20% of the N remaining after the first year mineralizes during the second year, 10% of the N remaining after the second year mineralizes during the third year, and 5% of the N remaining after the third year mineralizes during the fourth year. The N remaining after the first year that will become available for plant use in subsequent years is called residual N.

Pratt et al. (1973, 1976) expressed the available N as a fraction of total (organic plus inorganic) N. Because the immediately available inorganic N content varies greatly among manures, the first year's available N fractions proposed by Pratt were highly variable. For example, in the year of application, a liquid manure containing high amounts of dissolved ammoniacal N may be expected to release 75% of its N for plant uptake, but a dried manure whose ammoniacal N has been largely lost by volatilization may have a first year decay constant of only 35%.

Kolenbrander (1981) and Sluijsmans and Kolenbrander (1977) further refined the decay series concept by developing a system based on a truer interpretation of the physical, chemical and biological processes governing N release and availability from manure. Three manure N fractions that could be measured chemically (viz., inorganic N, quickly mineralizable organic N, and organic N that decomposes slowly over several years) were incorporated into equations that describe how manure N becomes plant available over time. In this scheme, decomposition (mineralization) factors are applied only to the organic N fraction.

Mathers and Goss (1979), using data from Pratt et al. (1973) and Willrich at al. (1974), demonstrated how the annual application of various animal manures for providing a constant rate of plant available N should be reduced to account for the build-up of residual soil N

(Table 3). The data show the effects of residual N in reducing the amount of subsequent N applications in manure to maintain a constant supply of 100 lbs N.

Table 3. Total N in manure calculated (using decay constants) to supply 100 lbs of available N to crops each year for years 1, 2, 5, 10, and 20.

Source	N	Decay constants[1]	Year 1	2	5	10	20
	%		----------lbs----------				
Poultry (hens)	4.5	0.9, 0.1, 0.05	111	109	108	106	103
Poultry (broiler)	3.8	0.75, 0.05, 0.05	133	131	125	117	108
Dairy, fresh	3.5	0.5, 0.15, 0.05	200	170	154	133	113
Dairy, anaerobic	2.0	0.3, 0.08, 0.07, 0.05	333	271	199	145	109
Swine	2.8	0.9, 0.04, 0.02	111	110	110	109	108
Bovine, fresh	3.5	0.75, 0.15, 0.1, 0.05	133	126	120	114	107
Dry corral	2.5	0.4, 0.25, 0.06, 0.03	250	156	157	134	113
Dry corral	1.5	0.35, 0.15, 0.1, 0.05	286	206	172	140	112
Dry corral	1.0	0.2, 0.1, 0.05	500	300	217	138	104

[1] Fractions of residual manure decaying each successive year. Last value in each series is the decay constant for each year thereafter.

The parameters of the decay series developed by Sluijsmans and Kolenbrander (1977) were based on a minimum of experimental data, scientific guesses, and only applied to relatively specific conditions under which they were developed. More recently, Klausner et al. (1994) developed a decay series for organic N in dairy manure (0.21, 0.09, 0.03, and 0.02) based on crop N uptake. There are no universal decay series across the United States because the rate of microbial breakdown depends primarily upon soil characteristics and climatic conditions. The decay series concept and its elaborations exceed the experimental data available for their quantitative verification; however, they have provided a framework for more experimental work and for summarizing available data.

N Mineralization Coefficients Employed in the Chesapeake Bay Region

As of the early 1990's, Delaware, Maryland, Pennsylvania and Virginia employed similar guidelines for determining the availability of N from manures (Evanylo, 1994). Each state recommended fractionating N into organic and inorganic components and using separate N availability rates for each form. Delaware (for beef and dairy cattle and swine), Pennsylvania (for beef and dairy cattle, swine, and poultry), and Virginia (for dairy cattle and poultry) also approximated availability based on total N where N fractionation had not been performed. Ammonia volatilization plus organic N mineralization estimates were combined to provide availability coefficients for manures whose analyses were performed on a total N basis.

In general, the N availability coefficients employed by Delaware, Maryland, Pennsylvania and Virginia for various manures during the year following application decreased in the order poultry>swine>beef=dairy (Evanylo, 1994). Residual (after year 1) N availability coefficients were generally similar among animal types. Larger fractions of plant available N due to mineralization were employed for Virginia than Pennsylvania, which was probably an empirical accounting of warmer climate effects on the N transformation process.

The N availability coefficients used by the Virginia Department of Conservation and Recreation (VDCR) in their nutrient management planning operations is presented in Table 4 (Virginia Department of Conservation and Recreation, 1995). It is interesting to note that the VDCR (and Pennsylvania) uses these coefficients as fractions of the original organic N that will become available with time, not as decay constants as employed in Table 3. That is, the VDCR coefficients are multiplied by the original amount of organic N applied, not by the amount of organic N remaining after the previous year's organic N fraction has mineralized.

Table 4. Available fractions of remaining organic nitrogen during the three years after application of various animal manures. (Source: Virginia Department of Conservation and Recreation, 1995)

Animal Type	Year 1	Year 2	Year 3	Year 4
	--------------------Fraction of available N--------------------			
Dairy & Beef	0.35	0.12	0.05	0.02
Swine	0.50	0.12	0.05	0.02
Poultry	0.60	0.12	0.05	0.02

The VDCR further simplified the calculations of the expected amounts of residual N from long term manure applications by adopting from Pennsylvania (Pennsylvania State University, 1990) an approach that employs coefficients based on the frequency of manure application instead of performing the residual N calculations using past years' application rates and residual availability coefficients.

The factors used in Pennsylvania vary according to manure type (i.e., poultry, all others), crop type (e.g., corn and summer annuals, small grains) and application timing (e.g., previous winter or fall, spring) to account for the effect of season on mineralization rate and ammonia volatilization, where TKN has not been separated into inorganic and organic N. Residual N coefficients are classified into three groups – fields that: 1) rarely received manure in past (<4 out of 10 years), 2) frequently received manure (4-8 out of 10 years), and 3) continuously received manure (>8 out of 10 years). No residual N is credited for rare application, 0.07 (poultry) and 0.15 (other manures) availability fractions are used for frequent application, and 0.12 (poultry) and 0.25 (other manures) availability fractions are used for continuous application. The VDCR uses 0, 0.1, and 0.2 as the availability factor for mineralization of residual N for rare, frequent, and continuous applications without regard to manure type. An example of calculating the amounts of N available from long term manure applications in Virginia based on decay rate constants, availability fractions, and frequency factors follows.

Example: A broiler litter is spread on a farm at 4 tons per acre and incorporated within 3-4 days. The manure contains 60 lbs total N (TKN) and 10 lbs NH_4-N per ton. Approximately the same amounts of manure have been applied to the field frequently (5 out of the last 10 years). Ammonium is expected to be 65% available (Virginia Department of Conservation and Recreation, 1995).

The data in Table 5 illustrate the amounts of plant available N estimated in the year of application using the three methods of calculation (i.e., decay rate constants, availability fractions, and frequency factors). It is important to note when selecting an approach for estimating plant available mineralizable N that the most commonly used methods will not influence the amounts of NH_4-N or first year mineralizable N, the forms of N that contributed 70% to 90% of the PAN. The highest estimates of PAN will be obtained when using decay rate constants to calculate the fraction of original organic N that will become available in subsequent years. Using frequency factors to estimate residual N was a compromise between decay rate constants and availability factors.

Table 5. Plant available N from long term broiler litter calculated using decay rate constants, availability factors, and frequency factors.

Fraction of available N	Decay rate constants[1]	Availability factors[2]	Frequency factors
	----------------------------lbs N/acre----------------------------		
NH_4-N	26	26	26
Mineralized N (Yr 1)	120	120	120
Mineralized N (Yr 2)	9.6	24	—
Mineralized N (Yr 3)	2.3	10	—
Mineralized N (Yr 4)	0.7	4	—
Residual N[3]	—	—	24
Total	159	184	170

[1] Based on organic N remaining after previous years mineralization.

[2] Based on original organic N applied.

[3] From years 2-4, based on frequency of application.

Conclusions Regarding an Approach for Estimating Manure N Mineralization

It is more important to identify and quantify the effects of the factors that contribute the greatest variability of manure organic N availability than to ensure that all of the factors are included in the calculations. The factors that should be considered in developing an approach for estimating mineralizable PAN are summarized below.

1. Manure analysis

The earliest decay constants were based on total N, but a more precise accounting of available N can be developed by analyzing manure for ammonium and organic N fractions for estimating the portions of each that will be available for plant uptake. The importance of the C:N ratio has been exaggerated owing to a lack of understanding of the differences in the decay of various C and N fractions. The C:N ratio and the forms of C and N in manure influence both decomposability and mineralization-immobilization, but it will be difficult to improve the predictability of N mineralization until the key C and N fractions for developing relationships with available N are identified. Future breakthroughs in this area may involve the development of models/computer simulations that predict N availability from short term incubation studies or chemical fractionation (Gilmour, 1998; Gordillo and Cabrera, 1997a; Hadas and Portnoy, 1994).

2. Animal type

Because the C:N ratio tends to be fairly constant in manures of the same species whose diets are managed and whose manure is handled similarly, combining animal type and N concentration has been recommended as a reliable indicator of the net amount of N that will mineralize. The exact amount of mineralizable N varies with many factors, but is most strongly correlated with C and N content and forms in the waste, which are correlated to animal feeding and manure handling programs. The variability observed in N mineralization rates for different animal types is high, but states will continue to use similar first year mineralization rates for poultry (50-70%), swine (40-50%), beef (30-45%) and dairy (25-35%) until a simple, more accurate tool is developed.

3. Waste handling

Beyond the initial properties of the waste, type of waste treatment (i.e., handling, storage and processing) is probably the next most important factor in determining the total amount of N that will mineralize. Reductions in available mineralizable N occur with increasing degree of waste processing. The N mineralization coefficients for composting reflect the feedstock recipes and decomposition process more than the animal type. Manure mineralization rate estimates should include a consideration of the storage, handling and processing methods, especially when final product has been greatly altered compared to the raw manure (e.g., compost, litter).

4. Residual N predictability

To determine residual N availability, some laboratories employ availability factors to be used with initial waste application rates and some base residual N on the fraction of organic N remaining following the previous year's mineralization. Although all predicted mineralization rates are similar, they do not always result in the same portion of N being mineralized. Because residual N contributions are generally small and N mineralization is so variable, it is questionable whether there is a need to be concerned about residual N mineralization beyond the second or third year. The discrepancies in the manner in which residual N is estimated are more important from a philosophical than from an N balance perspective; thus, the most important reason for demanding more uniformity in the manner that residual N is determined is to elicit greater public confidence. A simpler method for calculating residual N that does not greatly impact the precision of estimating decay rates is the system based on frequency of application developed by Pennsylvania and adopted by Virginia.

5. Environmental effects

Soil properties (texture, organic matter content) and climate (moisture amount and distribution, soil temperature regime) affect the rate of N mineralization more than cumulative mineralization. The total amount of N mineralized during a particular growing season is influenced by soil type and climate, crop type, and the length of the growing season. If mineralization rates do not reflect these differences, crop N insufficiency or nitrate loading to groundwater may result. The use of environmental factors to predict N availability may be more appropriately employed in a model that could be used to determine post-target crop residual N and to develop a site-specific strategy for reducing N losses and water contamination. The effects of the environment on mineralization have been accounted for to some extent by the adoption of regional mineralization coefficients.

References

Barker, J.C. 1980. Livestock manure production rates and approximate fertilizer content. Raleigh, NC: North Carolina Agr. Extension Service Leaflet 198 (revised).

Bitzer, C.C., and J.T. Sims. 1988. Estimating the availability of nitrogen in poultry manure through laboratory and field studies. J. Environ. Qual. 17:47-54.

Bouldin, D.R., S.D. Klausner, and W.S. Reid. 1984. Use of nitrogen from manure. *In* R.D. Hauck (ed.) Nitrogen in crop production. Madison, WI: American Society of Agronomy, Crop Science Society of America, and Soil Science Society of America, 221-45.

Brady, N.C. 1974. The nature and properties of soils. 8th ed.: Macmillan Publ. Company, New York.

Castellanos, J.Z. 1980. Nitrogen mineralization in manures. M.S. thesis, University of California, Riverside, CA .

Castellanos, J.Z., and P.F. Pratt. 1981. Mineralization of manure nitrogen. Correlation with laboratory indexes. Soil Sci. Soc. Am. J. 45:354-57.

Chae, Y.M. and M.A. Tabatabai. 1985. Mineralization of nitrogen in soils amended with organic wastes. J. Environ. Qual. 15:193-98.

Chescheir, G.M., III.; P.W. Westerman, and L.M. Safley, Jr. 1986. Laboratory methods for estimating available nitrogen in manures and sludges. Agric. Wastes 18:175-95.

Douglas, B.F. and F.R. Magdoff. 1991. An evaluation of nitrogen mineralization indices for organic residues. J. Environ. Qual. 20:368-72.

Eghball, B., J.F. Power, J.E. Gilley, and J.W. Doran. 1997. Nutrient, carbon, and mass loss of beef cattle feedlot manure during composting. J. Environ. Qual. 26:189-193.

Evanylo, G.K. 1994. Mineralization and availability of nitrogen in organic waste-amended mid-Atlantic soils. P.77-103. *In* S. Nelson and P. Elliott (ed.) Perspectives on Chesapeake Bay, 1994: Advances in estuarine sciences. CRC Publication No. 147. Chesapeake Research Consortium, Inc., Edgewater, MD.

Gilmour, J.T. 1998. Carbon and nitrogen mineralization during co-utilization of biosolids and compost. P. 89-112. *In* S. Brown, J.S. Angle, and L. Jacobs (ed.) Beneficial co-utilization of agricultural, municipal and industrial by-products. Kluwer Academic Publishers, Dordrecht/ Boston/ London.

Gordillo, R.M. and M.L. Cabrera. 1997a. Mineralizable nitrogen in broiler litter: I. Effect of selected litter chemical characteristics. J. Environ. Qual. 26:1672-1679.

Gordillo, R.M. and M.L. Cabrera. 1997b. Mineralizable nitrogen in broiler litter: II. Effect of selected soil characteristics. J. Environ. Qual. 26:1679-1686.

Hadas, A. and R. Portnoy. 1994. Nitrogen and carbon mineralization rates of composted manures incubated in soil. J. Environ. Qual. 23:1184-1189.

Han, X.G., and D.C. Wolf. 1992. Availability of N and P in two soils amended with swine lagoon effluent. p. 41. *In* Agronomy abstracts. Amer. Soc. of Agron., Madison, WI.

Jansson, S.L. and J. Persson. 1982. Mineralization and immobilization of soil nitrogen. *In* F.J. Stevenson (ed.) Nitrogen in agricultural soils. Agronomy 22:229-252. Amer. Soc. of Agron., Madison, WI.

Killorn, R. 1993. Crediting manure in soil fertility programs. Solutions, February, pp. 32-35.

Klausner, S.D., V.R. Kanneganti, and D. Bouldin. 1994. An approach for estimating a decay series for organic nitrogen in animal manure. Agron. J. 86:897-903.

Kolenbrander G.J. 1981. Limits to the spreading of animal excrement on agricultural land. *In* J.C. Brogan (ed.) Nitrogen losses and surface runoff from landspreading of manures. Martinus Nijhoff, The Hague/Boston/London.

Mathers, A.C. and D.W. Goss. 1979. Estimating animal waste applications to supply crop nitrogen requirements. Soil Sci. Soc. Am. J. 43:364-366.

Pennsylvania State University. 1990. The agronomy guide, 1991-1992. *In* R.D. Chambers (ed.) Agricultural information services. Pennsylvania State University, University Park, PA.

Pratt, P.F., F.E. Broadbent, and J.P. Martin. 1973. Using organic wastes as nitrogen fertilizers. Calif. Agric. 27:10-13.

Pratt, P.F., S. Davis, and R.G. Sharpless. 1976. A four-year field trial with animal manures. Hilgardia 44:99-125.

Reddy, K.R., R. Khaleel, and M.R. Overcash. 1980. Nitrogen, phosphorus, and carbon transformations in a coastal plain soil treated with animal manures. Agric. Wastes 2:225-38.

Reddy, K.R., R. Khaleel, M.R. Overcash, and P.M. Westerman. 1979. A non-point source model for land areas receiving animal wastes: I. Mineralization of organic nitrogen. Trans. ASAE, 863-72.

Schepers, J.S., and A.R. Mosier. 1991. Accounting for nitrogen in nonequilibrium soil-crop systems. P. 125-138. *In* R.F. Follett, D.R. Keeney, and R.M. Cruse (ed.) Managing nitrogen for groundwater quality and farm profitability. Soil Sci. Soc. of Amer., Madison, WI.

Sluijsmans, C.M.J. and G.J. Kolenbrander. 1977. The significance of animal manure as a source of nitrogen in agricultural soils. *In* Proc. Int. Seminar on Soil Fertility and Fertility Management in Intensive Agriculture. The Soc. of the Sci. of Soil and Manure, Tokyo, Japan.

Tyson, S.C. and M.L. Cabrera. 1993. Nitrogen mineralization in soils amended with composted and uncomposted poultry litter. Commun. Soil Sci. Plant Anal. 24:2361-74.

Virginia Department of Conservation and Recreation. 1995. Virginia nutrient management standards and criteria. Richmond, VA. 64 p.

Wilkinson, S.R. 1979. Plant nutrient and economic value of animal manures. J. An. Sci. 48:121-33.

Willrich, T.L., D.O. Turner, and V.V. Volk. 1974. Manure application guidelines for the Pacific Northwest. St. Joseph, MI: American Society of Agricultural Engineers ASAE Paper no. 74-4601.

Ammonia Volatilization
from Dairy and Poultry Manure

J. J. Meisinger, Ph.D
Soil Scientist, USDA-ARS
Environmental Chemistry Laboratory
Beltsville Agricultural Research Center
Beltsville, Maryland

W. E. Jokela, Ph.D.
Extension Soil Scientist
Plant and Soil Science Department
University of Vermont
Burlington, Vermont

Biographies for most speakers are in alphabetical order after the last paper.

Introduction

Ammonia volatilization is a major N loss process for surface-applied manures and urea fertilizers. The lost ammonia is important for both agricultural and non-agricultural ecosystems because it: i) is a direct loss of plant available N to the farmer, ii) reduces the N:P ratio in manure, which accelerates P build-up in soils, and iii) contributes to eutrophication in aquatic and low-N input ecosystems through atmospheric transport and deposition (Asman, et al. 1994; Asman et al., 1998; Sharpley et al., 1998). Atmospheric ammonia originating from agricultural activities has been implicated in widespread damage to natural ecosystems in Europe (Asman et al. 1998; Hacker & Du, 1993). Similarly, there is growing public concern in the US that current manure management practices may be promoting ammonia enrichment of streams, estuaries, and coastal waters.

Agriculture is the major source of ammonia emissions to the atmosphere, contributing about 90% of the total in Western Europe according to recent estimates (Kirchmann et al., 1998; Stevens & Laughlin, 1997; Bussink & Oenema 1998). Most ammonia emissions are from livestock production with cattle farming, especially dairy, regarded as the largest source (Bussink

& Oenema 1998). Land application of manure contributes close to half (46%) of the ammonia emissions from livestock in the UK, animal housing about one-third, and waste storage and grazing the remaining 20% (Phillips & Pain, 1998). Smaller ammonia emissions are attributed to non-animal agricultural, such as fertilizer and crops (Sommer & Hutchings, 1995). Most efforts to reduce agricultural ammonia losses have focused on land application, the single largest source. This paper will therefore focus on land application of dairy and poultry manures, which are two major livestock enterprises in the Northeast.

Ammonia volatilization occurs because ammonium-N in manure or solution is converted to dissolved ammonia gas, by the reaction:

$$NH_4^+\text{-}N \ \rightleftharpoons \ NH_{3g} + H^+ \qquad \qquad \text{(Eqn. 1)}$$

The reaction produces more NH_{3g} as pH or temperature increases, and as the NH_4-N concentration increases. The rate of ammonia release to the atmosphere is a function of the difference in NH_{3g} concentration in the manure and the air (Lauer et al., 1976; Freney et al., 1983). The details of ammonia volatilization are complex, being affected by the level of dissolved vs. clay adsorbed ammonium-N, the chemical conversion of ammonium-N to dissolved ammonia gas, and the physical transport of the ammonia gas into the atmosphere. A large number of environmental and management factors influence ammonia loss under field conditions (Freney et al., 1983). The dominant factors influencing losses can be categorized as: manure composition, application method, soil factors, and environmental conditions (Meisinger & Randall, 1991; Sharpley et al. 1998).

The above economic and environmental concerns emphasize the necessity for developing improved management practices for conserving ammonia N in manures. The goals of this paper are: i) to examine the major factors affecting ammonia loss by reviewing relevant ammonia volatilization data, ii) to examine ammonia volatilization estimates used in the Northeast, and iii) to provide suggestions for improving ammonia volatilization estimates used in nutrient management planning.

General Magnitude and Pattern of Field Losses

Ammonia volatilization losses vary greatly depending on environmental conditions and management. Losses can range from close to 100% for surface application with optimal conditions for volatilization, to only a few percent when manure is injected or incorporated immediately into the soil. Ammonia losses are usually expressed as a percentage of the total ammoniacal N (TAN, ammonium-N plus ammonia-N) in the manure or slurry, because it is that portion that is immediately susceptible to loss. Typical results of studies on the application of liquid cattle manure to grassland (incorporation not possible) lie in the range of 40 to 70% loss (Stevens and Laughlin, 1997). Losses of dairy slurries applied in the spring to land tilled the previous fall in Ontario were 24 to 33% of TAN (Beauchamp et al., 1982), while losses from solid dairy manure (about 20% solids) in several New York experiments ranged from 61 to 99% (Lauer et al., 1976). Ammonia losses from surface applied poultry litter in Europe are commonly

15 to 45% of TAN plus uric acid N (Jarvis & Pain, 1990; Moss, et al., 1995; Chambers, et al., 1997). Ammonia losses from spring surface-applied poultry litter to fescue pastures in the Southeast ranged from 28 to 46% of the NH_4-N (Marshall et.al., 1998). These data illustrate that ammonia losses from poultry litter are commonly 20-45% of TAN, which is considerably less than cattle slurry losses which are frequently 35-70% of TAN. Most of the research on ammonia volatilization from manures has been conducted in Europe. While the specific circumstances or conditions may be somewhat different from those in the Northeast, the general principles and conclusions derived from these studies should be relevant to Northeastern agriculture.

The temporal pattern of slurry ammonia emissions is a very rapid loss during the first 6 to 12 hours after application, and a prominent reduction in the rate during the next few days (Fig. 1). Poultry litter, because of its drier condition, has a slower initial rate of loss than slurries, but has significant losses extending over several days or weeks (Fig. 2).

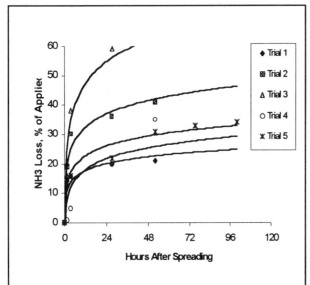

Fig. 1. Cumulative NH_3-N loss as a percent of surface-applied dairy slurry NH_4-N in five trials in Vermont during 1995-96.

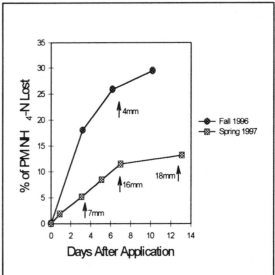

Fig. 2. Cumulative NH_3-N loss as percent of surface-applied poultry litter NH_4-N in Maryland in fall 1996 or spring 1997.

Results from a review of 10 studies with cattle slurry applied to grassland (Stevens and Laughlin, 1997) found that 30 to 70% of the total ammonia loss occurred in the first four to six hours, and 50 to 90% in the first day. One reason for the rapid losses from slurries is the slurry matrix, which is a well-mixed liquid abundantly supplied with urease. This matrix "sets the stage" for rapid ammonia losses once gas exchange is readily available. An example of a typical pattern of ammonia loss from surface broadcast dairy slurries is shown in Fig. 1 where 35 to 95% of the loss occurring in the first two to five hours. By contrast, the typical pattern of ammonia loss from surface-applied poultry litter in Maryland (Fig. 2) illustrates the high losses the first day after application, followed by continued substantial loss through day seven. Some investigators have even observed linear rates of volatilization from poultry litter for up to three weeks after application (Chambers et al., 1997). Volatilization losses from six poultry litter studies in the

Southeast (Marshall et al., 1998) show that an average of 25% of the total loss occurred on day one, 17% on day two, 15% on day 3, and 22% of the total loss over days four through seven. Thus, the time-course of ammonia loss is quite different for the liquid slurries (90-95% moisture) than for the drier poultry litters (20-40% moisture).

The pattern described above for slurries can be explained by a combination of manure and soil properties that change over time. Immediately after spreading the pH of slurry typically increases substantially, e.g., from the 7.6 to 8.4 (Sommer et al.,1991). The increased pH results from urea hydrolysis (Lauer et al. 1976) and loss of CO_2 by degassing. The initial concentration of TAN is usually high (1,000 to 2,000 mg TAN/l) and drying of the manure increases the TAN concentration further due to a decrease in the volume of water. After this initial high pH and high NH_4-N period, the length of which varies with environmental conditions, the rate of volatilization decreases dramatically due to: i) lower NH_4-N levels resulting from NH_3 losses, adsorption of NH_4-N onto soil colloids, and nitrification, ii) a lowering of the pH due to removal of the basic NH_3 molecule and release of H^+ (Eqn. 1), iii) infiltration of dissolved NH_4-N into the soil which decreases TAN at the air interface, and iv) formation of surface crusts which restrict gas exchange (Beauchamp et al., 1982; Brunke, et al., 1988; Sommer et al., 1991).

A review of ammonia volatilization literature quickly reveals that it is a highly variable process. But hidden beneath this variability are the major governing factors which affect ammonia volatilization in the field. Therefore, rather than focus on a case-by-case literature review and the variability of the process, we have chosen to emphasize the main factors affecting ammonia losses with resultant focus on techniques to improve manure N management.

Factors Affecting Ammonia Volatilization

Understanding the main factors affecting ammonia volatilization will delineate practices to reduce ammonia losses, will improve the prediction of these losses, and will aid in developing more efficient farm nutrient management plans. The factors can be categorized in four groups: i) manure characteristics (dry matter content, pH, NH_4-N content), ii) application management (incorporation, zone application, timing), iii) soil conditions (soil moisture, soil properties, plant/residue cover), and iv) environmental factors (temperature, wind speed, rainfall). The categories are ordered from the most practical factors to the least manageable factors.

Manure Characteristics

It is well known that manure is a highly variable commodity. Other papers at this workshop have focused on manure analysis; it is sufficient to state that sound manure management should begin with an analysis of the manure. Management of ammonia volatilization should include analysis of ammonium-N, total N, and dry matter (DM). Knowledge of the ammonium-N content is essential to set the upper limit on ammonia losses and gain better estimates of plant available N. Knowledge of dry matter can be useful in estimating ammonia loss rates.

The content of solids, or dry matter, in slurries has been shown to be an important factor in determining the ammonia volatilization potential in Europe (Sommer & Olesen 1991; Smith and

Chambers, 1995; Lorenz and Steffens, 1997; Pain & Misselbrook 1997). The general observation is that slurries with higher dry matter content show greater ammonia loss. For example, Sommer and Olesen (1991) showed a linear relationship between cattle slurry dry matter content and ammonia emission for slurries between 4 and 12% DM, however DM had little effect above or below those values. This relationship is due to the fact that slurries with lower solids tend to have greater fluidity and, therefore, infiltrate more readily into the soil where ammonium is protected from volatilization by adsorption onto soil colloids. Where vegetation is present, more fluid slurries make more direct contact with the soil, rather than adhering to plant material. The effect of dry matter content has been most pronounced in the short-term period immediately after application.

The 'fluidity and soil contact' concept explains why solid manures tend to volatilize a higher percentage of the TAN than dilute slurries, although solid manures lose less N the first day. For example, Menzi et al. (1997) found that, per unit of TAN applied, total emissions from solid manure were 30% higher than liquid manure in side-by-side comparisons. Researchers in Europe have used dry matter content to explain differences among manure of different species, e.g. more dilute pig slurri vs. thicker cattle slurry (Pain & Thompson, 1988; Brunke et al., 1988). The UK manure model "MANNER" employs a DM variable to predict losses by increasing NH_3 loss by about 5% of applied NH_4-N for each 1% increase in DM (Chambers et al., 1999). This principle has led to examination of dilution of manure with additional water as a management practice to reduce ammonia volatilization. A combination of solids separation and dilution to reduce dry matter content from 11.3 to 5.6% resulted in a 50% reduction in ammonia emissions (Stevens et al., 1992). Preliminary results from Vermont (Jokela et al, unpublished) are consistent with European results, showing about one-third less ammonia volatilization from liquid cattle manure diluted to reduce DM from 9 to 3%.

Dry matter content is not as dominant a factor for poultry litter, because most modern poultry units produce relatively dry litter, containing 55 to 75% DM. In fact, it is the low moisture level of poultry litter which is likely contributing to a lower potential for ammonia loss. It is often useful to think of ammonia volatilization as comparable to water evaporation. Thus, a drier poultry litter would loose less water and ammonia than a dairy slurry. Few studies have evaluated variations in potential ammonia loss among poultry manures. In a laboratory study of 18 poultry litters in Delaware, Schilke-Gartley and Sims (1993) found potential ammonia volatilization to vary from 4 to 31% of manure total N, but only weak correlations between ammonia loss and individual manure composition parameters. The multiple correlation using manure total N and pH produced an R^2 of 0.77 - if four manure samples which produced anomalous ammonia losses were omitted. Schilke-Gartley and Sims (1993) concluded that a manure test to estimate potential ammonia volatilization would be very useful, especially considering the wide range in potential ammonia loss among manures.

Measurement of manure N characteristics other than NH_4-N, total N, and DM are not in general use in the U.S. Manure pH is not regularly measured, even though a higher initial manure pH can increase the rate of ammonia volatilization (Sommer et al., 1991). However, initial manure pH has often not had a significant effect on slurry NH_3 emissions because of the rapid increase in slurry pH after application (Sommer & Hutchings, 1997; Sommer & Sherlock, 1996). Adding nitric or sulfuric acid to slurries before spreading to lower the pH to 6.5 has been effective at reducing ammonia volatilization (Stevens et al., 1992), but safety and other practical issues have limited adoption of the practice. Consideration of parameters such as pH, soluble Al, or soluble Fe may become useful if manure amendments such as alum $(AL_2(SO_4)_3)$ or ferrous sulfate $(FeSO_4)$ come into use. Both alum and ferrous sulfate have an acidifying effect on manures which could markedly decrease potential ammonia volatilization (Moore et al., 1995), because ammonia losses are minimal below pH 7. Consideration of the uric acid content of poultry litter may also be an important, as it is in the "MANNER" model, especially if the trend toward drier litters continues.

Application Management

Nitrogen losses during application can be grouped into losses during spreading and losses incurred after application. Unique problems and opportunities exist for reducing ammonia losses from slurries and solid manures for annual cropping systems, both tilled and non-tilled soils, and for grasslands. Management opportunities also exist for adjusting the time and rate of application.

<u>Annual Cropping Systems</u>

Volatilization losses during the spreading operation itself have generally been found to contribute little to total ammonia loss, usually less than 1% (Pain & Thompson, 1988; Phillips et al., 1990). The exception is irrigation of slurry, where ammonia losses can be much higher than conventional application methods (Phillips et al., 1990). Sharpe and Harper (1997) reported that 13% of slurry TAN was lost during irrigation in Georgia, while another 69% was volatilized from the sandy loam soil within 24 hours after application.

It is a well-known fact that soils are a good sink for ammonia, which leads to the corollary that incorporation of manure is a good method to reduce ammonia losses. There are a number of classic papers which clearly illustrate the importance of incorporation soon after application to achieve maximum agronomic response (Salter & Schollenberger 1939; Heck 1931). The rapid loss of ammonia from dairy slurries (Fig. 1) exemplifies the need to immediately incorporate these sources. In fact, even a one day delay in incorporating slurries can lead to loss of 50 to 90% to the TAN. Disking cattle manure reduced ammonia losses by 85 to 90% in a Canadian study (Brunke et al., 1988). Cultivating before slurry application can also reduce ammonia emissions because of increased infiltration into the soil (Bless, 1991; Sommer & Ersbøll, 1994). The literature is abounding with comparisons of tillage equipment to reduce ammonia loss (Amberger 1990; Klarenbeek & Bruins 1990; Dohler 1990). The general observation is that the more thorough and deep the tillage implement mixes the manure with the soil, the better it prevents ammonia losses, e.g., moldboard plows are more effective than fixed tines (Klarenbeek & Bruins, 1991). For solid manures (DM above 20%), direct tillage into the soil is the main avenue for incorporation, but slurries have many application options for conserving ammonia.

Various equipment options are available for injection or direct incorporation of liquid manure in annual row crop systems (Fig. 3).Deep injection with a knife or chisel (6 to 12 inches deep) has produced large reductions in ammonia emissions from slurries applied to corn in the US (Hoff, 1981). The reduced ammonia volatilization is generally reflected in improved N utilization and increased yields. Beauchamp (1983) obtained increased corn yields and approximately twice the N efficiency from liquid cattle manure when it was injected at either pre-plant or sidedress time compared to surface application. Klausner and Guest (1981) obtained increased corn yields from sidedressed injected dairy manure in New York. In recent years a horizontal sweep injector that operates at a shallower depth (4 to 6 inches; Fig. 3b) has become more popular because it provides more even distribution of manure, improves N availability, and requires less power (Schmitt et al., 1995).

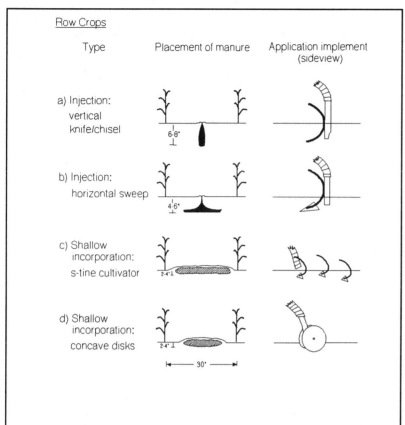

Fig. 3. Equipment options for injection or direct incorporation in row crops or on bare ground (Jokela & Côté, 1994).

A relatively new design, now available commercially from a few companies in Canada and the U.S., does not actually inject the manure but mixes and covers it with soil using either "s-tine" cultivator shanks or pairs of concave covering disks. (Figs. 3 c, d) These shallow incorporation methods require less power than injection options and can be operated at a faster ground speed and with less problem on stony soils. A long-term study with liquid swine manure as a sidedress application on corn (Côté et al., 1999) showed better utilization of N from manure applied between rows with "s-tine" incorporation than with deep injection. Results from a study in Vermont (Jokela et al., 1996) showed equal or slightly greater corn silage yields from 5000 gal/acre liquid dairy manure (68 lbs/acre NH_4-N and 135 lbs total N/acre) sidedressed with "s-tine" incorporation than from sidedressed fertilizer N at a 65 lb/acre. The above discussion shows that there are several good options to reduce ammonia losses from slurry and therefore improve N use efficiency for annual cropping systems.

Perennial Forage Systems

There are situations where injection or incorporation is not possible, e.g., manure applied to grasslands or manure applied to a no-till culture. In these situations modified application equipment is needed. Deep injection (6 to 12 inches) can effectivly reduce ammonia losses on grassland, but the practice has not been well accepted because of root damage and occasional yield reductions (Thompson et al., 1987). As a result, shallow injection systems (2-inch depth) have been developed (Fig. 4d) which still reduce ammonia emissions but produce less soil disturbance and crop damage (Pain & Misselbrook, 1997), although some yield reductions have been observed (Misselbrook et al., 1996). Ammonia volatilization has been reduced by 40 to 95% by shallow injection in various trials in the Netherlands and the UK (Frost, 1994; Misselbrook et al., 1996; Huijsmans et al., 1997). In some cases increased denitrification losses have been associated with reductions in ammonia emissions from injection, due to the localized high concentrations of carbon (which drives denitrification) and nitrogen (Thompson et al., 1987; Pain & Thompson, 1988).

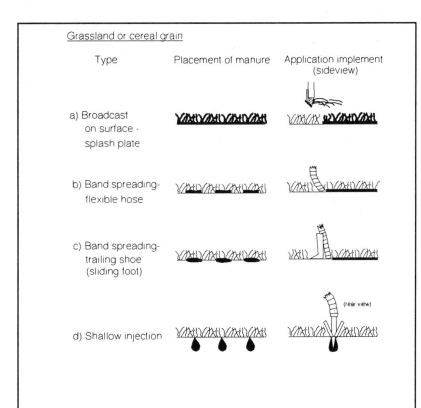

Fig.4. Equipment options for injection or direct incorporation in grassland or no-till crops (Jokela & Côté, 1994)

An approach that avoids soil disturbance entirely, while still reducing ammonia losses, is application of slurry in narrow bands either directly from the spreader hose or through a sliding shoe that rides along the soil surface (Fig. 4 b, c). The intent is to place the manure in a band close to the ground below the crop canopy, providing less surface exposure and some wind protection and preventing contamination of foliage with slurry. This equipment reduces ammonia volatilization, especially in the first few hours after application, though not as effectively as with injection. Most studies in Europe have reported volatilization reductions of 30 to 70% compared to surface application (Huijsmans et al., 1997; Frost, 1994; Pain & Misselbrook, 1997).

However, Thompson et al. (1990) reported a total reduction of only 17% over five days, a result of a slightly greater emission rates from the banded treatment during the last three days. This low effectiveness may have been because the bands were wider than in other studies and covered 35 to 40% of the ground surface.

Research with a trailing foot application system (Fig. 4c) in Vermont gave ammonia loss reductions of 30 to 90% compared to broadcast application, most of the difference occurring in the first several hours (Jokela et al., 1996). Small, but significant, yield increases of 6 to 14% resulted from band application in two of four site-years (Carter et al., 1998). A three-year study in British Columbia showed greater grass yields and N recovery from a sliding shoe system (Bittman et al., 1999), attributed to reductions in ammonia emissions although measurements were not made.

Timing

Another potential element for managing ammonia volatilization is time of application, considering either a seasonal scale (e.g. fall vs. spring) or a daily scale. If manure is immediately incorporated, timing issues center on applying the manure as close to the time of crop need as possible. If incorporation is not possible, timing should try to balance the objectives of applying close to crop need, yet avoid high ammonia loss seasons. Higher ammonia losses were reported from slurry on grassland in summer than in cooler seasons in the UK (Pain & Misselbrook, 1997) and in other Western European research (Amberger, 1990; Dohler, 1990). In other work in the UK (Smith & Chambers, 1995) ammonia losses decreased with each month's delay in application from September until January, and N efficiency was greater from spring than from fall-applied slurry. Smaller losses in cooler seasons are a result of lower temperatures, which provide less energy for volatilization, as demonstrated by the data in Fig. 2. Fall applications are not generally recommended in the Northeastern states due to the high susceptibility of loss through volatilization plus leaching. However, limited manure storage, soil trafficability issues, and time constraints have frequently contributed to significant fall-applications of manure in the region.

On a daily time scale, manure could potentially be applied in the late afternoon or evening to take advantage of the marked diurnal trend in ammonia losses, which consist of high daytime losses and lower losses at night (Beauchamp et al., 1982; Brunke, et al., 1988). However, evening applications have not always successfully reduced losses (Klarenbeek & Bruins 1991). Time and operational restraints greatly limit this approach to small operations. In any case, this short-term measure does not eliminate the need for incorporation the next day to minimize further losses.

Application Rate

Several researchers have found that total ammonia emissions were proportional to the application rate of manure TAN (Brunke, et al., 1988; Menzi et al., 1997; Svensson, 1994; Hoff, 1981). However, others (Thompson et al., 1990; Frost, 1994; Lauer et al., 1976) found a decreasing volatilization rate, per unit of slurry, as the application rate increased. The conflicting results are probably due to the competing factors of infiltration vs. volatilization. An explanation for these findings would be that a thinner layer of manure (lower rates) can lose a high percentage of its NH_4-N if adsorption or infiltration is small. In this situation higher rates increase the diffusion path length of NH_{3g} (deeper manure) and give more time for adsorption or infiltration to occur. However, if higher rates do not increase adsorption then more of the manure NH_4-N could be lost. Thus, the effect of application rates depends on the competing forces of adsorption vs. volatilization. In any event, application rates are generally governed by crop N needs and manure composition, rather than a desire to manage ammonia loss.

Soil Conditions

Soil conditions, such as moisture content, cation exchange capacity (CEC), pH, and plant or residue cover can also impact ammonia losses. The analogy between water loss and ammonia loss is useful for soil moisture because dissolved ammonia gas moves to the surface via the soil water, where it is subject of gaseous exchange with the atmosphere. A study of 32 soils showed a two- to three-fold increase in ammonia emissions from moistened soils compared to those in an air-dry condition (Kemppainen 1989), the increase ascribed to a lower absorption of the liquid fraction into the wetter soils (Kemppainen 1989; Pain et al., 1989; Sommer & Christensen, 1991).

Soil Chemical Properties

Soil chemical properties of pH, CEC, and texture can also impact ammonia loss. High soil pH increases ammonia losses by increasing concentrations of NH_3. For example, the percentage of TAN which is NH_3 is about 0.1, 1, 10, and 50% at pH values of 6, 7, 8 and 9, respectively (Court, et al., 1964). Ammonia volatilization from cattle slurry surface-applied to a fine-sand soil increased linearly with soil pH ($CaCl_2$) in the range of 5.4 to 6.9 (Kemppainen 1989). Factors which increase the change in pH will also increase potential ammonia loss. The buffering capacity of a soil is determined from its CEC, texture, soil minerals, and organic matter content. The CEC decreases with decreasing clay content (coarse textured sandy soils), decreasing organic matter content, and highly-weathered clay minerals (1:1 clays). A high CEC can impact ammonia loss by restricting the pH change associated with adding manures. In a study of 63 Finnish soils volatilization of NH_3 from surface-applied cattle slurry decreased with increasing CEC and, particularly, with increasing clay content (Kemppainen 1989). Thus, a low CEC sandy soil is susceptible to higher pH's and larger ammonia losses than a silt loam. Soil pH is readily managed, but since most Northeastern soils are acidic the pH factor is not a major option to control ammonia losses. The other soil properties related to CEC are not easily changed by management, so the best scenario for integrating soil properties into ammonia volatilization management is to use soil properties as a category variable to adjust estimates of ammonia loss.

Soil Cover

The presence of vegetative cover, the nature of the vegetation, and crop residues can also affect ammonia volatilization by restricting contact between manures and soil colloids. Thompson et al. (1990) reported 50% higher ammonia emissions from grassland than from a bare soil, most of the difference occurring in the first 24 hours. The explanation was that the grass served as a barrier and prevented much of the slurry from making contact with soil, and that slurry adhering to the grass created a larger surface area for volatilization. Likewise, in a French ammonia volatilization study with pig slurry there was about 30% greater losses from grassland than from wheat stubble (Moal et al. 1995).

Environmental Factors

Environmental factors can also impact ammonia losses because weather elements provide the energy and the driving force for the soil-air gas exchange. In general weather elements that increase the evaporative demand will also increase ammonia volatilization. Thus, ammonia volatilization is increased by higher temperatures and by increased wind speeds.

Temperature

The rate of ammonia volatilization increases with increasing temperature (Sommer et al. 1991; Svensson, 1994; Moal et al., 1995) with a greater effect observed in the first several hours after application (Sommer et al. 1991). Higher temperatures increase ammonia losses by decreasing the solubility of NH_3 gas in the soil solution and by increasing the proportion of TAN as NH_3 gas. Physical chemistry predicts that higher temperatures should cause ammonia losses to increase by a factor of about 3 for every 18°F (10°C) rise in temperature (Denmead et al., 1982). For example, a slurry containing 1500 mg NH_4-N/l at pH 7.8 would support equilibrium gaseous ammonia pressures of about 7, 23, and 69 mbars at temperatures of 50, 68, and 86°F (10, 20, and 30°C), respectively. The seasonal ammonia loss differences in Fig. 2 can be partially attributed to temperature because the average fall temperature was 64°F (18°C) while the spring temperature was 48°F (9°C). Thus, temperature can potentially have a considerable impact on ammonia losses. Temperature effects on ammonia loss have also been reported by others (e.g. Beauchamp et al., 1982; Harper et al., 1983; Nathan & Malzer, 1994; Sommer & Olesen, 1991; Sommer et al., 1991) but all the temperature effects have been less dominant than theory would suggest. This is because ammonia concentrations are seldom at equilibrium and because losses are also influenced by gaseous transport factors (tortuous air paths in soil, boundary layers, crusts, etc.). Ammonia losses do not stop at near-freezing temperatures. Laboratory studies with cattle manure in Vermont (Midgely & Weiser, 1937) and in New York (Steenhuis et al., 1979) reported losses of 50% of the TAN in two days at near-freezing temperatures. Losses near freezing can occur because a lower, but still substantial rate, of volatilization continues for a longer period of time (Sommer et al., 1991) and because freezing can have the same NH_4-N concentrating effect as drying (Midgely & Weiser, 1937; Lauer et al., 1976).

Temperature is not a universal driving variable, however. In a series of 11 experiments with swine slurries and solid dairy manure, Brunke et al. (1988) found that variations in ammonia flux were not well correlated with temperature. Brunke et al. (1988) attributed the results to interactions and correlations among meteorological parameters which affected the ammonia loss process. They suggested use of composite parameter, such as the hay drying index, which quantifies potential evaporation based on temperature, wind, and humidity, as a indicator of potential ammonia volatilization.

Wind speed

Higher winds contribute to higher ammonia losses by increasing the mass transfer and air exchange between the manured surface and the atmosphere. Most investigators have found a linear relation between wind speeds up to about 6 mph (2.5 m/s) and ammonia volatilization (Brunke et al., 1988; Sommer et al., 1991; Thompson et al., 1990). The greatest effect of wind speed is in the early phase of volatilization, before drying and surface depletion of NH_4-N occur. The precise impact of wind speed is difficult to assess from field data because wind increases are often confounded with changes in temperature and solar radiation.

Rainfall

Significant rainfall soon after slurry application can reduce ammonia volatilization by moving ammonium into the soil where it is held by soil colloids. The end result is an effect similar to shallow incorporation by tillage. Pain and Misselbrook (1997) reported ammonia reductions of about one-third from a 0.7 inch (18 mm) rainfall after application of cattle slurry. Significant reductions after rainfall was also reported by Beauchamp et al. (1982) in three Canadian studies with cattle slurry. Ammonia losses from urea fertilizers have suggested that only 0.3 inches (7-9 mm) of rainfall are needed to reduce ammonia losses and cause a significant yield response from grasses (Bussink & Oenema, 1996). Rainfall doesn't always stop ammonia losses, e.g., Chambers et al. (1997) noted an increase in NH_3 volatilization rate immediately following rainfall events several days after application of solid pig manure, perhaps due to re-wetting and subsequent re-drying of the solid manures. One management option to benefit from the rainfall effect is to irrigate soon after application. Work in Sweden (Malgeryd, 1998) reported a 70% reduction in ammonia losses from 1.2 inches (30 mm) of irrigation applied right after a surface broadcast application of pig slurry.

The above weather elements, of course, cannot be directly managed to control ammonia loss. Although some investigators have proposed applying manure before possible rainfall, or the use of irrigation on freshly manured fields. However, it is possible to include environmental conditions within a comprehensive ammonia management scheme. For example, ammonia emission values could be varied by categories based on average temperatures, drying conditions, or rainfall for the first day or two after manure application. Such an approach should improve ammonia loss estimates with attendant improvements in N availability estimates.

Estimation of Ammonia Volatilization

Ammonium-N is the fraction of manure most readily available to plants, but it is also the portion most easily lost via volatilization and most affected by field management and environmental conditions. Therefore, accurate estimates of ammonia loss are critical for improving the crop recovery of manure N and for reducing environmental losses of ammonia. Every US State and Canadian Province in the Northeast incorporates some type of estimate of ammonia volatilization into their manure N recommendation process (Table 1).

Table 1. Ammonia loss estimates for spring-applied manure in various Northeastern US States or Canadian Provinces (F.J. Coale, pers. comm., 2000; Penn. St. Coop. Ext., 1999; Klausner, 1995; Jokela et al., 1998; OMAFRA, 1999).

Location	Manure Type or Weather Condition	Injected or Immed. Incorp.	First Day Losses	Losses for non-incorporated
		Ammonia Loss, % of Applied NH_4-N		
Maryland	All Manures	0	20	100
Pennsylvania	Dairy	0	35	100
	Poultry [1]	0	20	80
New York	All Manures (spring)	35	47	100
	All Manures (sidedress in summer)	0	--	--
Vermont	Dairy <5% DM	5	30	40
	Dairy 5-10% DM	5	45	60
	Dairy 5-10% DM	10	60	80
	Dairy Solid	5	40	90
	Poultry [1]	10	20	80
Ontario [2]	All, Cool, Moist	0	10	40
	All, Cool, Dry	0	15	50
	All, Warm, Moist	0	25	75
	All, Warm, Dry	0	50	90

[1] Values from Univ. of Delaware recommendations.
[2] Nonincorporated is for bare soil condition.

The concept of separating manure total N into the ammonium N and organic N fractions, and adjusting the availability of ammonium-N for time of incorporation was implemented in New York about 20 years ago (Klausner & Bouldin, 1983). This approach was based on research done earlier in New York by Lauer et al, (1976) which utilized solid manures and estimated ammonia losses by difference. The original recommendations have undergone some revision over the years, but the current New York recommendations are not greatly different (Klausner, 1995, Table 2).

Table 2. Ammonia loss and N availability estimates for manure applications in New York (Klausner, 1995).

Time of application/incorporation		% of NH$_4$-N Lost	% of NH$_4$-N Available
During growing season as sidedress injection for row crops		0	100
Spring season	Immediate incorporation	35	65
	1 day	47	53
	2 days	59	41
	3 days +	Increase number by 12 for each day incorporation is delayed.	Reduce number by 12 for each day incorporation is delayed.
All other conditions		100	0+

Other Northeastern states (PA, VT, MD, etc.) adopted this approach along with further refinements. A common feature in most ammonia loss estimates is the predicted zero loss (or close to it) for manures immediately incorporated by tillage or by significant rainfall, commonly defined as > 0.5 inches (12 mm) of rain. However, ammonia loss estimates for all other situations vary greatly among States or Provinces because of differences in the assumed ammonia-loss vs. time relationship. In addition, some States employ manure type (animal species) as classification variables, while others use manure composition variables such as manure dry matter content, to predict ammonia losses (Table 1). One Province utilizes soil and weather conditions, e.g., temperature, moisture, and soil cover, to estimate ammonia emissions.

Manure recommendations by the University of Vermont initially utilized an approach similar to New York. However, a recent revision was undertaken to incorporate dry matter content and a different N-loss vs. time relationship (Jokela et al, 1998). These changes were based on recent Vermont research (Fig. 1; Jokela et al., 1996; Carter et al., 1998) and a number of European studies discussed above in the 'manure composition' section. The research results, and resulting modifications in ammonia loss estimates, incorporate the following points: i) the rate of ammonia loss from slurries is much greater the first few hours after application than recognized in the older recommendations, but the losses declines dramatically after a day or two, ii) ammonia loss is a function of slurry dry matter content (more accurately fluidity; Svensson, 1994), with losses being lower in dilute slurries because of greater soil infiltration, iii) under most circumstances there is significant utilization of some manure NH$_4$-N, especially from slurries, even when manures are left on the surface, i.e. there is not 100% loss of NH$_4$-N from nonincorporated manure. The precise estimates of loss and availability of NH$_4$-N are calculated from a series of equations similar to those used in the "MANNER" model (Chambers et al., 1999; see Fig. 5a,

5b, and Table 1).

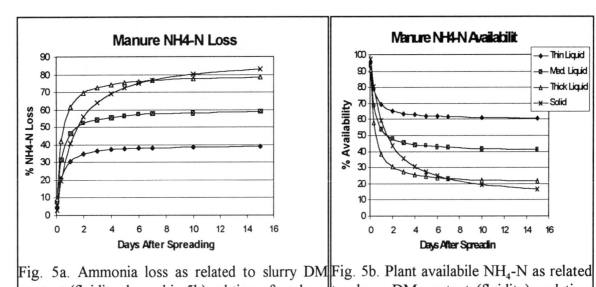

Fig. 5a. Ammonia loss as related to slurry DM content (fluidity, legend in 5b) and time after slurry application (Jokela et al., 1998).

Fig. 5b. Plant availabile NH₄-N as related to slurry DM content (fluidity) and time after spreading (Jokela et al., 1998).

Manure ammonia loss estimates in Ontario employ a weather and soil related approach. These estimates are based on interpretations of Canadian and European research which utilizes several of the elements discussed above in the 'soil conditions' and 'environmental factors' sections.

Table 3. Ammonia loss estimates from spring or summer manure applications in Ontario due to different weather and soil conditions. (OMAFRA, 1999).

Days from Application to Incorporation, Soil Condition	Cool Temps.		Warm Temps.	
	Wet Cond.	Dry Cond.	Wet Cond.	Dry Cond.
	Ammonia Loss, % of Applied NH₄-N			
Not Incorpor., Bare Soil	40	50	75	90
Not Incorpor., Standing Crop	20	25	40	50
Incorp. w/in 1 day, Bare Soil	10	15	25	50
Incorp. w/in 2 days, Bare Soil	13	19	31	57
Incorp. w/in 3 days, Bare Soil	15	22	38	65
Incorp. w/in 4 days, Bare Soil	17	26	44	73
Incorp. w/in 5 days, Bare Soil	20	30	50	80

The Ontario approach utilizes soil condition classes of: bare soil, crop residues, or the presence of a standing crop; plus the environmental factors of: season of year, temperature, and evaporative demand/soil moisture level. All of these factors form a multi-class ammonia estimation scheme which allows greater site-specificity (Table 3; OMAFRA, 1999). The Ontario system forecasts high losses when days are sunny and warm and soils are drying, and lower losses under cool, cloudy, rainy conditions when soils are moist. Estimated losses are highest for bare soil conditions and lower for a standing crop where the formation of an internal layer of calm air within a crop canopy can reduce gas exchange (Harper, et al., 1983; Freney, 1982).

Overview

Ammonia volatilization is a major N loss process for surface applied manures. Ammonia volatilization losses vary greatly depending on management practices and environmental conditions. The major factors affecting manure ammonia loss were categorized and discussed, namely: i) manure characteristics (dry matter content, pH, NH_4-N content), ii) application management (incorporation, zone application, timing), iii) soil conditions (soil moisture, soil properties, plant/residue cover), and iv) environmental factors (temperature, wind speed, rainfall).

The current ammonia loss recommendations in the Northeast region illustrate both the problems and opportunities that face researchers seeking to improve the management of manure N. The problems arise from the range of manures being applied in the region, the range of application equipment, and the range of soil and weather conditions commonly encountered. The most urgent items required to resolve these problems are reliable field data on ammonia losses under the soil, climate, and application regimes of the individual state. Fortunately there are a number of simplified field methods to measure ammonia volatilization, such as: the dynamic chamber methods (Svensson, 1994), wind-tunnel methods (Lockyer, 1984; Klarenbeek & Bruins, 1991; Thompson et al., 1990), and micrometeorological methods employing either multi-level or one-point passive samplers (Denmead, 1983; Ryden & McNeill, 1984; Wilson et al., 1983). Each of these methods can contribute valuable data on field ammonia loss that is needed to revise volatilization estimates. The collection of current data, the sharing of data into common databases, and the improved understanding of the factors affecting ammonia loss should all contribute to the realization of improved estimates for ammonia volatilization for the Northeast. These improved ammonia loss estimates then need to be combined with crop yield response to obtain an estimate of the manure 'fertilizer N equivalents', which also incorporates factors such as mineralization of organic N, leaching and denitrification losses, and manure N efficiency (timing, etc). The manure 'fertilizer N equivalent' can then be compared to the crop N requirement to determine the recommended rate of manure application and possible need for supplemental fertilizer N. Improved estimates and management techniques for recovering manure ammonium N will conserve a major plant nutrient, will improve the N:P ratio in manures, and will decrease the impacts of agricultural ammonia on low-N input ecosystems.

References

Amberger, A. 1990. "Ammonia emissions during and after land spreading of slurry". *Odour and Ammonia Emissions from Livestock Farming*. (Eds) V.C.Nielsen et al. pp. 126-131. Elsevier, London.

Asman, W.A.H., R.M. Harrison and C.J. Ottley. 1994. "Estimation of the net air-sea flux of ammonia over the southern bight of the North Sea". *Atmos. Environ.* 28:3647-3654.

Asman, W.A.H., M.A. Sutton and J.K. Schjørring. 1998. "Ammonia: emission, atmospheric transport and deposition". *New Phytol.* 139: 27-48.

Beauchamp, E.G., G.E. Kidd and G. Thurtell. 1982. "Ammonia volatilization from liquid dairy cattle manure in the field". *Can. J. Soil Sci.* 62:11-19.

Beauchamp, A.E. 1983. "Response of corn to nitrogen in preplant and sidedress applications of liquid cattle manure". *Can. J. Soil Sci.* 63:377-386.

Bittman, S., C.G. Kowalenko, D.E. Hunt and O. Schmidt. 1999. "Surface-banded and broadcast dairy manure effects on tall fescue yield and nitrogen uptake". *Agron. J.* 91:826-833.

Bless, H.-G., R. Beinhauer and B. Sattelmacher, 1991. "Ammonia emissions from slurry applied to wheat stubble and rape in North Germany". *J. Agric. Sci. (Camb.)* 117:225-231.

Brunke, R., P. Alvo, P. Schuepp and R. Gordon. 1988. "Effect of meteorological parameters on ammonia loss from manure in the field". *J. Environ. Qual.* 17:431-436.

Bussink, D.W., and O. Oenema. 1998. "Ammonia volatilization from dairy farming systems in temperate areas: a review". *Nut. Cyc. in Agroecosystems* 51:19-33.

Bussink, D.W. and O. Oenema. 1996. "Differences in rainfall and temperature define the use of different types of nitrogen fertilizer on managed grassland in UK, NL and Erie". *Neth. J. Agric. Sci.* 44:317-339.

Carter, J.E., W.E. Jokela, S.C. Bosworth, J.J. Rankin, and P. Pfluke. 1998. "Broadcast and band-applied manure effects on grass yield and N uptake". *Agronomy abstracts.* p. 317. Amer. Soc. Agron. Madison, WI.

Chambers, B.J., E.I. Lord, F.A. Nicholson and K. A. Smith. 1999. "Predicting nitrogen availability and losses following application of organic manures to arable land: MANNER". *Soil Use and Management* 15:137-143.

Chambers, B.J., K.A. Smith and T.J. van der Weerden. 1997. "Ammonia emissions following the land spreading of solid manures". *Gaseous nitrogen emissions from grasslands.* (Eds) S.C. Jarvis and B.F. Pain. pp.275-280. CAB Internat. Oxon, UK.

Côté, D., A. Michaud, T.S. Tran and C. Bernard. 1999. "Slurry sidedressing and topdressing can improve soil and water quality in the Lake Champlain Basin". *Water Science and Application* 1:225-238.

Court, M.N., R.C. Stephen and J.S. Waid. 1964. "Toxicity as a cause of the inefficiency of urea as a fertilizer". *J. Soil Sci.* 15:42-48.

Denmead, O.T., J.R. Freney and J.R. Simpson. 1982. "Dynamics of ammonia volatilization during furrow irrigation of maize". *Soil Sci. Soc. Am. J.* 46:149-155.

Denmead, O.T. 1983. "Micrometeorological methods for measuring gaseous losses of nitrogen in the field". *Gaseous loss of nitrogen from plant soil systems.* (Eds) J.R. Freney and J.R. Simpson. pp.133-157. Martinus Nijhoff, The Hague.

Dohler, H. 1990. "Laboratory and field experiments for estimating ammonia losses from pig and cattle slurry following application". *Odour and Ammonia Emissions from Livestock Farming.* (Eds) V.C.Nielsen et al. pp. 132-140. Elsevier, London.

Freney, J.R., J.R. Simpson and O.T. Denmead. 1983. "Volatilization of ammonia". *Gaseous loss of nitrogen from plant-soil systems.* (Eds) J.R. Freney and J.R. Simpson. pp.1-32. Martinus Nijhoff, The Hague.

Frost, J. 1994. "Effect of spreading method, application rate and dilution on ammonia volatilization from cattle slurry". *Grass and Forage Science* 49:391-400.

Hacker, R. R. and Du Z. 1993. "Livestock pollution and politics". *Nitrogen flow in pig production and environmental consequences.* (Eds) M.W.A. Verstegen, et al. pp. 3-21. Purdoc Sci. Pub. Wageningen, The Netherlands.

Harper, L.A., V.R. Catchpoole, R. Davis and K.L. Weir. 1983. "Ammonia volatilization: soil, plant, and microclimate effects on diurnal and seasonal fluctuations". *Agron. J.* 75:212-218.

Heck, A.F. 1931. "The availability of the nitrogen in farm manure under field conditions". *Soil Sci.* 31:467-481.

Hoff, J. D., D.W. Nelson and A.L. Sutton. 1981. "Ammonia volatilizatioin from liquid swine manure applied to cropland". *J. Environ. Qual.* 10: 90-94.

Huijsmans, J.F.M., J.M.G. Hol and D.W. Bussink. 1997. "Reduction of ammonia emission by new slurry application techniques on grassland". *Gaseous nitrogen emissions from grasslands.* (Eds) S.C. Jarvis and B.F. Pain. pp.281-285. CAB Internat. Oxon, UK.

Jarvis, S.C. and B.F. Pain. 1990. "Ammonia volatilisation from agricultural land". *The fertiliser society proceedings.* No. 298. pp.1-35. The Fertiliser Soc., London.

Jokela, W.E. and D. Côté. 1994. "Options for direction incorporation of liquid manure". *Liquid manure application systems.* p. 201-215. Northeast Reg. Agr. Engin. Serv. Cornell Univ., Ithaca, NY.

Jokela, W. E., S.C. Bosworth, P.D. Pfluke, J. J. Rankin and J.E. Carter. 1996. "Ammonia volatilization from broadcast and band-applied liquid dairy manure on grass hay". *Agronomy abstracts.* p. 315. Amer. Soc. Agronomy, Madison, WI.

Jokela, W. E., F. Magdoff, R. Bartlett, S. Bosworth and D. Ross. 1998. "Nutrient recommendations for field crops in Vermont". *Univ. Vermont Ext. Serv. Pub.* BR1390., Web: http://ctr.uvm.edu/pubs/nutrientrec/, Burlington, VT.

Kemppainen, E. 1989. "Nutrient content and fertilizer value of livestock manure with special reference to cow manure". *Annales Agriculturae Fenniae* 28:163-284.

Kirchmann, H., M. Esala, J. Morken, M. Ferm, W. Bussink, J. Gustavsson and C. Jakobsson. 1998. "Ammonia emissions from agriculture". *Nut. Cyc. in Agroecosystems* 51:1-3.

Klarenbeek J.V. and M.A. Bruins. 1991. "Ammonia emissions after land spreading of animal slurries". *Odour and Ammonia Emissions from Livestock Farming.* (Eds) V.C.Nielsen et al. pp. 107-115. Elsevier, London.

Klausner, S.D. 1995. "Nutrient management: crop production and water quality". *Cornell Univ. Whole Farm Planning Pub. Ser. no. 95CUWFP1.* Coop. Ext. Serv. Cornell Univ., Ithaca, NY.

Klausner, S. D. and R. W. Guest. 1981. "Influence of NH3 conservation from dairy manure on the yield of corn". *Agron. J.* 73: 720-723.

Klausner, S.D. and D.R. Bouldin. 1983. "Managing animal manure as a resource part I: basic principles". *Cornell Univ. Soil Fertility Series, p. 100.00, date 3-83.* Coop. Ext. Serv. Cornell Univ. Ithaca, NY.

Lauer, D.A., D.R. Bouldin and S.D. Klausner. 1976. "Ammonia volatilization from dairy manure spread on the soil surface". *J. Environ. Qual.* 5:134-141.

Lockyer, D.R. 1984. "A system for the measurement in the field of losses of ammonia through volatilization". *J. Sci. Food Agric.* 35:837-848.

Lorenz, F. and G. Steffens. 1997. "Effect of application techniques on ammonia losses and herbage yield following slurry application to grassland." *Gaseous nitrogen emissions from grasslands.* (Eds) S.C. Jarvis and B.F. Pain. pp.287-292. CAB Internat. Oxon, UK.

Malgeryd, J. 1998. "Technical measures to reduce ammonia losses after spreading of animal manure". *Nut. Cyc. in Agroecosystems* 51:51-57.

Marshall, S.B., C.W. Wood, L.C. Braun, M.L. Cabrers, M.D. Mullen and E.A. Guertal. 1998. "Ammonia volatilization from tall fescue pastures fertilized with broiler litter". *J. Environ. Qual.* 27:1125-1129.

Meisinger, J.J. and G.W. Randall. 1991. "Estimating nitrogen budgets for soil-crop systems". *Managing nitrogen for groundwater quality and farm profitability.* Proc. Sym. Am. Soc. Agron. Nov. 30, 1988 Anheim CA. pp. 85-124. Soil Sc. Soc. Am., Madison, WI.

Menzi, H., P. Katz, R. Frick, M. Fahrni and M. Keller. 1997. "Ammonia emissions following the application of solid manure to grassland". *Gaseous nitrogen emissions from grasslands.* (Eds) S.C. Jarvis and B.F. Pain. pp.265-274. CAB Internat. Oxon, UK.

Midgley, A. R. and V.L. Weiser. 1937. "Effect of superphosphates in conserving nitrogen in cow manure". *Vermont Agr. Expt. Stn. Bull. No.419.* Univ. of Vermont, Burlington, VT.

Misselbrook, T., J. Laws, and B. Pain. 1996. "Surface application and shallow injection of cattle slurry on grassland: nitrogen losses, herbage yields and nitrogen recoveries". *Grass and forage science.* 51:270-277.

Moal, J.F., J. Martinez, F. Guiziou and C.M. Coste. 1995. "Ammonia volatilization following surface-applied pig and cattle slurry in France". *J. Agri. Sci.(Camb.)* 125:245-252.

Moore, P. A., T. C. Daniel, D. R. Edwards and D. M. Miller. 1995. "Effect of chemical amendments on ammonia volatilization from poultry litter". *J. Environ. Qual.* 24:293-300.

Moss, D.P., B.J. Chambers and T.J. Van Der Weerden. 1995. "Measurement of ammonia emissions from land application of organic manures". *Aspects of Appl. Biology* 43:221-228.

Nathan, M.V. and G.L. Malzer. 1994. "Dynamics of ammonia volatilization from turkey manure and urea applied to soil". *Soil Sci. Soc. Am. J.* 58:985-990.

Ontario Ministry Agric. Food & Rural Affairs (OMAFRA). 1999. "Field crop recommendations 1999-2000". *Ont. Min. Agr. Food Rural Aff. Pub. 296.,* Ontario, Canada.

Pain B.F. and R.B. Thompson. 1988. "Ammonia volatilization from livestock slurries applied to land". *Nitrogen in organic wastes applied to soils.* (Eds) J.A. Hansen and K. Henricksen. pp. 202-211. Academic Press, London.

Pain, B.F., V.R. Phillips, C.R. Clarkson, and J.V. Klarenbeek. 1989. "Loss of nitrogen through ammonia volatilisation during and following the application of pig or cattle slurry to grassland". *J. of the Sci. of Food and Agric.* 47:1-12.

Pain, B.F. and T.H. Misselbrook. 1997. "Sources of variation in ammonia emission factors for manure applications to grassland". *Gaseous nitrogen emissions from grasslands.* (Eds) S.C. Jarvis and B.F. Pain. pp.293-301. CAB Internat. Oxon, UK.

Pennsylvania State Univ. Coop. Ext. Serv. 1999. "The agronomy Guide, 1999-2000". *College of Agr. Penn. St. Univ.,* University Pk., PA.

Phillips, V.R., B.F. Pain and J.V. Klarenbeek. 1990. "Factors influencing the odour and ammonia emissions during and after the land spreading of animal slurries". *Odour and Ammonia Emissions from Livestock Farming.* (Eds) V.C.Nielsen et al. pp. 98-106. Elsevier, London.

Phillips, V.R. and B.F. Pain, 1998. "Gaseous emissions from the different stages of European livestock farming". *Environmentally friendly management of farm animal waste.* (Ed.) T. Matsunaka pp. 67-72 Kikanshi Insatsu Co. Ltd., Sapporo, Japan.

Ryden, J.C. and J.E. McNeill. 1984. "Application of the micrometeorological mass balance method to the determination of ammonia loss from a grazed sward". *J. Sci. Food Agric.* 35:1297-1310.

Salter, R.M. and C.J. Schollenberger. 1939. "Farm manure". *Ohio Agr. Expt. Stn. Bull. No. 605.* Dept. of Agron. Ohio St. Univ., Columbus, OH.

Schilke-Gartley, K.L. and J.T. Sims. 1993. "Ammonia volatilization from poultry manure-amended soil". *Biol. Fertil. Soils* 16:5-10.

Schmitt, M.A., S.D. Evans and G.W. Randall. 1995. "Effect of liquid manure application methods on soil nitrogen and corn grain yields". *J. Prod. Agric.* 8:186-189.

Sharpe, R.R. and L.A. Harper. 1997. "Ammonia and nitrous oxide emissions from sprinkler irrigation applications of swine effluent". *J. Environ. Qual.* 26:1703-1706.

Sharpley, A.N., J.J. Meisinger, A. Breeuwsma, J.T. Sims, T.C. Daniel and J.S. Schepers. 1998. "Impacts of animal manure management on ground and surface water quality". *Animal waste utilization: effective use of manure as a soil resource.* (Ed) J.L. Hatfield. pp. 173-242. Ann Arbor Press, Chelsea, MI.

Smith, K. A. and B.J. Chambers. 1995. "Muck: from waste to resource utilisation: the impacts and implications." *Agricultural Engineer,* Autumn 1995: 33-38.

Sommer, S.G. and A.K. Ersbøll. 1994. "Soil tillage effects on ammonia volatilization from surface-appplied or injected animal slurry". *J. Environ. Qual.* 23:493-498.

Sommer, S.G. and N. Hutchings. 1995. "Techniques and strategies for the reduction of ammonia emission from agriculture". *Water, Air and Soil Pollution* 85: 237-248.

Sommer, S. G. and J. E. Olesen. 1991. "Effects of dry matter content and temperature on ammonia loss from surface-applied cattle slurry". *J. Environ. Qual.* 20:679-683.

Sommer, S.G. and B.T. Christensen. 1991. "Effect of dry matter content on ammonia loss from surface applied cattle slurry". *Odour and Ammonia Emissions from Livestock Farming.* (Eds) V.C.Nielsen et al. pp. 141-147. Elsevier, London.

Sommer, S.G. and R.R. Sherlock. 1996. "pH and buffer component dynamics in the surface layer of animal slurries". *J. Agri. Sci. (Camb.)* 127:109-116.

Sommer, S.G., J.E. Olesen and B.T. Christensen. 1991. "Effects of temperature, wind speed and air humidity on ammonia volatilization from surface applied cattle slurry". *J. Agri. Sci. (Camb.)* 117:91-100.

Steenhuis, T.S., G.D. Budenzer and J.C. Converse. 1979. "Ammonia volatilization of winter spread manure". *Trans. Am. Soc. Agr. Engin.* 22:152-157,161.

Stevens, R.J. and R.J. Laughlin. 1997. "The impact of cattle slurries and their management on ammonia and nitrous oxide emissions from grassland. *Gaseous nitrogen emissions from grasslands*. (Eds) S.C. Jarvis and B.F. Pain. pp.233-256. CAB Internat. Oxon, UK.

Stevens, R.J., R.J. Laughlin and J.P. Frost. 1992. "Effects of separation, dilution, washing and acidification on ammonia volatilization from surface-applied cattle slurry". *J. Agri. Sci. (Camb.)* 113:383-389.

Svensson, L. 1994. "A new dynamic chamber technique for measuring ammonia emissions from land-spread manure and fertilizers". *Act Agric. Scand. Sect. B Soil Plt. Sci.* 44:35-46.

Svensson, L. 1994. "Ammonia volatilization following application of livestock manure to arable land". *J. Agric. Engin. Res.* 58:241-260.

Thompson, R.B., B.F. Pain and Y.J. Rees. 1990. "Ammonia volatilization from cattle slurry following surface application to grassland. II. Influence of application rate, wind speed and applying slurry in narrow bands". *Plant & Soil* 125:119-128.

Thompson, R.B., B.F. Pain and D.R. Lockyer. 1990. "Ammonia volatilization from cattle slurry following surface application to grassland". *Plant and Soil.* 125:109-117.

Thompson, R.B., J.C. Ryden and D.R. Lockyer. 1987. "Fate of nitrgogen in cattle slurry following surface application or injection to grassland". *J. Soil Sci.* 38:689-700.

Wilson, J.D., V.R. Catchpoole, O.T. Denmead, and G.W. Thurtell. 1983. "Verification of a simple micrometeorological method for estimating the rate of gaseous mass transfer from the ground to the atmosphere". *Agric. Meteorology* 29:183-189.

Diagnostic Nitrogen Tests
for Manure-Amended Soils:
Current Status and Future Outlook

Gregory D. Binford
Assistant Professor of Soil and Water Quality
Department of Plant and Soil Science
University of Delaware, Newark, Delaware

David J. Hansen
Assistant Professor of Soil and Water Quality
Department of Plant and Soil Science
University of Delaware, Georgetown, Delaware

Biographies for most speakers are in alphabetical order after the last paper.

INTRODUCTION

One of the greatest costs associated with crop production is nutrient management. The application of too little nutrient can result in large yield reductions and correspondingly large losses in potential income. At the same time, however, the application of too much nutrient can result in wasted money spent on these nutrients. Another concern with the application of too much nutrient that has received tremendous attention for many years is the environmental cost associated with the application of too much fertilizer, in particular nitrogen (N) and phosphorus (P). These environmental concerns have escalated in recent years, especially in areas with intensive animal production. This concern is focused on the large volume of manure that is produced during the production of these animals and what happens to this manure once it leaves the livestock facility.

Because animal manures contain nutrients, they are typically applied to agricultural fields to supply nutrients for crops. Concerns occur, however, when the amounts of nutrients generated in the manure are much greater than the nutrient requirements of the crops to be grown on the farm. Over-application of nutrients from manure, increases potential risk of these nutrients moving from the land into nearby water supplies. Excessive nutrient levels in water supplies can cause water quality problems. As a result, there is great need for accurate diagnostic tools for determining optimal rates of nutrients to apply during the production of crops when animal manures have been applied.

Soil testing has been considered the standard practice for determining how much P and potassium (K) should be applied to a crop. However, determining optimal rates of N needed to optimize crop production has been a much more difficult task. This difficulty occurs because N is more dynamic in the soil environment than P and K. As a result, plant available N can change drastically due to weather or management factors. Many studies over the last several decades have focused on developing diagnostic tools for managing N during crop production. Because of the large number of acres of corn that are grown each year, much of this research has focused on corn, and therefore, much of this paper will focus on N management in corn. The overall objective of this paper is to summarize the research results of diagnostic tools that have been proposed for improving N management in corn. All of these tools will work in both manured and nonmanured conditions; however, some tools may have greater potential value in manured situations, which will be addressed in this paper.

Presidedress Soil Nitrate Test (PSNT):

Soil testing to predict the N needs of corn has been a practice commonly used for decades in the western part of the Corn Belt (i.e., Nebraska and Kansas) and in the High Plains where rainfall amounts during the winter are not usually sufficient to cause significant losses of nitrate from soils. In the humid corn growing areas of the United States, however, soil testing has not been utilized to predict the N needs of corn until the development of the presidedress soil nitrate test (PSNT). This test was originally proposed by Magdoff et al. (1984). Since this initial study, there have been numerous research projects that have evaluated the merits of using the PSNT as a nitrogen management tool in corn.

Figure 1. Example of when PSNT samples are taken.

The basic premise behind the PSNT is to sample the surface 12-inch layer of soil when corn plants are between 6 and 12 inches tall (measured from the soil surface to the center of the whorl, see Figure 1). Sampling at this time will account for the potentially large increases in amounts of available N that can occur due to mineralization from organic sources of N in the soil (e.g., animal manures, plant residues, soil organic matter) or the potentially large decreases that can occur due to leaching and/or denitrification. Most research with the PSNT has shown that the greatest value from this test is in identifying those fields where additional N is not needed to produce optimal yields; however, the time of sampling is early enough that additional N can be added as a sidedress before the corn is too tall. The typical relationship between relative corn grain yields and soil nitrate concentrations at the time of sampling for the

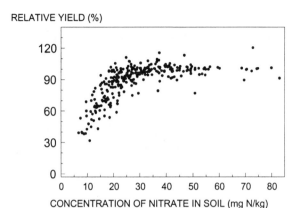

RELATIVE YIELD (%)

CONCENTRATION OF NITRATE IN SOIL (mg N/kg)

Figure 2. Typical PSNT relationship showing relative yields vs. soil nitrate-N concentration (Binford et al., 1992a).

PSNT is shown in Figure 2.

The consistency of the proposed critical concentration for the PSNT across numerous research studies and many states suggests that this test is a reliable indicator of when additional N is not needed for producing optimal corn yields. The optimal concentration of soil nitrate found in Iowa ranged from 21 to 26 ppm N (Blackmer et al., 1989; Binford et al., 1992a), while work in Pennsylvania showed an optimal concentration of 21 to 25 ppm nitrate-N (Fox et al., 1989). Research in both Maryland (Meisinger et al., 1992) and New Jersey (Heckman et al., 1996) found that 22 ppm nitrate-N was the critical concentration for the PSNT. A large study in Delaware suggested that the critical concentration for the PSNT on soils of the Atlantic Coastal Plain should be between 20 to 25 ppm nitrate-N (Sims et al., 1995). A recent report by Evanylo and Alley (1997) suggested a critical concentration of 18 ppm nitrate-N for the PSNT in Virginia, which is slightly less than the critical concentration of other studies.

The depth of sampling for residual soil nitrate testing in the West is usually at least 24 inches and often as deep as 48 inches. Samples are taken to these depths because of the mobility of nitrate in the soil profile. As a result, it would seem likely that a sampling depth of greater than 12 inches could improve the ability of the PSNT to predict the N status of corn. Studies in both Iowa (Binford et al., 1992a) and Delaware (Sims et al., 1995) evaluated the merits of sampling deeper than 12 inches. Both of these studies showed slight increases in the predictability of the PSNT, but the basic conclusions were that the slight increases in predictability were not worth the added efforts and costs associated with the deeper sampling depths.

Use of the PSNT in the Eastern United States is usually recommended only on soils that have received animal manures or similar organic amendments. It should be noted, however, that correlations of the PSNT that were developed in Iowa (Blackmer et al., 1989; Binford et al., 1992a) were from studies where broadcast applications of inorganic fertilizers were made prior to PSNT sampling. This would suggest that the PSNT can be utilized on fields that have received either organic or inorganic additions of N prior to sampling. Random sampling, however, should be avoided if banded applications of N have occurred during the spring prior to taking PSNT samples. It is important to note, however, that the overall value of taking PSNT samples is limited on soils low in organic matter and that have no history of animal manure applications; this is because soil nitrate concentrations at this time are often less than the PSNT critical concentration.

It is obvious that there have been a tremendous number of research projects that have demonstrated the value of the PSNT as a diagnostic tool for improving N management during corn production. Nonetheless, a recent survey showed that the use of the PSNT on corn acres in the United States is quite small (Fox et al., 1999). There are a number of possible reasons for the limited adoption of the PSNT. These reasons include: the time of sampling is an extremely busy time of the year on farms, the perceived value of the test does not seem to justify the time and costs associated with sampling, lack of knowledge and understanding of how the test works, fear of experiencing a N deficiency by using less N than normal, a perception that the test is not practical due to timing and the need to apply sidedress N before the corn gets too tall, and overall resistance to change normal habits of operation.

The value of the PSNT seems to represent a tremendous opportunity on manured soils (or soils that are relatively high in organic matter) for improving the overall profitability of corn production. In fact, an economic analysis by Musser et al. (1995) showed that the PSNT increased average profits in Pennsylvania corn production from $3.78 to $13.65 per acre when compared to standard N fertilization practices. Considering the factors that can influence the potential amounts of plant available N that will be present for the growing corn crop, it seems that the PSNT should be used on many more acres. This is especially true for soils that have received applications of animal manures. The potential sources of variation in amounts of plant available N on manured soils include: the inability to apply an accurate rate of manure, the inability to predict spring weather, the inability to predict the mineralization rate of N from the manure, the inability to get a representative sample of manure to determine the N content, and the inability to quantify amounts of N volatilized from manures after soil application. By utilizing the PSNT, the ability to predict or determine these potential variability factors becomes much less important because the PSNT can quantify all these sources of variation.

Research on other crops has found that the PSNT is a tool that can likely provide value for any crop that requires the application of N. Multiple location studies in New Jersey, Delaware, New York, and Connecticut showed value of the PSNT as a N management tool in fall cabbage (Heckman et al., 1999). In another study by Heckman et al. (1995), the PSNT was shown to be a reliable indicator of the N status in sweet corn. Binford et al. (1996) found that the PSNT test could be a useful N management tool in sugar beets. It is interesting to note that the critical concentration for all of these crops tends to be somewhere near 20 to 25 ppm nitrate-N. Recommended rates of sidedress N for PSNT concentrations below the critical concentration tend to vary slightly from state to state, so it is recommended to consult your local extension office for specific sidedress N recommendations with the PSNT.

End-of-season Stalk Nitrate Test

It has been known for decades that corn will accumulate nitrogen when excessive levels of N are available in the soil (Hoffer, 1926; Hanway et al., 1963). The end-of-season stalk nitrate test is based on this concept of nitrate accumulating in the lower portion of cornstalks and was first proposed by Binford et al. (1990). Figure 3 shows the typical relationship that is observed between relative grain yields and concentration of nitrate in cornstalks at maturity. The sharp break in this relationship clearly shows those stalk nitrate concentrations that were not adequate and those that were adequate for obtaining optimal yields. This test is useful for identifying fields that have much greater amounts of available N during the growing season than what is needed to attain optimal corn growth.

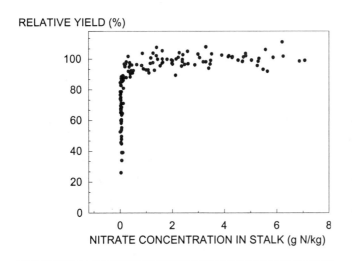

Figure 3. Relationship of relative corn grain yields vs. stalk nitrate concentrations at maturity (Binford et al., 1990).

Figure 4. Example of where the corn plant is cut for the end-of-season stalk nitrate test.

Because samples are taken at the end of the season, this test should not be used to guide fertilization practices after only one season of testing. The value of this test should come from using this test after a period of years. This test should be used to fine-tune N management practices. For example, if this test shows excessive levels of N for several years, it is likely that N management practices could be adjusted to reduce amounts of available N during future seasons. When this test is used in conjunction with other diagnostic tools, it becomes possible to have a much greater understanding of how current management practices are impacting the overall N status of corn during production and the potential impact of these practices on the environment.

Additional work by Binford et al. (1992b) identified a critical concentration range for the end-of-season stalk nitrate test of 700 to 2000 ppm nitrate-N. Sims et al. (1995) found that this critical concentration was essentially the same for corn production in Delaware. A study in Nebraska indicated that end-of-season stalk nitrate concentration was a more sensitive indicator of N treatment effects on plant N status than corn grain yield (Murphy and Ferguson, 1997). Another Nebraska study showed that stalk nitrate concentrations were useful for improving future N fertilization recommendations (Varvel et al., 1997a).

The procedure for taking stalk samples is explained by Hansen et al. (1999) and Blackmer and Mallarino (1997). Basically, corn stalks should be sampled at least one week after black layers (i.e., physiological maturity) have formed on about 80% of the kernels of most ears. Sampling can be performed up to grain harvest. Collect corn stalk samples by cutting the 8-inch segment of stalk found between 6 and 14 inches above the soil (Figure 4). Stalk nitrate concentrations less than 250 ppm N are considered low, 250 to 700 ppm N are considered marginal, 700 to 2000 ppm N are considered optimal, and greater than 2000 ppm N are considered excessive. Again, it is important to remember that this is an end-of-season assessment and drastic changes in nutrient management practices should not occur after only one year of testing.

Chlorophyll Meters

It is well known that plants that are deficient in N will have a yellowish appearance when compared to plants with adequate N. Because N is a major constituent of chlorophyll, this yellowing is a result of reduced chlorophyll levels in the plant. In fact, studies have demonstrated positive correlations between chlorophyll concentration and corn leaf N concentration (Wolfe et al., 1988; Wood et al., 1992). Yadava (1986) showed that a hand-held meter developed by Minolta Corporation could be used to provide a non-destructive measurement of the relative chlorophyll level of intact leaves. Piekielek and Fox (1992) and Schepers et al. (1992) first proposed the use of chlorophyll meters as a tool for monitoring the N status of corn. Since these initial reports, numerous studies have evaluated the relative merits of using the SPAD-502 chlorophyll meter as a diagnostic tool for monitoring the N status of corn (Blackmer and Schepers, 1994; Sims et al., 1995; Jemison and Lytle, 1996; Waskom et al., 1996).

The SPAD-502 chlorophyll meter works by clamping the meter onto a plant leaf and providing an immediate value of the relative chlorophyll content of the leaf; this value given by the meter is unitless (Figure 5). The meter measures relative chlorophyll concentration by measuring light transmittance through the leaf at 650 and 940 nm. The transmittance at 940 nm is used as a reference to compensate for factors such as leaf moisture content and thickness, while the 650-nm source is sensitive to chlorophyll concentration (Blackmer and Schepers, 1995). Even with this compensation, Piekielek and Fox (1992) and Blackmer and Schepers (1995) found that the use of an actual value from the chlorophyll meter was of questionable value when used across multiple locations because factors other than N status (e.g., hybrids, rotation, etc.) can impact meter readings. In addition, Blackmer et al. (1993) found that corn plant spacing can impact chlorophyll meter readings.

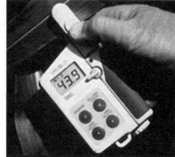

Figure 5. Example of chlorophyll meter on leaf.

Piekielek and Fox (1992) found that the chlorophyll meter readings at V6 (six fully emerged leaves) were capable of identifying sites that would respond to sidedress N applications. However, they suggested that this tool was probably not useful for making sidedress N recommendations. A Delaware study

(Sims et al, 1995) found that chlorophyll meter readings at V6 were about as good as the PSNT for predicting if additional N was needed at sidedressing; however, when data from multiple sites were pooled, the predictive ability of the chlorophyll meter was not as good as individual sites. They indicated that a much larger database was needed to determine the impact of soil, plant, and environmental factors on chlorophyll meter readings. Blackmer and Schepers (1995) found that grain yields did tend to increase as V6 chlorophyll meter readings increased within a location, but when data were pooled, site-to-site variation was much greater than variation observed within sites. They concluded that early in the season (i.e., V6 stage) chlorophyll meter readings have limited potential as a diagnostic tool for determining the N status of corn, unless the readings can be adjusted for environmental factors and hybrid differences. When considering the use of the chlorophyll meter as an early-season diagnostic tool for managing N in corn, it is important to note that when the chlorophyll meter showed severe N stress at the V8 stage (i.e., eight fully emerged leaves), Varvel et al. (1997b) were unable to obtain maximum yields from adding additional fertilizer N to corn.

The use of chlorophyll meter readings later in the season seems to have greater ability to predict N status than early season measurements. Blackmer and Schepers (1995) found better relationships between grain yields and chlorophyll meter readings taken at either R4 (dough stage) or R5 (dent stage) than when readings were taken early in the season at V6. Nonetheless, because of site-to-site variation they suggested that these readings at R4 or R5 should be adjusted to account for this site-to-site variation. Schepers et al. (1992) adjusted chlorophyll meter readings for this site-to-site variation by calculating an N sufficiency index. The N sufficiency index is simply the chlorophyll meter reading for a given field or treatment divided by the chlorophyll meter reading of an area within the field that is known to have adequate N. In an N response study, this adequately fertilized area would be the treatment with the highest N rate. Figure 6 shows an example of how strips that are known to have adequate levels

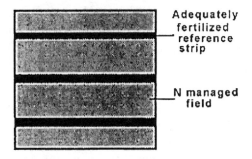

Figure 6. Diagram of using N reference strips with adequate N (Peterson et al., 1993).

of N can be established in a field. This type of normalization of the chlorophyll meter readings should reduce the impact of factors that influence meter readings other than N status. Research in Pennsylvania indicated that early-season (i.e., V6) chlorophyll meter readings should be referenced to a high-N reference strip (i.e., N sufficiency index); however, the need for a high-N reference strip was not as necessary for meter readings taken at the early dent stage (Piekielek and Fox, 1992; Piekielek et al., 1995).

The greatest value of using the chlorophyll meter in corn appears to be for evaluating the need for applying additional N during the later stages of vegetative growth through irrigation water (i.e., fertigation). Blackmer and Schepers (1995) used the chlorophyll meter to monitor the N status of corn in Nebraska fields. When the N sufficiency index (as defined above) dropped below 0.95 for two consecutive weeks, N fertilizer was applied in the irrigation water at a rate of 30 lb N/acre. They concluded that this technique of using the chlorophyll meter to schedule fertigation could be a valuable tool for improving the profitability of N fertilization practices and reducing potential environmental concerns from excess application of N. Varvel et al. (1997b) evaluated the potential of using this N sufficiency index approach in a five-year study. Their results indicated that a 95% sufficiency was adequate for maintaining optimal yields of corn. They did note, however, that when corn at the V8 stage dropped below the 90% sufficiency level, grain yield potential had been irreversibly reduced and multiple applications of N could not increase grain yields to the level of corn that had been adequately fertilized throughout the season. A similar study by Shapiro (1999) found the chlorophyll meter to be an adequate tool for managing N in situations where high levels of nitrate were present in waters used for irrigation. Specific methodology on how to use the chlorophyll meter in corn is explained by Peterson et al. (1993).

Leaf N Test

One of the first diagnostic tools developed for evaluating the N status of corn was based on the N concentration of the leaf opposite and below the primary ear when plants are silking (Tyner, 1946; Tyner and Webb, 1946). Since the initial study, numerous research projects have evaluated the value of using this tissue test as a diagnostic N test for corn (Bennett et al., 1953; El-Hout and Blackmer, 1990; Melsted et al., 1969). Work by Cerrato and Blackmer (1991) demonstrated that the sensitivity of leaf N analysis when evaluated across numerous environments was poor and that the use of a single critical concentration should be discouraged.

Although leaf N analysis is used extensively in corn research studies that contain various levels of N sufficiency, this test is not commonly used in production agriculture as a diagnostic tool. To achieve valuable information in a production agriculture field from leaf N analysis, it is necessary to use an approach similar to the N sufficiency index approach that is suggested when using a chlorophyll meter. In other words, compare the leaf N concentration from samples taken in the area of concern with the leaf N concentration from samples in an area that appears healthy. This approach can be used to diagnose most suspected nutrient deficiencies.

Visual Evaluation of N Status

Nitrogen deficiency symptoms in corn are associated with yellowing of the bottom leaves of the plant. Agronomists often refer to this yellowing of lower leaves as firing, and they sometimes use the degree of firing as an indicator of N status. As the degree of N deficiency increases, the number of lower leaves that turn yellow will increase. In fact, Viets et al. (1954) reported a strong correlation between corn grain yields and the number of N-deficient leaves on the plant.

Binford and Blackmer (1993) conducted an extensive evaluation of using visual evaluations as a tool to quantify N status in corn. They measured the number of green leaves below the ear. The reason for counting green leaves instead of dead leaves is because if a dead leaf has already fallen from the plant, it would be impossible to detect when counting. Also, they only counted leaves below the ear, because N deficiency symptoms (i.e., leaf firing) are rarely, if ever, visually apparent above the ear. They found that an absolute optimum number of green leaves below the ear could not be determined because of variation across locations in total number of leaves below the ear. Adjusted leaf ratings, however, could be used as an indicator to determine degree of N stress. Adjusted leaf ratings were calculated by subtracting the actual number of green leaves in an N-deficient area from the number of green leaves in an area with adequate N. Use of this adjusted leaf rating is also necessary because, regardless of N status, the actual number of green leaves decreased with time. Their findings suggested that during the later stages of reproductive growth (i.e., early R3 through early R5) grain yields were reduced by about 11% for each one-leaf increase in adjusted leaf ratings. In other words, for each leaf that died due to N deficiency, there was a corresponding 11% decrease in grain yields.

Remote Sensing

Remote sensing involves measurement of the amount of light (or energy) reflected from an object. In the case of nutrient management, remote sensing is a measurement of light reflectance from a crop canopy or soil surface. Remote sensing has been used in Nebraska and Kansas as indirect measurements of soil organic matter (Schepers, personal communication). Walberg et al. (1982) showed that remote sensing could be used to measure differences in the N status by measuring light reflectance of corn canopies. The advantage of remote sensing is that it can be used to measure the health of a collection of plants or an entire field, whereas chlorophyll meter measurements are from single plants with the assumption that they represent the entire field. The basic concept of using reflectance as a measure of N status is that a plant that is deficient in N will have reduced levels of chlorophyll. As a result, with reduced levels of chlorophyll, plants will not be able to absorb as much of the incoming light and more reflectance will occur.

It must be remembered, however, that increased light reflectance can be a result of many factors other than N stress. Examples of factors that can cause changes in light reflectance include: nutrient deficiencies other than N, amount of crop biomass, soil cover, plant height, canopy architecture, and about any factor that causes stress on a crop. With this in mind, Blackmer et al. (1996a) proposed the use of in-field reference strips that are known to contain non-limiting levels of N. Basically, this is the same idea as using the N sufficiency index with the chlorophyll meter (see Figure 6). In their study, Blackmer at al. (1996a) measured light reflectance from corn canopies that had received various levels of fertilizer N. They found highly significant differences among N treatments in amounts of light reflected at certain wavelengths. The amount of light reflected did vary with hybrid, but when data were normalized relative to non-limiting N strips, good correlations were found between grain yields and relative reflectance. They suggested by using relatively inexpensive sensors, light reflectance at specific wavelengths could be measured and this information could be used to monitor the N status of corn.

Aerial photography is another method of using remote sensing to monitor N status of crops. Many studies have evaluated the potential of using aerial photographs to measure various crop stresses; such as diseases, insect damage, water stress, and nutrient deficiencies (Colwell, 1974; Wildman, 1982). In fact, some crop consultants traditionally use aerial photography to make better recommendations to their clients. Blackmer et al. (1996b) evaluated numerous types of aerial images and basically concluded that this technique offers a relatively inexpensive method of monitoring N status in corn. It is important to note that this technique would require non-limiting strips of N in a field, if it were used as a diagnostic tool for N management.

SUMMARY

In the last several months, some states have developed new laws that regulate the use of nutrients on farms. These laws have heightened the awareness of the need for proper management of nutrients. With this increased awareness, it is expected that diagnostic tools for improving nutrient management will be used on a more regular basis than in the past. Areas with intensive animal agriculture seem to be of greatest concern due to the abundant supply of organic nutrients (i.e., manure). As noted in this paper, the greatest value from using the PSNT is on soils where organic N has been applied. Therefore, we expect increased awareness and use of the PSNT in these areas. Also, because of the unknown N availability of animal manures, it is likely that use of the end-of-season stalk test could become a more popular tool on soils where manures have been applied; that is, at least until application rates have been fine-tuned to provide optimal nutrient rates.

Because the chlorophyll meter and remote sensing both seem to work best on corn that is in the later stages of vegetative growth, the greatest value from these tools will likely occur on fields with irrigation capabilities. This is because the injection of N into irrigation water offers a convenient low-cost method for applying additional N to corn that is too tall for most types of fertilizer application equipment. Currently, the value of these tools does not justify the expense and labor of applying N to corn that is taller than most application equipment can operate without damaging the corn, unless this N is injected into irrigation water.

It is also expected that the use of yield monitors will show the true value of diagnostic tools in production agriculture situations. All of these tools have been developed in controlled small-plot environments. The value of these tools has been hard for growers to measure with traditional combines and methods of harvest, unless the value was dramatic. The yield monitor now offers the grower a method to measure relatively small differences that impact grain yields and overall profitability. Therefore, if a diagnostic tool can demonstrate an increase in profits through the use of yield monitors, the adoption rate of that tool will be more rapid than in the past.

REFERENCES

Bennett, W.F., G. Stanford, and L. Dumenil. 1953. Nitrogen, phosphorus, and potassium content of the corn leaf and grain as related to nitrogen fertilization and yield. Soil. Sci. Soc. Am. Proc. 17:252-258.

Binford, G.D. and A.M. Blackmer. 1993. Visually rating the nitrogen status of corn. J. Prod. Agric. 6:41-46.

Binford, G.D., A.M. Blackmer, and N.M. El-Hout. 1990. Tissue test for excess nitrogen during corn production. Agron. J. 82:124-129.

Binford, G.D., A.M. Blackmer, and M.E. Cerrato. 1992a. Relationships between corn yields and soil nitrate in late spring. Agron. J. 84:53-59.

Binford, G.D., A.M. Blackmer, and B.G. Meese. 1992b. Optimal concentrations of nitrate in cornstalks at maturity. Agron. J. 84:881-887.

Binford, G.D., D.D. Baltensperger, and A.D. Blaylock. 1996. In-season soil testing for nitrogen management in sugar beets. p. 313. *In* Agronomy Abstracts. ASA, Madison, WI.

Blackmer, A.M. and A.P. Mallarino. 1997. Cornstalk testing to evaluate nitrogen management. Iowa State Univ. Coop. Ext., PM-1584. Ames, IA.

Blackmer, A.M., D. Pottker, M.E. Cerrato, and J. Webb. 1989. Correlations between soil nitrate concentrations in late spring and corn yields in Iowa. J. Prod. Agric. 2:103-109.

Blackmer, T.M. and J.S. Schepers. 1994. Techniques for monitoring crop nitrogen status in corn. Commun. Soil Sci. Plant Anal. 25:1791-1800.

Blackmer, T.M. and Schepers. 1995. Use of a chlorophyll meter to monitor N status and schedule fertigation of corn. J. Prod. Agric. 8:56-60.

Blackmer, T.M., J.S. Schepers, and M.F. Vigil. 1993. Chlorophyll meter readings in corn as affected by plant spacing. Commun. Soil Sci. Plant Anal. 24:2507-2516.

Blackmer, T.M., J.S. Schepers, G.E. Varvel, and E.A. Walter-Shea. 1996a. Nitrogen deficiency detection using reflected shortwave radiation from irrigated corn canopies. Agron. J. 88:1-5.

Blackmer, T.M., J.S. Schepers, G.E. Varvel, and G.E. Meyer. 1996b. Analysis of aerial photography for nitrogen stress within corn fields. Agron. J. 88:729-733.

Cerrato, M.E. and A.M. Blackmer. 1991. Relationships between leaf nitrogen concentrations and the nitrogen status of corn. J. Prod. Agric. 4:525-531.

Colwell, J.E. 1974. Vegetation canopy reflectance. Remote Sens. Environ. 3:175-183.

El-Hout, N.M. and A.M. Blackmer. 1990. Changes in nitrogen concentrations of corn leaves near silking time. Commun. Soil Sci. Plant Anal. 21:169-178.

Evanylo, G.K. and M.M. Alley. 1997. Presidedress soil nitrogen test for corn in Virginia. Commun. Soil Sci. Plant Anal. 28:1285-1301.

Fox, R.H., G.W. Roth, K.V. Iversen, and W.P. Piekielek. 1989. Soil and tissue nitrate tests compared for predicting soil nitrogen availability to corn. Agron. J. 81:971-974.

Fox, R.H., W.P. Piekielek, and L.G. Bundy. 1999. Status of diagnostic tests for nitrogen availability to corn. p. 242. *In* Agronomy Abstracts. ASA, Madison, WI.

Hansen, D.J., G.D. Binford, and J.T. Sims. 1999. End-of-Season Corn Stalk Nitrate Testing to Optimize Nitrogen Management. Univ. of Delaware, College of Ag. and Nat. Res., NM-03, Newark, DE.

Hanway, J.J., J.B. Herrick, T.L. Willrich, P.C. Bennett, and J.T. McCall. 1963. The Nitrate Problem. Iowa State Univ. Ext. Serv. Spec. Report 34.

Heckman, J.R., W.T. Hlubik, D.J. Prostak, J.W. Patterson. 1995. Pre-sidedress soil nitrate test for sweet corn. HortScience 30:1033-1036.

Heckman, J.R., R. Govindasamy, D.J. Prostak, E.A. Chamberlain, W.T. Hlubik, R.C. Mickel, E.P. Prostko. 1996. Corn response to sidedress nitrogen in relation to soil nitrate concentrations. Commun. Soil Sci. Plant Anal. 27:575-583.

Heckman, J.R., U. Krogmann, P.J. Nitzsche, T.F. Morris, R.A. Ashley, J.T. Sims, and S.B. Sieczka. 1999. The presidedress soil nitrate test for fall cabbage. p. 242. *In* Agronomy Abstracts. ASA, Madison, WI.

Hoffer, G.N. 1926. Testing corn stalks chemically to aid in determining their plant food needs. Purdue Univ. Agric. Exp. Stn. Bull. 298.

Jemison, J.M. and D.E. Lytle. 1996. Field evaluation of two nitrogen testing methods in Maine. J. Prod. Agric. 9:108-113.

Magdoff, F.R., D. Ross, and J. Amadon. 1984. A soil test for nitrogen availability to corn. Soil Sci. Soc. Am. J. 48:1301-1304.

Meisinger, J.J., V.A. Bandel, J.S. Angle, B.E. O'Keefe, and C.M. Reynolds. 1992. Pre-sidedress soil nitrate test evaluation in Maryland. Soil Sci. Soc. Am. J. 56:1527-1532.

Melsted, S.W., H.L. Motto, and T.R. Peck. 1969. Critical plant nutrient composition values useful in interpreting plant analysis data. Agron. J. 61:17-20.

Murphy, T.L. and R.B. Ferguson. 1997. Ridge-till corn and urea hydrolysis response to NBPT. J. Prod. Agric. 10:271-282.

Musser, W.N., J.S. Shortle, K. Kreahling, B. Roach, W.C. Huang, D.B. Beegle, and R.H. Fox. 1995. An economic analysis of the pre-sidedress nitrogen test for Pennsylvania corn production. Review Agric. Econ. 17:25-35.

Peterson, T.A., T.M. Blackmer, D.D. Francis, and J.S. Schepers. 1993. Using a chlorophyll meter to improve N management. Univ. of Nebraska, NebGuide G93-1171-A, Lincoln, NE.

Piekielek, W.P. and R.H. Fox. 1992. Use of a chlorophyll meter to predict sidedress nitrogen requirements for maize. Agron. J. 84:59-65.

Piekielek, W.P., R.H. Fox, J.D. Toth, and K.E. Macneal. 1995. Use of chlorophyll meter at the early dent stage of corn to evaluate nitrogen sufficiency. Agron. J. 87:403-408.

Schepers, J.S., D.D. Francis, M.Vigil, and F.E. Below. 1992. Comparison of corn leaf nitrogen concentration and chlorophyll meter readings. Commun. Soil Sci. Plant Anal. 23:2173-2187.

Shapiro, C.A. 1999. Using a chlorophyll meter to manage nitrogen applications to corn with high nitrate irrigation water. Commun. Soil Sci. Plant Anal. 30:1037-1049.

Sims, J.T., B.L. Vasilas, K.L. Gartley, B. Milliken, and V. Green. 1995. Evaluation of soil and plant nitrogen tests for maize on manured soils of the Atlantic Coastal Plain. Agron. J. 87:213-222.

Tyner, E.H. 1946. The relation of corn yields to leaf nitrogen, phosphorus, and potassium content. Soil Sci. Soc. Am. J. 11:317-333.

Tyner, E.H. and J.R. Webb. 1946. The relation of corn yields to nutrient balance as revealed by leaf analysis. Agron. J. 38:173-185.

Varvel, G.E., J.S. Schepers, and D.D. Francis. 1997a. Chlorophyll meter and stalk nitrate techniques as complementary indices for residual nitrogen. J. Prod. Ag. 10:147-151.

Varvel, G.E., J. S. Schepers, and D.D. Francis. 1997b. Ability for in-season correction of nitrogen deficiency in corn using chlorophyll meters. Soil Sci. Soc. Am. J. 61:1233-1239.

Viets, F.G., C.E. Nelson, and C.L. Crawford. 1954. The relationships among corn yields, leaf composition and fertilizers applied. Soil Sci. Soc. Am. Proc. 18-297-301.

Walberg, G., M.E. Bauer, C.S.T. Daughtry, and T.L. Housley. 1982. Effects of nitrogen nutrition on the growth, yield, and reflectance characteristics of corn canopies. Agron. J. 74:677-683.

Waskom, R.M., D.G. Westfall, D.E. Spellman, and P.N. Soltanpour. 1996. Monitoring nitrogen status of corn with a portable chlorophyll meter. Commun. Soil Sci. Plant Anal. 27:545-560.

Wildman, W.E. 1982. Detection and management of soil, irrigation, and drainage problems. p. 387-401. *In* C.J. Johannsen and J.L. Sanders (ed.) Remote sensing for resource management. Soil Conserv. Soc. Am., Ankeny, IA.

Wolfe, D.W., D.W. Henderson, T.C. Hsiao, and A. Alvino. 1988. Interactive water and nitrogen effects on senescence of maize: II. Photosynthetic decline and longevity of individual leaves. Agron. J. 80:865-870.

Wood, C.W., D.W. Reeves, R.R. Duffield, and K.L. Edmisten. 1992. Field chlorophyll measurements for evaluation of corn nitrogen status. J. Plant Nutr. 15:487-500.

Yadava, U.L. 1986. A rapid and nondestructive method to determine chlorophyll in intact leaves. HortScience 21:1449-1450.

Manure Sampling and Testing

John B. Peters
Director
Soil and Forage Analysis Laboratory
Soil Science Department
University of Wisconsin–Madison

Biographies for most speakers are in alphabetical order after the last paper.

Background

In September 1996, a joint NCR-13 and SERA-6 soil testing work group meeting was held in Raleigh, North Carolina. Earlier in that year, a manure sample exchange was conducted with NCR-13, SERA-6 and NEC-67 member laboratories. Results from that sample exchange were presented at the Raleigh meetings and sparked interest in joining efforts to develop a manure-testing manual, which could be used in all regions. A multi-regional committee was formed to share expertise in the development of a common manual. Members of the multi-regional committee working on the publication include John Peters and Sherry Combs (NCR-13, WI), Ann Wolf and Doug Beegle (NCR-13 and NEC-67, PA), Maurice Watson (NCR-13, OH), Jan Jarman (MN Dept. of Agriculture), Nancy Wolf (SERA-6, AR), John Kovar (SERA-6, LA), and Bruce Hoskins (NEC-67, ME). The author wishes to acknowledge the significant editorial contributions of members of this committee in the sampling section of this paper and to Sherry Combs for her assistance in the laboratory methods portion of this paper.

Introduction

There are essential pieces of information required to determine the proper application rate and nutrient credits for livestock waste to meet crop needs. These include the acreage of the field, capacity of the spreader and nutrient concentration of the manure. Nutrient concentration can be assigned by using estimated "book" or average available N, P_2O_5 and K_2O concentrations. However, testing manure may better indicate how factors such as animal management affect manure nutrient content. Using good sampling technique is critical for maintaining confidence in manure nutrient analysis results. Appropriate sample handling and laboratory methods are also important to ensure accurate results.

Sampling Livestock Waste

Data in the livestock waste facilities handbook (MWPS, 1985) provides "typical" or average nutrient contents for manures of several animal types. These values probably give an acceptable estimate for "typical" producers, especially if current sampling methods used do not represent the pit, pack or gutter adequately. However, an analysis of a well-sampled system may give a better estimate of manure nutrient concentrations for individual farms than book values, especially if herd and manure management are not "typical". The MWPS total nutrient estimates are compared in Table 1 to actual manure analysis of 51 farms in Minnesota (Wagar et al., 1994) and from 1128 manure samples submitted to the University of Wisconsin Soil Testing Labs between 1986-1991 (Combs, 1991). On average, the actual farm values compare well to the MWPS estimates. Note however that the actual analysis values range widely from the MWPS estimates, indicating poor sampling, management or other on-farm differences. Lindley et al., (1988) also found actual manure analysis values to be highly variable and ranged from 50 to 100% of published values.

Table 1. Comparison of analyzed manure total nutrient concentrations to "typical" nutrient concentrations.

Animal type	System	Nutrient	Minnesota *		Wisconsin**			MWPS ***
			Avg.	Range	Avg.	s.d.	Range	Avg.
			-- lbs/1000 gal --					
Dairy	Liquid	N	29	10-47	28	11	1-71	24
		P_2O_5	15	6-28	13	9	1-118	18
		K_2O	24	11-38	29	16	1-171	29
			-- lbs/t --					
Dairy	Solid	N	13	7-25	10	3	3-33	9
		P_2O_5	6	3-13	6	3	0.2-35	4
		K_2O	8	2-18	11	5	0.2-24	10
			-- lbs/1000 gal --					
Swine	Liquid	N	48	7-107	41	40	1-281	36
		P_2O_5	28	3-64	17	19	1-141	27
		K_2O	21	7-51	21	15	2-83	22

*	Nutrient levels in manure samples taken from 51 farms.
**	Nutrient levels in 388 solid/semi-solid dairy, 380 liquid dairy and 260 liquid swine manure samples submitted to the University of Wisconsin Soil Testing Labs 1986-1991.
***	Livestock Waste Facilities Handbook (MWPS, 1985).

Technique

In virtually any type of agricultural analytical work the results are greatly influenced by sampling. For solid manure, it is generally recommended to sample from loaded spreaders rather than from the actual manure pack. A Wisconsin study (Peters and Combs, 1998) showed that even when well trained professionals sampled dairy manure, variability was much higher when samples were collected directly from the barnyard and pack compared to those collected from the loaded spreader (Table 2). The data also indicated that taking several samples would help minimize potential variability.

In this same study, several samples of liquid manure were taken from a thoroughly agitated lagoon while being pumped into a spreader tank. The results of multiple samples taken by different individuals from a well-agitated liquid dairy manure lagoon indicate that variability is much lower than in the solid manure/barnyard system.

Table 2. Variability in dairy manure nutrient content when sampled by multiple individuals.* Marshfield 1997.

Material	Sampling Method	No. of Samples	DM	N	P	K
			----------------------- % -----------------------			
Dairy - Solid	Barnyard- hand	6				
		Mean	35.02	1.87	0.42	2.48
		SD	2.81	0.22	0.04	0.27
	Barnyard- shovel	7				
		Mean	31.37	2.10	0.50	3.45
		SD	4.50	0.40	0.09	1.16
	Spreader- hand	6				
		Mean	34.35	1.98	0.42	2.60
		SD	1.41	0.17	0.03	0.39
	Spreader- shovel	6				
		Mean	34.60	1.98	0.41	2.30
		SD	4.82	0.31	0.04	0.31
Dairy - liquid	Pump-direct	8				
		Mean	5.11	4.66	1.27	5.23
		SD	0.08	0.32	0.09	0.66
	Pail- subsample	4				
		Mean	5.20	4.80	1.30	5.15
		SD	0.06	0.10	0.03	0.23

* Wisconsin Farm Training Instructions used in the study.

Further, variability can exist among different samplings even when they are taken by the same individual under ideal conditions. This occurred when samples of liquid and semi-solid dairy manure were collected. In this Wisconsin study, five-gallon samples were mixed as thoroughly as possible before being split into twenty-four subsamples. The results indicate that the variability between liquid samples was quite low, but higher with semi-solid dairy samples. This was particularly apparent with total N and dry matter measurements (Peters and Combs, 1998).

Time

An evaluation of long-term sampling of solid/semi-solid manure showed little variability occurred in nutrient concentration over a three-year period at the University of Wisconsin Arlington Agricultural Research Station (Combs, 1991). Sampling from a stanchion barn pack periodically for three years showed that all samples had similar total nutrient values. The least variation occurred for N while the greatest variation was associated with K. These results seem to indicate that with good representative sampling and no significant change in herd management, consistent results, even for solid manure, are possible.

On the other hand, results from sampling solid manure in a poultry-laying barn at the University of Wisconsin Arlington Agricultural Research Station indicated inconsistent results over time (Peters and Combs, 1998). These poultry manure samples taken from the same barn approximately five months apart show a significant difference in all parameters measured. This could be partially a result of seasonal changes in the feed ration, feed contamination or differences in individual sampling technique. Commonly, 5-6 batches of birds are grown out before the litter is removed. Poultry houses are normally sampled when the last batch of birds is removed from the house. Sampling earlier is not recommended, since the nutrient content in poultry litter will change over time.

Due to these variations over time, manure nutrient concentration values used to determine field nutrient credits should ideally be based on long-term farm averages, assuming herd and manure management practices have not changed significantly.

Storage Management

The segregation of manure that occurs in liquid storage requires that special care be taken to ensure that a homogeneous mix is sampled. In a Minnesota study, manure agitated for 2-4 hours before application had highly consistent results for total N, P, K concentrations and percent solids when individual tanks (first to last) were analyzed (Wagar et al., 1994). Samples taken at various stages during the storage system emptying process at Wisconsin also showed very little variability providing the material was thoroughly agitated (Peters and Combs, 1998).

Sampling Recommendations

The number of manure samples tested by public and private labs has increased from approximately 6,220 in 1988 to almost 16,000 in 1996 (Soil, Plant and Animal Waste Analysis Status Report, 1992-96). However, the majority of animal producers still do not sample manure. Reasons for not doing so include sample heterogeneity and the inherent difficulty of taking a representative sample.

Several states have developed guidelines for sampling manure to minimize the sample heterogeneity problem. This information was used to help develop the sampling guidelines presented here. It is generally not recommended to attempt to sample bedded packs or unagitated liquid manure storage facilities. In fact, using nutrient analysis results from poorly sampled systems will not improve the accuracy in estimating N or P loading to a field and may in fact be detrimental.

Taking an adequate number of subsamples is critical for getting a good estimate of nutrient value. In order to characterize N content of a beef manure stockpile within 10%, it took a Colorado State researcher 17 subsamples (Successful Farming, August 1998). However, getting that level of accuracy for P required 20 subsamples and for K it required 30. As a minimum, take a composite sample three independent times and subsample at least five times per composite.

Recommended Procedures for Sampling Livestock Waste for Analysis

Recommended procedures for sampling liquid and solids waste are given below. Producers may choose from these methods as appropriate.

Solid manure – Dairy, Beef, Swine, Poultry
Obtain a composite sample by following one of the procedures listed below. Thoroughly mixing the composite sample is important. One method is to pile the manure and then shovel from the outside to the inside of the pile until well mixed. Fill a one-gallon plastic heavy-duty ziplock bag approximately one-half full with the composite sample, squeeze out excess air, close and seal. Store sample in freezer if not delivered to the lab immediately.

Sampling while loading – *Recommended method for sampling from a stack or bedded pack.* Take at least five samples while loading several spreader loads and combine to form one composite sample. Thoroughly mix the composite sample and take an approximately one pound subsample using a one-gallon ziplock plastic bag. *Sampling directly from a stack or bedded pack is not recommended.*

Sampling during spreading – Spread tarp in field and catch the manure from one pass. Sample from several locations and create a composite sample. Thoroughly mix the composite sample and take a one-pound subsample using a one-gallon ziplock plastic bag.

Sampling daily haul – Place a five-gallon pail under the barn cleaner 4-5 times while loading a spreader. Thoroughly mix the composite sample and take a one-pound subsample using a one-gallon ziplock plastic bag. Repeat sampling 2-3 times over a period of time and test separately to determine variability.

Sampling Poultry In-house - Collect ten samples from throughout the house to the depth the litter will be removed. Samples near feeders and waterers may not be indicative of the entire house and subsamples taken near here should be proportionate to their space occupied in the whole house. Mix the samples well in a five-gallon pail and take a one-pound subsample, using a one-gallon ziplock bag.

Sampling Stockpiled litter – Take ten subsamples from different locations around the pile at least 18 inches below the surface. Mix in a five-gallon pail and take a one-pound subsample using a one-gallon plastic ziplock bag.

Liquid Manure - Dairy, Beef, Swine

Obtain a composite following one of the procedures listed below and mix thoroughly. Using a plunger, an up-and-down action works well for mixing liquid manure in a five-gallon pail. Fill a one-quart plastic bottle not more than three-quarters full with the composite sample. Store sample in freezer if not delivered to the lab immediately.

Sampling from storage – Agitate storage facility thoroughly before sampling. Collect at least five samples from storage facility or during loading using a five-gallon pail and mix. Place subsample of the composite sample in a one-quart plastic container. *Sampling a liquid manure storage facility without proper agitation (2-4 hrs. minimum) is not recommended.*

Sampling during application – Place buckets around field to catch manure from spreader or irrigation equipment. Combine and mix samples into one composite sample. Place the subsample of the composite sample in a one-quart plastic container.

Sample identification and delivery

Identify the sample container with information regarding the farm, animal species and date. This information should also be included on the sample information sheet along with application method, which is important in determining first year availability of nitrogen.

Keep all manure samples frozen until shipped or delivered to a laboratory. Ship early in the week (Mon.-Wed.) and avoid holidays and weekends.

Analyzing Livestock Waste

The ability of a laboratory, or individual using a quick test, to accurately analyze the nutrient value depends not only on sampling technique and handling but how a sample is prepared for analysis and which method is selected. In many cases, different methods can be used depending on instrumentation available and staff expertise. Even when only one method is available, how an individual executes its steps can be important to the overall results.

Laboratory Methods

Methods used to determine manure nutrient content were surveyed among public and private soil testing labs in order to develop suggested standardized methods for manure and other organic waste analysis.

Twenty-five of the 32 labs surveyed reported using the Kjeldahl technique for total N analysis. However, to arrive at 'total N' the nitrate-N must be determined and added to the Kjeldahl N value or the Kjeldahl test must be modified for total N analysis to include nitrate-N by use of salicylic acid. Nitrogen was also determined by the Dumas type of dry combustion analyzers. The Dumas type of analyzer measures total N in one single value that includes all forms of N (organic-N, ammonium-N, nitrate-N and nitrite-N). Most labs used distillation if NH$_4$-N was reported, but the Nessler color technique or an ion selective electrode were also identified.

Total P and K were determined in 19 labs by wet digestion techniques using a variety of acid combinations. Several used a microwave system instead of the traditional beaker /tube and hot plate. Eight labs dry ashed samples. The Inductively Coupled Plasma or Atomic Absorption Spectrometers were the most common detection systems but some labs reported using a colorimeter or Directly Coupled Plasma Spectrometer.

The choice of drying times and temperatures used to determine dry matter content varied substantially among all labs. Samples were dried anywhere from 4 hrs to 4 days at temperatures ranging from 50°C to 140°C. Overnight (16-24 hrs) was the most common drying time (15 labs). Most public labs used temperatures < 100°C while many private labs dried at temperatures ≥ 100°C.

Method Performance

Because of the variety of methods found to be used by soil testing laboratories to determine nutrient content, the NCR-13 Waste Analysis Subcommittee conducted a manure sample exchange to evaluate performance of various laboratory analytical methods. This exchange included 18 public laboratories in the NCR-13, NEC-67 and SERA-6 working groups. Solid and semi-solid manures were spread on a concrete floor and thoroughly mixed with manure forks and then subsampled. Liquid manure samples were taken from the lagoon at the Marshfield Agricultural Research Station while it was being agitated and field spread. Two sets of liquid samples were taken two days apart when the lagoon was approximately one-quarter and three-quarters empty.

Variability

To measure the variability present in the samples, four liquid and eight solid subsamples were analyzed by the University of Wisconsin Soil and Forage Analysis Laboratory. The variability in the liquid samples was quite low, but was 2 to 20 times higher for total N and DM in poultry and dairy semi-solid samples (Table 3). Because of this variability between samples, despite extensive mixing, solid manures were dried and ground prior to the NCR-13, NEC-67 and SERA-6 distribution. Bulk samples of poultry and sheep manure were dried in large forced air forage dryers at 55°C and ground through a 4-mm Wiley mill. The ground material was then thoroughly mixed in a large pail and subsampled into zip lock bags.

Table 3. Effect of in-lab variability on total nutrient content of liquid and solid/semi-solid manure.

Material	No. of Analyses	Nutrient* (dry weight basis)			
		DM	N	P	K
		------------------------ % ------------------------			
Liquid Dairy Manure #3	4				
mean		7.13	4.25	1.04	3.63
SD		0.08	0.09	0.03	0.04
Liquid Dairy Manure #4	4				
mean		6.05	4.65	1.28	4.07
SD		0.09	0.05	0.05	0.04
Poultry (fresh)	8				
mean		28.14	6.31	1.76	3.08
SD		0.15	1.12	0.04	0.05
Dairy Semi-Solid (fresh)	8				
mean		14.14	3.75	0.83	3.27
SD		0.14	0.26	0.02	0.03

* University of Wisconsin Soil and Forage Analysis Lab - Marshfield

To assess the in-lab variability for all labs in the exchange program, a duplicate sample of the sheep manure was included in the five samples distributed. The mean and standard deviations for the individual samples were in close agreement with the overall values indicating that there was, generally, relatively low in-lab variability.

Five manure samples were sent to soil testing labs in the NCR-13, NEC-67 and SERA-6 regions and included:

Exchange Sample	Material
1	Poultry manure, no bedding, dried & ground
2	Sheep manure, bedded pack (same as #5)
3	Liquid dairy manure, lagoon ¾ empty
4	Liquid dairy manure, lagoon ¼ empty
5	Same as #2

The methods used by soil testing laboratories in the NCR-13, NEC-67 and SERA-6 regions generally fall into two categories for N, P and K. For total nitrogen, either Kjeldahl (micro and macro) or combustion were used. Total P and K were determined by either a dry ash (5 labs) or wet digestion (6 labs) using varying mixes of acids and detection systems (ICP, AA, colorimeter). Sample dry matter was determined by weight loss by all labs with drying temperatures ranging from 50 to 110°C and times from overnight to 72 hours.

Analytical results were received from 14 of the 18 labs who received samples. Slightly lower standard deviations from the mean were shown by using the Kjeldahl method for total N analysis compared to combustion, however, differences between methods were small. The liquid dairy manure samples showed higher variabilities by both methods than when samples were analyzed after drying and grinding. The combustion method means were higher than the overall mean, except sample #1, whereas Kjeldahl means showed no consistent pattern. Both methods seemed equally capable of producing reasonable results, especially for dried/ground manure.

Phosphorus analysis was extremely consistent between the dry ash and wet digestion methods used. This low variability occurred between methods in spite of differing acid mixes for ash dissolution or digestion and detection systems.

Potassium analysis was more variable than phosphorus. Samples prepared by dry ashing tended to result in slightly lower total K contents than by wet digestion. Wet digestion methods had highest standard deviations for the liquid dairy manure (samples 3 and 4) perhaps because of subsampling difficulties.

Dry Matter

The variability associated with dry matter determinations depended on whether a single lab or multiple lab results were considered and if manure was solid or liquid. Standard deviation for liquid dairy manure was <0.09% with only somewhat higher levels found for solid fresh dairy and poultry manure when analyzed by a single lab. Colorado researchers found that they could achieve within 10% accuracy for dry matter when only 3 subsamples were tested by the same lab (Successful Farming, August, 1998). However, when results were pooled across the varied times and temperatures used by labs to determine dry matter, the associated standard deviations increased especially for solid manure. Consistent results were most difficult to achieve for pre-dried poultry and sheep manure. Liquid manure variability was less but not as consistent as the single lab results.

Rather small differences in dry matter determinations can have substantial effects on reported fresh weight nutrient contents. The values in table 4 show how a difference in DM from 12 to 18% can change 'as is' N, P_2O_5 and K_2O values. Since producers use the 'as is' values for determining fertilizer equivalence, this could translate into an inappropriate application rate. Therefore, a more consistent method for determining dry matter is needed.

Table 4: Dry matter effect on calculated manure nutrient content.

DM %	N	P_2O_5	K_2O
	------------ lbs/wet ton ------------		
12	7.2	5.5	11.5
15	9	6.9	14.5
18	10.8	8.3	17.3

*based on dry weight analysis of 3.00% N, 1.00% P and 4.00% K

An in-depth study was initiated in 1998, among public labs at Arkansas, Maine and Pennsylvania to explore the effects of drying time and temperature on manure dry matter content (Wolf, Hoskins, and Wolf, Multi-regional Committee, 1998). Three temperatures (50, 70 and 110°C) and 4 drying times (6, 16, 24 and 48 hrs) were chosen to be used on 3 or 4 manure samples routinely submitted for analysis. A representative subsample was suggested to be approximately 10 g fresh weight for solid manures (>85% DM) or 20 g fresh weight for liquid manure (<85% DM). Subsample size was found to have a significant effect on drying time to constant weight for some samples dried at lower temperatures. Dairy manure samples, larger that 10g, required 24 hrs at 50°C to reach constant weight but only 16 hrs for more moderately sized samples. This effect was not shown for poultry manure or for those samples dried at higher temperatures. Liquid and solid dairy manure samples dried to constant weight at the lower (50-70°C) temperatures had $\leq$ 1-2% absolute relative moisture remaining after 48 hrs compared to drying at 110°C. However, poultry manure did not follow this trend and had relatively high moisture content remaining at 50 and 70°C compared to 110°C.

Table 5: Suggested minimum drying times at various temperatures.

	50°C	70°C	110°C
	----------- hrs -----------		
solids (<85% H_2O)	24	16	6
liquids (>85% H_2O)	48	48	16

On-farm Testing

In addition to using data from a commercial laboratory or book values, a third option for determining manure nutrient credits is the use of rapid on-farm testing. There are several quick tests that are commercially available for on-farm nutrient analysis. These measure either one or more of the following: ammonium N, total N, total P, and total K. Analysis takes less than 10 minutes and the equipment is, typically, relatively simple to use. Some of the instrument methods, which were reviewed by Van Kessel, et. al., 1999, include the hydrometer, conductivity meter and pen, ammonia electrode, reflectometer, and hypochlorite reaction meters. The hydrometer is used to indirectly measure total N and total P in slurry samples. The conductivity meter and conductivity pen are commonly used to measure ammonium N and K in slurry samples. The other instruments are mainly used for measuring ammonium N only. In all cases, a properly taken, representative sample is required. The skill of the individual operator is critical to obtaining consistently accurate results. The on-farm or "quick" test should be calibrated to traditional laboratory methods. In most situations, more information will be necessary than can be obtained from the somewhat limited scope of this type of instrumentation.

These "quick" on-farm methods are less accurate and should not be considered as a replacement for traditional laboratory testing. If used in conjunction with standard laboratory testing, these quick tests may have a value in monitoring nutrients in well-agitated liquid manure systems.

Summary

Obtaining a representative sample is critical for any method to be of value. Results of on-farm or laboratory testing are limited by how well the sample represents the farm's manure. Recommended sampling protocol must be followed to ensure a representative sample is obtained.

Results of the NCR-13 exchange tend to emphasize that laboratory analysis method and execution of the method influence results. In view of the consistency of the public labs' results with varying methods, standardizing methods may not result in better analysis. However method standardization may be a good first step in gaining producer confidence in using laboratory results. It is difficult to identify a 'best' method and suggests that probably of greater importance is good execution of individual methods. The exception is dry matter. The importance an accurate estimate of dry matter has on determining fertilizer equivalence makes development and use of a standardized method important.

REFERENCES

Combs, S. M. 1991. Effects of herd management and manure handling on nutrient content: A lab summary. New Horizons in Soil Sci. No. 5-91. Dept. Soil Sci., University of Wisconsin-Madison, WI.

Lindley, J. A., D. W. Johnson and C. J. Clanton. 1988. Effects of handling and storage systems on manure value. ASAE. Applied Engr. Agric. 4(3), 246-252.

MWPS Livestock waste facilities handbook. Handbook #18, 2nd ed. 1985. Midwest Plan Service. Ames, Iowa.

Peters, J. B. and S. M. Combs. 1998. Variability in Manure Analysis as Influenced by Sampling and Management. New Horizons in Soil Sci. No. 6-98. Dept. Soil Sci., University of Wisconsin-Madison, WI.

Successful Farming Magazine. August, 1998. Make sense of manure sampling. Successful Farming, Des Moines, IA. Copyright Meredith Corporation.

Van Kessel, J. S., R. B. Thompson, and J. B. Reeves III. 1999. Rapid on-farm analysis of manure nutrients using quick tests. J. Prod. Agric. 12:215-224.

Wagar, T., M. Schmitt, C. Clanton and F. Bergsrua. 1994. Manure sampling and testing. FE-6423-B. University of Minnesota, Extension Service.

Wolf, N. B. Hoskins, and A. Wolf. 1998. Dry matter analysis methods for livestock manure. Unpublished data from the multi-regional manure testing manual development committee.

Choosing a Liquid Manure Application Method

Donald Hilborn, P.Eng., M.Sc. (Eng.)
By-Product Management Specialist
Ontario Ministry of Agriculture, Food, and Rural Affairs
Woodstock, Ontario

Biographies for most speakers are in alphabetical order after the last paper.

INTRODUCTION

Manure application equipment must be able to properly apply manure to undertake a nutrient management plan. The selection of the actual manure application system(s) depends on meeting a particular combination of requirements that can vary from farm to farm. Unfortunately, there is not the one ideal system that can meet all needs. This paper looks at the necessary requirements for a manure application system, reviews the combinations of systems available and describes particular systems or combinations of systems having the most potential.

Nutrient Management Plan

The application of manure no longer just involves "taking" manure to a field. It has evolved to a scientific practice to handle the manure in an environmentally acceptable and economically viable manner. The manure must be handled following a plan. This plan takes into account many factors such as soil type, soil slope, nutrient level, type of crop grown, yield of crop grown and type of manure. Most areas in North America have developed software or information packages to assist the farmer in determining a plan.

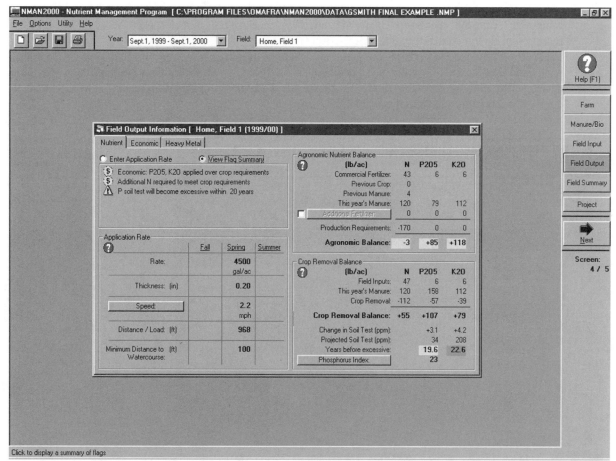

The above graphic shows a screen of Ontario's version nutrient management program. This field output screen shows the agronomic and crop removal balances which involves both commercial fertilizer and manure application of 4500 of gallons per acre of manure from a finishing hog barn. The program completes a rough calculation of application speed if a tanker system was used. The program also indicates the minimum distance to a watercourse.

Other sections of the software describe the different fields or sectors of fields making up the plan.

REQUIREMENTS FOR A MANURE APPLICATION SYSTEM

A manure application system must be able to undertake the specified plan. It must be able to evenly apply the manure at a specified application rate, maintain proper setback distance and be able to access all the fields indicated in the plan. As livestock operations become larger the need to access a remote landless increases. The system must be able to meet the following criteria....

- Achieve prescribed application rates

    The plan specifies a typical rate of 4500 gallons per acre. How difficult is it to achieve this rate? 4500 gallons per acre works out to an application thickness of 0.2 inches. With most application systems, this is near the lower practical rate. In fact, many injection systems on tankers or hard hose irrigation systems will not achieve this low rate.

    Manufacturers have been asked by farmers to develop systems that work faster to be able to handle the large amount of manure produced by farms. Manufacturers have responded by building larger tanks with larger pumps and pipes to unload the liquid. A 6000-gallon tanker, with a 30' wide application pattern and a 120 second unloading time, needs to travel at 11 miles per hour to achieve a rate of 4500 gallons per acre. Is this achievable? Can the operator increase the unloading time to reduce the speed? What is the effect of reduction of flow on the distribution characteristics?

Injection systems seem to present the greatest challenge. In most cases, injectors don't exceed 16 foot in width and can't exceed 4 miles per hour application speed. Systems incorporating injection should be developed to allow an emptying time of 1 minute per 1000 gallon of tanker capacity. Flows to all of the injection points must remain uniform at all flow rates.

Similarly, hard hose irrigation systems applying a strip 200' wide should be able to pull back at 1 ft per min per 1000 gallons per hour of capacity to achieve the lower rates.

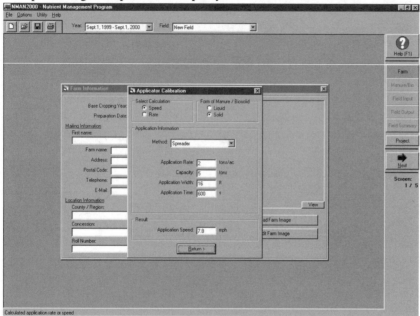

The same type of challenge holds true for solid manure spreaders, especially for poultry manure where rates can be as low as 2 tons per acre. The above graphic from the Nman program indicates that for a 16' width of application, a solid spreader holding 5 tons of manure requires an application time of 600 seconds to keep the speed below 8 mph. This works out to a minimum of 0.5 tons per minute to keep the application speed below 8 miles per hour.

With most application systems there is a flow range which works well. Application systems should be sold with specifications including the usable application ranges, calibration protocol and test pattern results. Perhaps a standard should be developed for manufacturers to follow when developing, testing and marketing equipment.

- Achieve Even Application Rates

   To maximize nutrient use, the manure must be evenly spread so that the farmer can have the confidence that the nutrients from the manure will be available for crop growth. Once the farmer has this confidence, he/she will reduce commercial fertilizer application rates. Application systems that are sensitive to wind drift or are difficult to calibrate will not give the necessary confidence.

   Manure application systems must be able to apply manure evenly over the width and length of application. Research by Greg Wall et al, 1996 have shown that the application rate down length of application for a tanker was quite consistent. This negates some of the concern regarding the variance in application during the unloading process. However, the study indicated greater concern across the width of application.

As shown in the above photo, a boom unit was developed that uses 5 delivery points. The boom covers a span of 12 narrow corn rows (30'). Wall et al. found a uniformity coefficient for this system that was significantly higher than the standard single point delivery system. Wall et al. used a series of trays was developed which measures the manure application rate at 3 different points during the application of a single load.

- Access proper landbase

  With larger livestock operations, a land base much larger than the traditional 100 or 200 acre farm surrounding the farmstead is required. This will increase the distance to transport the manure and will likely involve road crossings and travel. This enters a challenge for direct flow (pipeline) systems. Often a combination direct flow/ tanker/ spreader system will be considered. The more dilute manure is spread at the farmstead and the more concentrated liquid and/or solid manure component is spread at remote sites.

  Using a larger tanker to transfer manure from farm to field becomes competitive once the travel distance exceeds 3 to 5 miles. Since the main time requirements with the tanker relate to the loading and unloading procedures, travel distances up to 20 plus miles do not add a significant cost.

The above graphic shows a new system that allows the vacuum tanker to remove manure from the transport truck without the requirement for transfer tanks or plumbing systems.

- Minimize Possibility of Environmental Damage

  Manure application systems must be designed to avoid or minimize spills due to equipment failure. Due to the nature of the task, it is impossible to have a perfectly failproof system therefore the goal must be to make sure that if equipment failure occurs, the operator can quickly detect it and shut down the system. This is especially true for direct flow systems such as irrigation or drag hose units.

  A typical drag hose unit will operate at 20000 to 30000 gallons per hour. If a pipe disconnects, 500 plus gallons per minute will spill. The operator must be able to shut this down in a very short period of time. A radio link is a must. An automatic shutdown at the main pump is required if a second operator is not guaranteed to be around the main pump at all times.

  Processes need to be set up to ensure that proper setback from water sources is maintained. Pre Flagging setbacks should be considered to ensure setbacks are maintained.

- Minimize Field Damage.

  Field Damage occurs in two ways. Via compaction or via direct crop damage via wheels and application

384

This photo shows some of the challenges, which occur when taking wide wheels through narrow rows. Narrow, large diameter tires should be considered (these also are advantageous wrt compaction concerns). Headlands are always significantly damaged. Using grass strips for a portion of the headlands may be worth consideration.

Another form of damage occurs during the application of manure directly on a crop. In the tests completed in Ontario, damage caused by application as shown above has not been found to be significant as long as lower application rates are used, application during hot sunny days is minimized and more dilute manure is used (This may mean application of non-agitated manure).

Finally compaction damage. Ontario recommends that axle loads never exceed 10 tons. This limits size to approximately 3400 gallons tanker for a dual axle and 5500 gallons for a triple axle and 7500 for a quad axle. Larger tankers should not be used unless the land base is very resistant to compaction. To resist compaction a long narrow footprint causes less deep damage than a wide short foot print. Hence large diameter narrower tires with multiple axles are desired. This also works best for row crop access and road width concerns.

- Minimize road damage and safety concerns.

    Systems with 3 or more wheels need to have an individual axle turning system to minimize damage to roads etc during transportation. Municipal officials are starting to claim that damage to road etc by tankers is exceeding tax base revenue from a livestock operation.

    Safety is another major concern. Tankers carrying up to 7000 gallons of manure are sometimes operating without brakes. Couple this with the new tractors that operate at 25 to 30 miles per hour. Add untrained operators and you have a very dangerous combination. Rules should be made to require full brake systems and operator training especially for the larger tankers/fast tractor combination.

    Safety is not just an issue with tractor pulled or truck mounted equipment. Systems using high pressure/high flow rates are quite hazardous especially for untrained applicators. Using air compressors to flush lines is extremely hazardous unless proper equipment and training is used.

- Minimize Community Relations Problems.

    Odour Levels are reduced by...

    - applying the manure as quickly as possible.
    - reducing the air contact time between the point it is released from the application systems and hits the ground.
    - covering or incorporating the manure as quickly as possible.

    However, maintaining community relations is more than just reducing odour levels. Splashing or tracking manure onto a roadway can cause as many problems as having high odour levels from non-incorporated manure. Certain obvious problems such as droplet drift onto cars or houses must not occur.

- Minimize Cost

    The cost associated with handling the manure should be evaluated based on the undertaking of a total nutrient management plan for the farming operation. This plan includes the application of manure, the purchase and spreading of commercial fertilizer and anticipated yield gains (or losses) due to the use of manure. Cost must also take into account good relations. The additional cost of properly injecting manure may pay if a livestock operation is allowed to expand without intensive neighbourhood resistance.

- Minimize Frustrations

    At the best of times, the handling of manure is a difficult, challenging task. There is little tolerance for application systems that continually break or plug. Many injector units have been permanently parked behind the shed for this reason.

    To properly operate some equipment such as narrow injection or dribbler systems, pretreatment of the manure is necessary for some types of manure to remove solids that cause pluggage. Larger operations have a better opportunity to introduce manure treatments both on an economic and management basis.

COMBINATIONS OF MANURE APPLICATION SYSTEMS AVAILABLE

To choose a particular system you need to look at all the combinations of systems available.

To reduce the repetition in describing various systems available, systems have been split into the...

- transportation unit which moves the manure to a single point in the field.
- distribution unit which takes the manure from a single point and spreads it.

Description of Manure Transporters

Tanker Units

A) Tractor Pulled Tanker.

    The tanker is available in sizes from 1000 to 8000 gallons. Typical tankers in North America are filled via pumps and emptied by gravity feed to a pto pump which pressurizes flow to a splash plate or boom system giving effective application widths of 12' to 50'. These pumps have the capability to empty the tank very quickly having instantaneous rates equivalent to 50000 gallons per hour.

B) Truck Mounted Tanker

A tank is mounted on a dedicated truck or floatation unit such as a Terragator. Many units use vacuum systems to load and unload the tanks. This more expensive system eliminates the need of a filling pump and tends to keep a cleaner unit since spills during transfer are eliminated. Proper access to the manure tank and a separate agitation unit is required.

Direct Flow Systems

These systems transport the manure from the tank to the field via temporary surface or permanent underground pipelines. The fields generally need to be located within about 2 miles of the manure storage for this to work. Permission must be received to place a pipe across (or through) all properties, roads and streams between the farmstead and the receiving field.

C) Soft Hose Application System

A 3" to 5" soft hose is dragged around the field behind a tractor. The hose feeds an application system mounted on the tractor's 3 point hitch.

Application rates of 13000 to 50000 gallons per hour.

D) Hard Hose Application System

Using irrigation technology, a 3" to 6" hard hose applies manure while being slowly pulled back by the reel unit. Application rates of 10000 to 40000 gallons per hour.

A hard hose can also be used as a drag hose. The hoses appear to be having a reasonably long life span.

The major wear and tear appears to be right at the hose. For this reason a flex hose/chain combination is used for the first 20' as shown in this picture. Note the dobbler flow sensor attached at the top of the vertical pipe.

E) Centre Pivot Irrigation System

Uses irrigation technology, a long boom rotates around a fixed centre point. Multiple application points are used. Pretreatment of most manures will be necessary to avoid pluggage.

F) Small "Pulse Jet" Irrigator

This system uses small diameter pipe to operate an irrigation system that sends out a pulse of flow. This allows a wide application width yet allows smaller piping to be used. The system can advance on its own and continually monitors the pulses allowing automatic shutdown if problems occur.

Combined Systems

These systems typically involve the use to road tankers to transfer the manure from the barn to the field. A holding tank is used to transfer the manure to a direct flow or a field tanker unit for application.

Description of the Distributor Units

1) Single Point Splash Plate

Application widths between 10' to 50'.

The advantage to a plate system is that it is inexpensive and relatively plug resistant. However, the problem it is very difficult to design a plate that can evenly distribute the manure over the full width.

2) Single Point Oscillating Nozzle (or gun cart)

Application width between 40' to 400'.

This system evolving from irrigation technology has several advantages. All the manure flows through one point and therefore will resist plugging. The oscillating nozzle should allow for more uniform application. The main problem associated with this system is the wind and droplet drift occurring at the higher widths and/or higher pressures. With proper management, these units continue to have potential.

3) Multiple Point Splash Plates.

Total application width between 20' to 60'.

To avoid the evenness problems associated with a splash plate, several manufacturers are building a boom that carries several splash plates. Testing has shown that even application can occur. The main problem is the requirement to split the flow into several pipes. With thicker manure, a powered distribution unit is required which adds to cost. With manure containing very high amounts of debris, a powered distribution unit will not solve the plugging problems.

4) Multiple Point Vortex Nozzles

Total application width between 40' to 160'.

A vortex nozzles are mounted on a boom with each nozzle covering a width of 16' to 20'. The same problems associated with thicker manure limits the use of the multiple vortex nozzles to thinner manure.

5) Multiple Point Drop Hoses

Total application width between 12' to 40'.

The drop hose system will lay the manure down on the soil surface. Research has shown that this approach will greatly reduce odours and ammonia loss. The problem associated with this system is the high number of drop pipes required (one every 12" to 48"). This involves a larger number of splits with associated increased cost and plugging concerns. Limited to thinner manure or the liquid component of separated manure.

6) Multiple Point Injection Units

Total application width between 10' to 24'.

For odour control there is no better system than immediate injection. However, these units have the same plugging concerns as described above. Due to the increased power and strength required to pull, narrower widths are used, which tends to lead to overapplication problems since tankers have been designed to empty very quickly. Injecting at proper rates will slow down the overall amount of manure a farmer can handle.

One concern with injectors at wide spacing is the high concentration of manure in one zone. For example, if one is applying at average of 4000 gallons per acre through a 4" wide injector spaced 40" apart, he is applying the equivalent of 40000 gallons per acre in that zone. That works out to a band about 2" thick. Research has indicated that this concentration leads to increases in the problem of macropore flow of pollutants to tile drains.

7) Multiple Point Incorporation Units.

Total application width between 10' to 24'.

An incorporation unit works by laying the manure out on the surface and then covering this band with dirt. The incorporation will be shallower than injection and more plug resistant if inwards facing disk closers are used. It will tend to work better than injectors in between growing crops since less soil will be thrown out the side.

8) Centre Pivot Multiple Point Nozzles

This involves using a large number of small bore nozzles. Probably this system will only work with very thin manure or the liquid component of separated manure.

Combinations Available

Table I gives all the possible combinations available. 19 combinations can be purchased from manufacturers in North America.

Table I Possible Combinations of Manure Application Systems

	1) Single Point Slash Plate	2) Single Point Swinging Nozzle	3) Mult. Point Slash Plates	4) Mult. Point Vortex Nozzles	5) Mult. Point Drop Hoses	6) Mult. Point Injection Units	7) Mult. Point Incor Unit	8) Pivot Nozzle Units
Tractor Pulled Tanker	Stand.	Poss.	Opt.	Poss.	New	Opt.	Opt.	
Truck Mounted Tanker	Stand.	Poss.	Opt.	Poss.	New	Opt.	Opt.	
Soft/ Hard Hose Tractor Pulled	Stand.	Opt.	Opt.	Poss.	New	Opt.	Opt.	
Hard Hose Reel Return		Stand.		Opt.	Poss.			
Centre Pivot Irrigation System					Poss			Stand

Code
- Stand. -This is the standard distribution system that comes with the unit
- Opt. -This is an optional distribution system that is commercially available
- New -This is a combination that has just become available by one manufacturer and may still be in the prototype stage.
- Poss. -This is a possible system not currently commercially available.

SYSTEMS HAVING THE MOST POTENTIAL

To identify the best available systems, the farms utilizing the equipment were split into three sizes. Farms having less than 1 million gallons of manure per year (a "smaller operation"), farms having 1 to 3 million gallons per year (the typical operations currently constructed in Ontario) and farms with greater than 3 million gallons per year (a typical "corporate farm's" manure handling requirement).

Farms Handling Less Than 1 Million Gallons of Manure

Most of these farms will be handling their manure on farms within 1 mile of the farmstead. This is within the range of a tractor pulled tanker or a smaller drag hose or hard hose irrigation system. In most cases, a custom applicator that is available at the proper time will be less cost then the farm having it's own equipment.

Transport Unit Options

      Tractor Pulled Tanker

           Following Assumptions Used

               3500 gallon tanker

               25 minute load, transport, empty, transport cycle

               Operates 9 hrs/day

                    21 loads per day - 73000 gallons/day

                    Requires 13.6 days of operation per year for 1 million gallons.

This time seems reasonable for one farm. Horsepower requirements should match the tractor size already on the farm.

Costs will be about $20000 for tank and $10000 for pump. At a 25% depreciation rate this works out to $7500 per year without the tractor costs. At even 1 cent per gallon custom application is only $2500 more.

Compaction issues are a concern. Adequate storage is needed to allow a farmer to miss an early spring application if too wet.

Works in all liquid manure types.

Works on remote fields.

Small Drag Hose System
> Following assumptions used
>> Net application rate of 12000 gallons per hour
>> Operates 9 hrs/day
>> 108000 gallons/day
>> Requires 10 days of operation for 1 million gallons/yr.
> This system is underused. Best for larger or 2 to 3 farmers.
> Works on manure with less than 6% solids (will not work for thick Dairy Manure from tie stall barn).
> Cost depends on amount of pipe required however more costly than custom applicator for a single operation (if custom applicator is available at proper time).
> Minimal Compaction concerns.
> Will not work for intercrop application.
> Would use portable pipes.

Hard Hose Irrigation System
> Same as Drag Hose except costs are higher.
> Very minimal compaction.

Distribution Unit Options
> Tanker and Small Drag Hose
>> If close to neighbours use a boom or injector.
>> If neighbours are not a problem, consider surface application applicator with tillage right after.
> Irrigation
>> Boom system too expensive. Gun only works if neighbours are not a problem and you do not operate on windy days.

Farms Handling 1 to 3 Million Gallons of Manure

Most of these farms will be handling their manure within 3 miles of the farmstead. The tractor operated tanker units will only work if large tankers (5000 to 6000 gallons) and the land is resistant to compaction. Truck mounted tankers are worth considering especially if land can't be accessed by pipes. Direct flow systems especially with a permanently installed pipeline is the most feasible idea.

Transport Unit Options
> Tractor Pulled Tanker
>> Following Assumptions Used
>>> 5000 gallon tanker
>>> 35 minute cycle (over a 3 mile radius)
>>> Operates 9 hrs/day
>>>> 16 loads/day 80000 gallons/day
>>>> 13 days for 1 million gallons (practical)
>>>> 25 days for 2 million gallons (borderline)
>>>> 37 days for 3 million gallons (not practical)
>>> Could use 2 tanker units for the larger operation

> Truck Mounted Tanker
>> Following Assumptions Used
>>> 4500 gallon tanker
>>> 20 minute cycle
>>> Operates 9 hrs/day
>>>> 27 loads/day 120000 gallons/day
>>>> 17 days for 2 million gallons (practical)
>>>> 26 days for 3 million gallons (acceptable)
>>> Best case if fields not accessible by pipelines.

> Drag Hose System
>> Following Assumptions Used
>>> 6" or 8" diameter underground line to fields
>>> 4" or 5" drag hose unit
>>> Average Flow Rate of 30000 gallons/hr
>>> Operates 9 hrs/day
>>>> 270000 gallons/day
>>>> 7.5 days for 2 million gallons (more capacity than necessary)
>>>> 11 days for 3 million gallons (practical)
>>> Ideal to apply 50% prior to corn and 50% post corn or alfalfa plowdown

Minimal Compaction.
Large investment
Medium Irrigation System
Following Assumptions Used
6" underground line to fields
4" diameter hard hose
Average Flow Rate of 16000 gallons/hr
Operates 9 hrs/day
144000 gallons/day
14 days for 2 million gallons (practical)
21 days for 3 million gallons (practical)
Works if you apply prior to corn, intercrop application and post corn or alfalfa plowdown.

Distribution Options
For Tanker and Drag Hose Units
An injector system will likely slow you down too much for the tanker unit. Should consider a separate tractor/cultivator immediately incorporating. A boom surface applicator should be considered to allow operation in windy conditions.
An injector unit will work well in combination with a surface applicator option.

For Irrigation System
Would consider a vortex, boom applicator. This allows application in windy conditions and encourages good management. The other option is to convert to a hose hose drag unit.
A conventional gun will work, however proper management is essential. This system does not have a good reputation and may create neighbour problems even when managed properly. The good news is if there are problems, you could convert to a boom or a low trajectory applicator or a drag hose system.

Operations Handling over 3 Million Gallons of Manure

Operations this size will require a very significant landbase. For example a 6 million gallon operation would typically require 1000 acres (at 6000 gallons per acre).

Transport Unit Options
Tanker System
Conventional tractor tanker systems would generally not be considered for this size due to the large number of units required and the distance to spread.
Truck mounted tankers could be considered if proper crop rotations and soil types are available to allow extended periods of application.
Drag Hose Systems
A 30000 gallon per hour system operating 15 hrs/day (using 2 shifts and proper lighting) would handle...
450000 gallons per day
13 days for 6 million gallons
22 days for 10 million gallons
13 days is reasonable for corn application. 22 days may be reasonable for lighter soil types and wider crop rotations.
Main problem would be finding sufficient land base within 2 miles of farmstead.
Medium to Large Irrigation Systems
An irrigation system running at 20000 gallons per hour net running 12 hours per day would handle...
240000 gallons per day
25 days for 6 million gallons
42 days for 10 million gallons
25 days is possible if intercrop application is used.
42 days seems impractical. 2 systems could be considered.
Combination Systems
Large operations could consider using 2 systems.
For example, a 10 million gallon/year operation has a 5000 gallon floater and a drag hose system.
The truck would access remote fields having a cycle time of 45 minutes. Operating 14 hrs per day the truck would handle 95000 gallons per day. Over a 20 day period the truck would apply approximately 2 million gallons.
This leaves 8 million gallons for the drag hose requiring about 19 days.
This would require 3 operators if both systems working simultaneously.
Another choice would be to have road trucks transporting manure to a remote site. Costs will quickly rise using this system.
For example moving 5 million gallons would cost $50000 at 1 cent per gallon.

Distribution Options

An operation this large would have several options available to fit the situation. With the large landbase requirement, community relations will become a sizeable task and proper manure handling practices will be necessary.

Treatment of Manure

Consideration should be made to separate the solids from the manure.

A separator system may remove 20% of the manure by volume (and nutrient content). The solid portion could be composted and sold off the farm. This means that only 80% of the manure needs to be handled as a liquid proportionally reducing application time and landbase requirements. Also, due to the removal of solids, the liquid portion will be easier to handle and other systems such as the centre pivot irrigation system could be considered.

NEW IDEAS

Technology is evolving to measure flow rates and match this information to GPS technology. Equipment manufacturers have equipment available that can either apply manure at a variable rate according to the nutrient management plan. Probably the best feature is the ability to continuously calibrate application and produce a manure application map similar to a "yield map" except instead of tracking grain coming in a combine you would measure manure leaving a tanker or a direct flow system. This information could be used to fine-tune manure and commercial fertilizer applications. This could allow an operator to either optimize or maximize the manure application.

The application map can be used as a means to show to society that an operation is properly following a Nman plan.

This is a photo of a magnetic inductance flow unit used on a tanker system. This unit works accurately and gives a quick response. The dobbler flow unit was found to work on a continuous flow system but was not effective on a tanker system due to the slow response time.

This is a photo of a variable rate flow unit connected to the magnetic inductance system. Hydraulically controlled units are proposed to replace the 12 volt DC unit to improve response time.

Conclusion

There is not the one perfect system that works for all cases. The selection of an application system should take into account all factors including environmental acceptability, maximum nutrient utilization, societal assurance, human safety and economic viability. In the past, economic viability ruled however there is too much at risk with larger operations to select a system without looking at the whole picture.

Session 10

Site Management

Frost Incorporation and Injection of Manure

Harold M. van Es, Ph.D.
Associate Professor
Department of Crop and Soil Sciences
Cornell University

Robert R. Schindelbeck, M.S.
Research Support Specialist
Department of Crop and Soil Sciences
Cornell University

Biographies for most speakers are in alphabetical order after the last paper.

Introduction

In the Northeastern US, tillage is seldom possible during the winter due to frozen or excessively wet soil conditions. It may nevertheless be desirable to perform soil disturbance to improve water infiltration (Pikul et al., 1991), improve the beneficial effects of freeze-thaw cycles (Lehrsch et al., 1991), or incorporate animal manures or other soil amendments. The method of "frost tillage" (van Es and Schindelbeck, 1995, van Es et al. 1998) was developed as a tillage practice which is performed when a thin frozen layer exists at the soil surface and the underlying soil is tillable.

The physical changes in soils that lead to frost-tillable conditions are defined by a process referred to as freezing-induced water redistribution (Dirksen and Miller, 1966). It is generated by changes in the water potential gradient resulting from a lowering of the pore water pressure in the freezing zone, which in turn results from unstable air-ice interfaces when pore water freezes (Miller, 1980). When

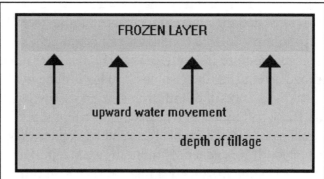

Figure 1.   Soil drying through freezing-induced water redistribution.

frost enters initially-unfrozen soils, this process results in the accumulation of water (as ice) in the frozen zone concurrently with water extraction from the unfrozen zone below (Fig. 1). Therefore, soils that are too wet for tillage in the unfrozen state become workable after frost has entered into the ground. It can be ripped by a tillage or injection tool as long as the frozen layer is thin enough. Until recently, little information was available on the effect of the freezing process on the underlying zone and the potential for the soil to become workable.

Figure 2. Frost tillage using a chisel implement

Conditions conducive to frost tillage include (van Es and Schindelbeck, 1995):

1.   A frozen soil layer of 25 to 100 mm depth, providing adequate equipment support, but shallow enough to be readily ripped by a tillage tool.
2.   Limited snow accumulation.
3.   Sub-zero temperatures at the soil surface during at least part of the day to maintain high-friction (nonslippery) surface conditions.

Figure 2 shows an example of frost tillage using a chisel implement, the preferred tool due to the large chunks of frozen soil material. Based on eight years of field experimentation, we have determined that frost tillage conditions always occurred when an initially thawed soil experienced two to three days of good freezing conditions (daily minimum temperatures generally below -8°C and maximum temperatures below 0°C). Frost tillage conditions typically persisted for at least two days, being terminated by either the extension of the frozen layer beyond 100 mm, or above-zero daytime temperatures causing supersaturated and slippery surface conditions. In some cases, snow accumulated soon after the initial development of frost tillage conditions, and they therefore persisted for up to eight days. It was concluded that if soil is unfrozen, frost tillable conditions can be reliably anticipated based on three to five-day weather forecasts. Before frost tillage is performed, it is advised to evaluate the consistency state of the unfrozen soil using the "ball test". This is done by digging through the frozen layer with a spade and taking a handful of unfrozen soil, and subsequently attempting to squeeze the material into a ball. If the soil molds and forms a firm ball, the soil is still in the plastic state and too wet. If it crumbles, it is in the friable state and workable. Also, the state of the frozen layer should be evaluated. If it is generally more than 100 mm thick, it will be difficult to work, especially with a plow. Also, the frozen layer may be very hard if the soil was saturated when the frost initially entered the ground. In some cases, this may reduce the ability of the implement to penetrate the

Figure 3. Frost-tilled soil

soil. It is also recommended to evaluate different parts of a field if it contains much topographic variability or the presence of other factors that affect the soil microclimate (e.g., tree belts).   In the majority of the cases, such issues will not be encountered

Frost tillage leaves a rough soil surface that is conducive to water infiltration due to high surface roughness (Fig. 3). After soil melt, the soil still maintains high surface storage capacity and facilitates infiltration (Fig. 4). Frost (chisel) tillage was agronomically

compared to conventional spring chisel tillage at Ithaca, NY during 1992 to 1995 and two other sites (Aurora, NY and Mt. Pleasant, NY) in 1994 and 1995, using plots ranging in size from 300 to 600 m^2 in a spatially-balanced randomized complete block design (van Es and van Es, 1993).

Mean maize (*Zea maize* L.) grain yields from frost-tilled plots were statistically similar (8870, 9608, and 5972 kg ha^{-1} for frost-tillage and 9003, 9474, and 5375 kg ha^{-1} for spring tillage for the Ithaca, Aurora and Mt. Pleasant sites, respectively). Crop residue cover

Figure 4.   High surface roughness after soil melting promotes water infiltration and reduces runoff.

was statistically higher for frost tillage compared to spring tillage, indicating a potential benefit for erosion control. Frost tilled soil was also found to increase spring drying compared to untilled soil.

Frost Tillage and Manure Management
One of the main advantages of frost tillage for livestock farms is the potential to incorporate or to

inject manure during the winter. Farms with limited storage capacity (<30 days) still apply manure during the winter time on frozen or wet soils. When frost tillage conditions occur, then such farmers may take advantage of this opportunity to incorporate the manure and reduce the potential for runoff losses during late-winter and early-spring rains. Farmers that have longer-term manure storage may use frost-tillage time windows to apply-and incorporate or to directly inject manure. If injection tools are used in frozen soil, the injector bar and knifes may need to be reinforced. Fig. 5 shows manure injection at Table Rock farms in Wyoming County, NY in February 1999. Frost injection conditions persisted for an eight day period and an estimated 200 acres were manured during this time period. The main advantages of using frost tillage and injection for land application of manure are:

Figure 5. Manure injection into frozen soil (photo by E. Jacobs)

- already-spread manure that is laying on the soil surface can be incorporated to prevent subsequent runoff
- application of stored manure can be moved away from the spring time when workloads are heavy
- compaction damage is nonexistent, because the application equipment is supported by the frost layer
- odor problems are reduced during frost conditions due to lower gas volatilization potential.

Disadvantages of this method are the higher power requirements (approximately 20%) and in some cases the need for sturdier injection equipment.

Seasonal Occurrence of Conditions for Frost Manure Incorporation
Although frost tillage conditions are predictable on the short term based on antecedent soil conditions and three-to-five-day weather forecasts, the *seasonal predictability* of conditions for frost tillage and injection appeared to be a critical assessment need. Measurements from the various sites indicated that soil type differences have little effect on freezing progressions and the potential for frost tillage. Apparently, soil thermal properties did not vary greatly among soil types and, with appropriate weather conditions, frost tillage will occur similarly at each location. An objective was to establish estimates on seasonal probabilities for frost manure incorporation

for the Northeastern USA through the use of a soil freezing model and regional climate information.

A reliable climate data set including daily values of minimum and maximum temperature, precipitation and snow depth with relatively high spatial resolution was available through the U.S. National Weather Service network of volunteer observers. A one-dimensional heat-flow model capable of estimating frost penetration depth was derived that uses data from this network as input. The model, described in detail by DeGaetano et al. (1996), is physically-based in as much as possible given the limitations of the network data. It simulates frost penetration based on the process of thermal diffusion. Details on the model assumptions and calibration efforts are discussed in van Es et al. (1998). The calibrated soil freezing model was used to predict the number of frost tillage/injection days per winter season for a 14 state area in the Northeastern USA. For each of the 275 weather observation sites, data from the period 1950 to 1995 were used to model soil freezing depth under bare ground and sod. The number of annual frost tillage/injection days was determined for all sites and averaged for the two soil cover types to simulate surface conditions with crop residue.

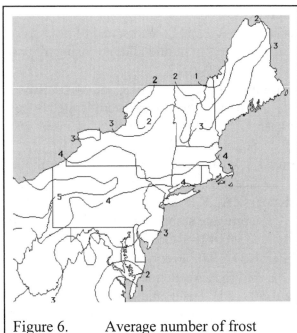

Figure 6.   Average number of frost tillage days in the northeastern USA

The results show that frost tillage/injection opportunities in the Northeast are generally greatest along the 40-42° latitude range including Southern New England, Southern New York, Pennsylvania, New Jersey, and Ohio (Fig. 6). Frost tillage opportunities in the northern part of the region are limited by more persistent snow cover and greater soil freezing depth. The southern part of the region (Virginia, West Virginia, Maryland, and Delaware) experiences less frost tillage opportunities, primarily as a result of milder climates. Further analyses of the simulation data for this best adapted area show that frost tillage conditions generally occur for at least one day per winter and may span 8 to 9 days in one out of four years (van Es et al., 1998)

In conclusion, time windows with shallow-frozen soil conditions provide opportunities for save application of manure on fields, or incorporation of earlier-applied manure that remained on the soil surface. Using these time windows that are conducive to frost injection and incorporation provides distinct advantages in reducing work loads and soil compaction damage in the spring, as well as decreasing the potential for runoff losses. Farmers are recommended to experiment with the use of this technique and evaluate its usefulness to their nutrient management system.

References

DeGaetano, A.T., D.S. Wilks, and M. McKay. 1996. A physically-based model of soil freezing in humid climates using air temperature and snow cover data. J. Appl. Meteor., 35, 1009-1027.

DeGaetano, A.T., D.S. Wilks, and M. McKay. Extreme-value statistics for frost penetration depths in the Northeastern United States. J. Geotechn. Engin. (in print).

Dirksen, C., and R.D. Miller. 1966. Closed-system freezing of unsaturated soil. Soil Sci. Soc. Am Proc. 30, 168-173.

Konrad, J.M. and C. Duquennoi. 1993. A model for water transport and ice lensing in freezing soils. Water Resour. Res. 29:3109-3124.

Lehrsch, G.A., R.E. Sojka, D.L. Carter, and P.M. Jolley. 1991. Freezing effects on aggregate stability affected by texture, mineralogy and organic matter. Soil Sci. Soc. Am. J. 55:1401-1406.

Miller, R.D. 1980. Freezing phenomena in soils. In: D. Hillel, ed. Applications of soil physics. Academic Press, San Diego. pp. 254-299.

Pikul, J.L., J.F. Zuzel, and D.E. Wilkens. 1991. Water infiltration into frozen soil: field measurements and simulation. In: Gish, T.J., and A. Shirmohammadi, eds. Preferential flow. Am. Soc. Agric. Engin., St. Joseph, MI. pp 357-366.

van Es, H.M., A.T. DeGaetano, and D.S. Wilks. 1998. Upscaling plot-based research information: Frost tillage. Nutrient Cycling in Agroecosystems 50:85-90.

van Es, H.M., and R.R. Schindelbeck. 1995. Frost tillage for soil management in the Northeastern USA. J. Minn. Acad. Sci. 59:37-39.

van Es, H.M. and C.L. van Es. 1993. The spatial nature of randomization and its effect on the outcome of field experiments. Agron. J. 85:420-428.

Impact of Fencing on Nutrients: A Case Study

Daniel G. Galeone
Hydrologist
U.S. Geological Survey
Water Resources Division
Lemoyne, Pennsylvania

Biographies for most speakers are in alphabetical order after the last paper.

Introduction

The Mill Creek basin within Lancaster County, Pennsylvania has been identified as an area needing control of nonpoint-source (NPS) pollution to improve water quality. A cooperative effort between the U.S. Geological Survey (USGS) and the Pennsylvania Department of Environmental Protection (PaDEP), this project supports an initiative by the U.S. Department of Agriculture to implement NPS-control management practices within the Mill Creek basin. The project is funded by PaDEP through the National Monitoring Program (NMP) of the U.S. Environmental Protection Agency. The NMP stems from Section 319 of the 1987 amendment to the Clean Water Act. The NMP was developed to document the effects of NPS pollution-control measures and associated land-use modifications on water quality (Osmond et al., 1995).

A common agricultural NPS pollution-control practice implemented in the Mill Creek basin is streambank fencing. To provide land managers information on the effectiveness of streambank fencing in controlling NPS pollution to water bodies, a 6-10 year study is being conducted in two small (approximately 1.5 square miles (mi^2)) paired basins within the Mill Creek basin. This paper describes the preliminary effects of streambank fencing on nutrient concentrations and yields in surface water for a small agricultural watershed in Lancaster county, Pennsylvania. The pretreatment data collected from October 1993 through June 1997 for surface water at the outlet of both the control and treatment watersheds is compared to the post-treatment data collected from July 1997 through November 1998.

Problem

Agriculture is the predominant land use in the Mill Creek basin of Lancaster county, Pennsylvania. Much of the area along streams is used to pasture animals, primarily dairy cattle. Streambank fencing to exclude animal access to streams is a best-management practice (BMP) that is targeted to reduce suspended-sediment and nutrient inputs to streams. This BMP reduces direct nutrient inputs from animals, eliminates streambank trampling, and promotes streambank revegetation. Livestock trampling of streambanks increases bank erosion (Kauffman et al., 1983). Livestock also can change physical soil properties in grazed areas by increasing soil compaction (Alderfer and Robinson, 1949; Orr, 1960; Bryant et al., 1972), which causes decreases in soil-infiltration rates (Rauzi and Hanson, 1966) and subsequent increases in overland flow. Development of a vegetative buffer along each side of the stream is used to stabilize streambanks, thereby reducing bank erosion (Rogers and Schumm, 1991) and potentially reducing the input of nutrients to the stream channel by filtration of overland flow (Cooper et al., 1987; Parsons et al., 1994; Pearce et al., 1997) and through the retention of nutrients in the subsurface of the riparian zone (Jacobs and Gilliam, 1985; Lowrance, 1992; Nelson et al., 1995).

Site Description

The two smaller study basins are adjacent to each other within the larger Big Spring Run basin (figure 1). Both the study basins have about 3 stream miles (mi) and 2 mi of pasture along streams. Big Spring Run flows north and discharges to Mill Creek about 0.6-0.7 mi from the outlets of the study basins. Mill Creek is located within the Susquehanna River basin. Annual precipitation averages 41 inches and the average annual temperature is 52 degrees Fahrenheit at a National Oceanic and Atmospheric Administration site about 1-2 mi northeast of the basins (National Oceanic and Atmospheric Administration, 1994).

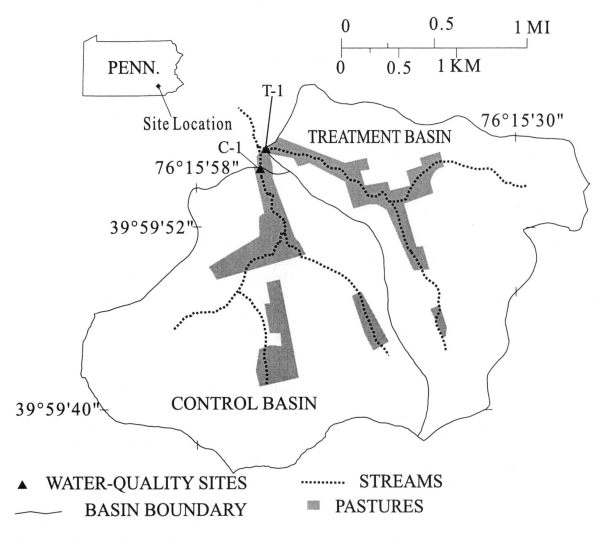

Figure 1. Map of study basin.

The basins are underlain by carbonate rock of the Conestoga Formation. This is an Ordovician-aged rock containing gray limestone with fine- to coarse-crystalline texture (Poth, 1977). The primary soils of the basins are of the Lehigh and Conestoga series (Custer, 1985). The Lehigh series is a fine-loamy, mixed, mesic Aquic Hapludalf and the Conestoga series is a fine-loamy, mixed, mesic Typic Hapludalf. Both soil series are well-drained and relatively deep and have slopes that range from 0 to 15 percent; most slopes in the basins range from 3 to 8 percent (Custer, 1985). Ground-water well drilling logs report the depth to bedrock in the basins ranges from 5 to 20 feet (ft).

Land use is primarily agricultural (about 80-90 percent) and urban. Agriculture in the two basins primarily involves crop production and dairy farming. Major crops are corn, soybeans, and alfalfa. From 1993 through 1996, the average annual nitrogen (N) and phosphorus (P) applications to the control basin (1.77 mi^2) were about 103,000 pounds (lb) of N and 22,000 lb of P per mi^2. The average annual N and P applications to the treatment basin (1.42 mi^2) were about 80,000 lb of N and 21,000 lb of P per mi^2. The number of dairy cattle in either basin at any one time over the study period has ranged from 200 to 400.

Annual surface-water yields of N, P, and suspended sediment that were computed for water years 1994 through 1996 at the outlets of both basins were similar. Yields were computed using least-squares regression equations. Annual yields of total N, total P, and suspended-sediment estimated at T-1 and C-1 were approximately 30,000, 15,000, and 1,500,000 lb/mi^2 (Galeone, 1999). For both basins, about 90 percent of the total-N yield was attributable to dissolved NO_3-N and 90 percent of the total-N yield occurred during non-stormflow; conversely, about 90 percent of the total-P yield was attributable to stormflow and 60-65 percent of the total-P yield was in suspended form. The nutrient and suspended-sediment yields are comparable to those reported by Lietman et al. (1983) and Unangst (1992) for other small agricultural drainage basins located in Lancaster county, Pennsylvania.

Study Approach

The primary approach used for this study to determine effects of streambank fencing on surface-water quality is a paired-watershed analysis (Galeone and Koerkle, 1996). This approach requires a calibration period prior to BMP implementation (or treatment). The calibration period is used to account for inherent differences between the two basins (Clausen et al., 1996). That is, if the basins respond differently to climatic and hydrologic variations, the calibration relation will account for the variations. Deviations from the calibration relation during the treatment period may be attributed to the treatment. The paired basins were calibrated (pretreatment period) from Oct. 1993 through June 1997. Fencing was installed in the treatment basin to exclude dairy animals (except for cattle crossings) and provide vegetated buffers of 10-12 ft width on either side of the streambank. Fencing was completed by June 30, 1997. Post-treatment data include all data collected after June 30, 1997. Generally, there was no other significant change in agricultural practices in either basin from October 1993 through November 1998 except for streambank fencing in the treatment basin. This paper discusses data collected through November 1998.

Although data were collected at five surface-water sites in the study area, results discussed herein include only data collected at the outlet of the treatment basin (T-1) and at the outlet of the control basin (C-1). Grab and stormflow samples were collected at each site and analyzed for total and dissolved forms of nitrogen and phosphorus and suspended sediment. Grab (fixed-time) samples were collected every ten days from April through November; otherwise, samples were collected monthly. The more intensive sampling from April through November coincided with the typical period when cows are pastured in south-central Pennsylvania. Fixed-time samples were collected by hand at the downstream side of the cement v-notch weir used to monitor flow at the outlets of both basins. Flow was continuously monitored at both sites. Storm samples were collected with an automated sampler having a 72-bottle capacity. Sample collection during a storm event was initiated by a float switch that turned the samplers on at a specific stage. After initialization, samples were collected every 15 minutes until either the 72 bottles were filled or the stage dropped below the point at which initialization occurred. Storm samples were retrieved within a day of the completion of the event and chilled prior to sample processing. After defining the storm interval so that similar time intervals and parts of the hydrograph were used for both C-1 and T-1, storm samples were composited into one storm sample per site. Aliquots pipeted from the bottles were flow weighted so that the composite sample represented the mean conditions for the storm event. Chemical and suspended-sediment analyses were done on the composited samples.

Chilled samples were shipped to the USGS National Water Quality Laboratory in Arvada, Colorado for nutrient analysis for both grab and storm samples. Analyses were performed according to techniques described in Fishman and Friedman (1989). Suspended-sediment concentration analyses were conducted by the USGS Sediment Laboratory in Pennsylvania through water year 1995 and thereafter at the USGS Sediment Laboratory in Kentucky. Both sediment laboratories used procedures described by Guy (1969) to determine suspended-sediment concentrations.

Statistical analyses were conducted on fixed-time and storm samples separately. Samples collected during the fixed-time schedule were not used in the statistical analyses if the stream flow at the time of sample collection exceeded the 90th percentile of flow for that station. Thus, from this point onward, fixed-time (grab) samples will be designated as **low-flow** samples.

Data collected before and after fence installation at the outlets is statistically compared in this paper using Wilcoxon rank-sum tests (Helsel and Hirsch, 1995) in order to determine if streambank fencing had a significant effect (alpha level equal to 0.05) on surface-water quality. This is a nonparametric test that requires matching the paired data from the basins, taking the difference in the paired data, and then ranking the differences. The data were then separated into pre- and post-treatment data to determine if there was a significant difference in the ranked data between the two periods.

Analysis of covariance (ANCOVA) could also be used to determine if the treatment had a significant effect on water quality. ANCOVA is a parametric procedure that requires a normal distribution; therefore, data transformations such as log_{10} may be required prior to testing (Clausen et al., 1996). Although this testing was conducted, this analysis will not be discussed, but, in general, the results parallelled those identified using the Wilcoxon rank-sum test.

Results

Variations in stream discharge from the calibration to treatment period were evident for both study basins (figure 2). Stream discharge, measured in units of cubic feet per second (cfs) was about 10 percent higher than the mean annual (the mean annual was calculated for data from water years 1993 through 1998) for both basins prior to fence installation (October 1993 through June 1997); however, stream discharges following fence installation (July 1997 through November 1998) were about 33 percent below the mean annual. The discharge relation between C-1 and T-1 did not change from the pre- to post-treatment period (figure 3). The mean annual discharges for T-1 and C-1 are 1.91 cfs (1.35 cfs/mi^2) and 3.20 cfs (1.81 cfs/mi^2), respectively. The discharge at C-1 is 33 percent greater than at T-1 on a per area basis. According to regional curves developed by Flippo (1982), the cfs per unit area relation for the area should be 1.27; thus, discharge measured at C-1 is higher than normal, and this could be caused by ground-water crossing surface-water boundaries. This has implications when comparing basins to detect changes in water quality due to streambank fencing.

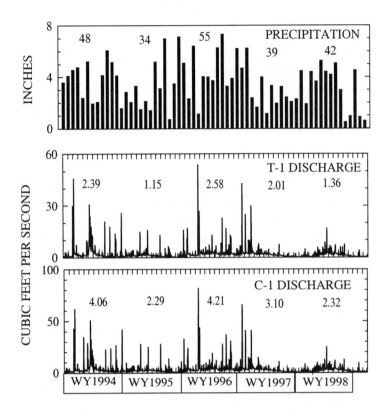

Figure 2. Monthly precipitation measured at T-1 and continuous stream flow at T-1 and C-1 from October 1993 through December 1998. The annual precipitation totals are given for each water year, as are the annual mean discharges.

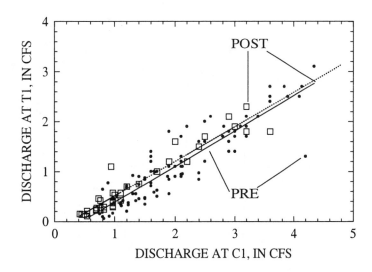

Figure 3. Relation between discharge at T-1 and C-1 at time of fixed-time sample collection from October 1993 through December 1998. The lines for the pre- and post-treatment period represent the predicted relation between the basins for both periods.

Concentrations of nutrients measured at T-1 and C-1 indicated a varied response to fencing depending on the constituent and sample type (low-flow or storm) (figures 4 and 5). There was no significant change in the total-P concentrations for low flow or stormflow at T-1 relative to C-1 during the post-treatment period. There were significant reductions (relative to the control basin) in total-N concentrations (20 percent) for low-flow samples and a significant reduction in suspended-sediment concentrations (35 percent) for storm samples collected at T-1 during the post-treatment period. The reduction in total N during low flow was attributable to decreased concentrations of NO_3-N. The mean concentration for total N for T-1 from the pre- to post-treatment period decreased from 11.7 to 9.0 milligrams per liter (mg/L). Approximately 95-97 percent of the total N in low flow for both basins was in the form of NO_3-N, and this was consistent from the pre- to post-treatment period. The concentration of suspended sediment in storm samples for T-1 decreased from a mean of 710 to 190 mg/L from the pre- to post-treatment period. This was a 73 percent decrease as opposed to a 27 percent decrease for the control basin.

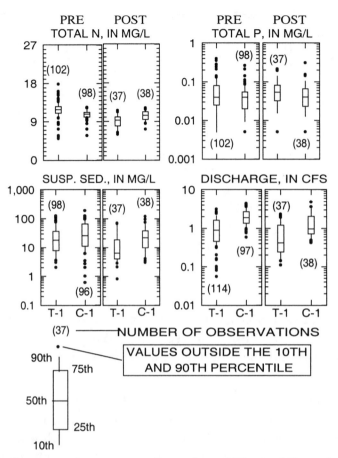

Figure 4. Ranges of discharge and concentrations of total N, total P, and suspended sediment for low-flow samples collected during the pre- and post-treatment period at T-1 and C-1.

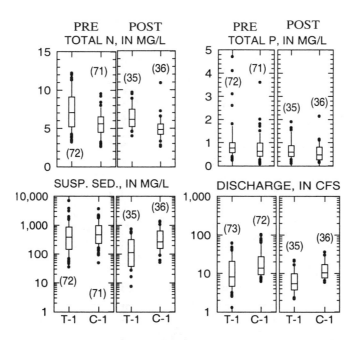

Figure 5. Ranges of discharge and concentrations of total N, total P, and suspended sediment for stormflow samples collected during the pre- and post-treatment period at T-1 and C-1.

There were no significant changes in the relations between the data collected at the outlet of the treatment and control basins for nutrient or suspended sediment yields from the pre- to the post-treatment period for low-flow or stormflow samples (figure 6). For example, the difference in the N yield for T-1 and C-1 for stormflow during the post-treatment period was not significantly different than the difference in the N yield for T-1 and C-1 during the pre-treatment period. It should be noted that the yields for low flow and stormflow in figure 6 are not comparable. The low-flow yields are daily yields calculated from the sample concentration multiplied by the discharge (cfs was multiplied by 86,400 to convert from seconds to days) at the time of sample collection and divided by the drainage area. The stormflow yields are the total yield for that particular storm; therefore, the mean discharge for the storm was multiplied by the sample concentration for the storm and the storm duration and divided by the area, resulting in pounds per square mile. The yields for N, P, and suspended sediment for both low-flow and stormflow samples decreased from the pre- to the post-treatment period in both basins. This was indicative of the decreased flow (figure 2) in both basins caused by the below normal precipitation during the post-treatment period. The discharge at the time grab (low-flow) samples were collected was 30 percent lower during the post-treatment relative to the pre-treatment period (figure 4). Similarly, the mean discharge of storms during the post-treatment period was about 50 percent of the mean discharge for the pre-treatment period (figure 5).

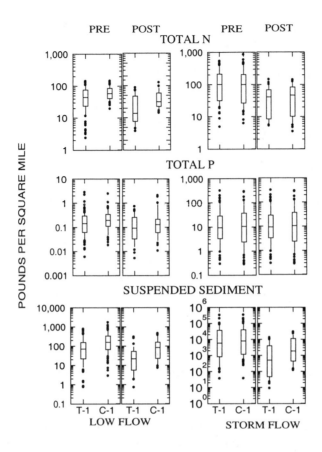

Figure 6. Ranges of daily yields for low flow and total yields for storm events for total N, total P, and suspended sediment at T-1 and C-1 during the pre- and post-treatment period.

Overall, even though paired comparisons for yields at T-1 and C-1 during the post-treatment period were not found to differ significantly from the paired comparisons during the pre-treatment period, there were percent reductions for concentrations and yields for T-1 during the post-treatment period that equaled or exceeded the reductions in N, P, and suspended sediment for C-1. This was true for both low-flow and stormflow samples.

Conclusions

This paper presents data from the first 17 months (July 1997 through November 1998) of the post-treatment period for a streambank fencing project in Lancaster county, Pennsylvania that will be collecting data at least through November 2000. Additional data and data from other aspects of the study, which includes monitoring effects on shallow ground-water quality and documenting changes in benthic-macroinvertebrate communities, will provide further information on streambank fencing effects in pastured land. The data from the project will eventually be used to quantify the potential effects of streambank fencing on nutrient loads from watersheds where pasture area is adjacent to waterways.

Additional Information

All water-quality and water-quantity data collected for this study are published in the USGS Annual Data Reports for water years 1994 through 1998 for the Susquehanna and Potomac River Basins. These reports or other information about the USGS activities in Pennsylvania can be accessed at the web site **---- *http://pa.water.usgs.gov*** or contact author.

References

Alderfer, R.B., and R.R. Robinson. 1949. Runoff from pastures in relation to grazing intensity and soil compaction. J. Amer. Soc. Agron. 39: 948-958.

Bryant, F.T., R.E. Blaser, and J.R. Peterson. 1972. Effect of trampling by cattle on bluegrass yield and soil compaction of a Meadowville Loam. Agron. J. 64: 331-334.

Clausen, J.C., W.E. Jokela, F.I. Potter III, and J.W. Williams. 1996. Paired watershed comparison of tillage effects on runoff, sediment, and pesticide losses. J. Environ. Qual. 25: 1000-1007.

Cooper, J.R., J.W. Gilliam, R.B. Daniels, and W.P. Robarge. 1987. Riparian areas as filters for agricultural sediment. Soil Sci. Soc. Am. J. 51: 416-420.

Custer, B.H. 1985. Soil survey of Lancaster county, Pennsylvania. U.S. Department of Agriculture, Soil Conservation Service. 152 p.

Fishman, M.J., and L.C. Friedman. 1989. Methods for determination of inorganic substances in water and fluvial sediments. U.S. Geological Survey Techniques of Water-Resources Investigations, book 5, chap. A1. 545 p.

Flippo, H.N. 1982. Evaluation of the streamflow–data program in Pennsylvania. U.S. Geological Survey Water-Resources Investigation 82-21. 56 p.

Galeone, D.G. 1999. Calibration of paired basins prior to streambank fencing of pasture land. Journal of Environmental Quality 28: 1853-1863.

Galeone, D.G. and E.H. Koerkle. 1996. Study design and preliminary data analysis for a streambank fencing project in the Mill Creek basin, Pennsylvania. U.S. Geological Fact Sheet 193-96. 4 p.

Guy, H.P. 1969. Laboratory theory and methods for sediment analysis. U.S. Geological Survey Techniques of Water-Resources investigations, book 5, chap. C1. 58 p.

Helsel, D.R., and R.M. Hirsch. 1995. Statistical methods in water resources. Elsevier Science Publishing Co. Amsterdam. 529 p.

Jacobs, T.C., and J.W. Gilliam. 1985. Riparian losses of nitrate from agricultural drainage waters. J. Environ. Qual. 14: 472-478.

Kauffman, J.B., W.C. Krueger, and M. Vavra. 1983. Impacts of cattle on streambanks in northeastern Oregon. J. Range Manage. 36: 683-685.

Lietman, P.L., J.R. Ward, and T.E. Behrendt. 1983. Effects of specific land uses on nonpoint sources of suspended sediment, nutrients, and herbicides - Pequea Creek Basin,

Pennsylvania, 1979-80. U.S. Geological Survey Water-Resources Investigations Report 83-4113. 88 p.

Lowrance, R. 1992. Groundwater nitrate and denitrification in a coastal plain riparian forest. J. Environ. Qual. 21: 401-405.

National Oceanic and Atmospheric Administration. 1994. Climatological Data - Annual Summary - Pennsylvania 1994. National Climate Data Center. Asheville, N.C. 35 p.

Nelson, W.M., A.J. Gold, and P.M. Groffman. 1995. Spatial and temporal variation in groundwater nitrate removal in a riparian forest. J. Environ. Qual. 24: 691-699.

Orr, H.K. 1960. Soil porosity and bulk density on grazed and protected Kentucky bluegrass range in the Black Hills. J. Range Manage. 13: 80-86.

Osmond, D.L., D.E. Line, and J. Spooner. 1995. Section 319 National Monitoring Program -- An overview: NCSU Water Quality Group, Biological and Agricultural Engineering Department, North Carolina State University, Raleigh, North Carolina, 13 p.

Parsons, J.E., R.B. Daniels, J.W. Gilliam, and T.A. Dillaha. 1994. Reduction in sediment and chemical load agricultural field runoff by vegetative filter strips. Report No. 286. Water Resources Research Institute of the University of North Carolina. Raleigh, North Carolina. 75 p.

Pearce, R.A., M.J. Trlica, W.C. Leininger, J.L. Smith, and G.W. Frasier. 1997. Efficiency of grass buffer strips and vegetation height on sediment filtration in laboratory rainfall simulations. J. Environ. Qual. 26: 139-144.

Poth, C.W. 1977. Summary ground-water resources of Lancaster county, Pennsylvania. Pennsylvania Geological Survey. Water Resources Report 43. 80 p.

Rauzi, F., and C.L. Hanson. 1966. Water intake and runoff as affected by intensity of grazing. J. Range Manage. 19: 351-356.

Rogers, R.D., and S.A. Schumm. 1991. The effect of sparse vegetative cover on erosion and sediment yield. Journal of Hydrology 123: 19-24.

Unangst, D., chair., Pennsylvania State Rural Clean Water Program Coordinating Committee. 1992. The Conestoga Headwaters Project 10-Year Report of the Rural Clean Water Program. U.S. Department of Agriculture, Agricultural Stabilization and Conservation Service. 329 p.

Overview of Federal Cost Sharing

Anthony J. Esser
National EQIP Program Manager
USDA-Natural Resources Conservation Service
Washington, DC 20013

Biographies for most speakers are in alphabetical order after the last paper.

INTRODUCTION

The focus of this presentation will be regarding the Environmental Quality Incentives Program (EQIP) although there are other programs that can be used to help producers implement portions of their conservation plan. It is targeted program with an environmental focus and a legislative mandate that at least fifty percent of the EQIP funds allocated be utilized for implementation of conservation systems relating to livestock production.

Background

The Environmental Quality Incentives Program (EQIP) provides in a single, voluntary program flexible technical, financial, and educational assistance to farmers and ranchers who face serious threats to soil, water, and related natural resources on agricultural land and other land, including grazing lands, wetlands, forest land, and wildlife habitat. Assistance is provided in a manner that maximizes environmental benefits per dollar expended, to help producers comply with Title XII of the Food Security Act of 1985, as amended, and Federal and State environmental laws. Assistance also encourages environmental enhancement. Producers are aided in making beneficial, cost-effective changes needed to conserve and improve soil, water, and related natural resources on their farm and ranch operations. A consolidated and simplified conservation planning process is used to reduce administrative burdens on producers.

Funds of the Commodity Credit Corporation are used to fund the assistance provided under EQIP. For fiscal year 1996, $130 million was made available to administer an interim program; $200 million is to be made available for each of fiscal years 1997 through 2002. However, Congress only approved $174 million in fiscal year 1999. Fifty percent of the funding available for the program must be targeted at practices relating to livestock production.

Needs Assessment and Selecting Priority Areas

The program is primarily available in priority conservation areas throughout the Nation. The priority areas are watersheds, regions, or areas of special environmental sensitivity or having significant soil, water, or related natural resource concerns. The process for selecting these priority areas begins with the local conservation district(s) convening local work groups to advise the Natural Resources Conservation Service (NRCS) in various conservation elements. The local work group is a partnership of the conservation district, NRCS, Farm Service Agency, Farm Service Agency county committee, Cooperative Extension Service, and other local government entities with an interest in natural resources conservation. They provide leadership in conducting a comprehensive conservation needs assessment of the natural resource conditions in a locality and identify program priorities and resources available. They also develop proposals for priority areas, develop ranking criteria used to prioritize producer's applications for EQIP, make program policy recommendations, and other related activities.

The local conservation needs assessment is incorporated into the State, regional, and national natural resources strategic plans by NRCS, thus aiding in program decision-making. The priority areas recommended to NRCS by the local work group are submitted to the NRCS State Conservationist. The State Conservationist, with the advice of the State Technical Committee, sets priorities for the program, including approval and funding levels of priority areas.

State Conservationists, with the advice of the State Technical Committee, may also determine that program assistance is needed by producers located outside of funded priority areas that are subject to environmental requirements, or who have other significant natural resource concerns.

The Chief of NRCS may determine the need for national conservation priority areas where eligible producers may receive enhanced program assistance in EQIP, Wetland Reserve Program, or Conservation Reserve Program.

Conservation plan and contract

Program participation is voluntary and initiated by the producer who makes an application for participation. Contract applications are accepted throughout the year. Ranking and selecting the offers of producers occur during designated periods. To rank and select the highest priority applicants, NRCS conducts an evaluation of the environmental benefits the producer offers to achieve by using the program. The evaluation uses ranking criteria developed with the advice of the local work group to give a higher priority to producer offers that maximize environmental benefits per dollar expended. The Farm Service Agency county committee approves for funding the highest ranking applications.

Approved applicants are responsible for developing and submitting a conservation plan encompassing the producer's farming or ranching unit of concern. The conservation plan, when implemented, must protect the soil, water, or related natural resources in a manner that meets the purposes of the program and is acceptable to NRCS and the conservation district. A conservation plan is developed by the producer, with the assistance of NRCS or other public and private natural resource professionals, in cooperation with the local conservation district. The plan is used to establish an EQIP contract.

The contract, developed and administered by Farm Service Agency, provides for cost-sharing and incentive payments to the producer for applying the needed conservation practices and land use adjustments within a specified time schedule. Payments are made to the producer when the conservation practices specified in the contract are satisfactorily established. Up to 75 percent cost-sharing may be provided for conservation practices. A person is limited to receiving up to $10,000 in a fiscal year, and a contract is limited to a total of $50,000. Contracts must be for 5 to 10 years in length.

Accomplishments

Since its inception in FY1997, USDA has entered into over 64,500 contracts with producer throughout the country to implement conservation systems. These contracts represent one-third of the requests that have been received. These contracts cover 26.9 million acres of farmland and 7 million acres of cropland and commit $367 million of USDA funds to assist producers implement over 618,000 conservation practices.

In keeping with the trends, fiscal years 2000 and 2001 should produce another 40,000 contracts from 120,000 producers and should result in the commitment to install an another 400,000 conservation practices. If the Administration's budget is approved, there could be a sizable increase in the EQIP budget for FY2001.

Summary

The EQIP program is just one of the conservation toolbox that is available to producers to implements conservation systems to resolve resource issues. Other programs include the Conservation Reserve Program (CRP), Wetlands Reserve Program (WRP), Forestry Incentives Program (FIP), Wildlife Habitat Incentives Program (WHIP), Farmland Protection Program (FPP), and Conservation Farm Option (CFO). Additionally there are other federal programs and many states funded conservation initiatives which can also be used to implement animal agriculture conservation systems. How these programs are blended together to accomplish the "Management of Nutrient and Pathogens from Animal Agriculture" is determined thought the locally-led process under the direction of the conservation districts.

Session 11

Nutrient Management Plans

Nutrient Management Plans — Poultry

Greg L. Mullins
Virginia Polytechnic Institute and State University
Department of Crop and Soil Environmental Sciences
418 Smyth Hall (0403)
Blacksburg, Virginia 24061

Biographies for most speakers are in alphabetical order after the last paper.

Introduction:

The poultry industry is an important component of the agricultural economy of the United States and production continues to increase. During 1982 to 1994, production in the U.S. increased by 44% (Bagley et al., 1996). The annual U.S. production in 1997 was 14 billion broilers and 301 million turkeys with receipts totaling $14 and $2.9 billion, respectively (USDA, 1999b). Trends in agriculture in the United States as well as throughout the world has been toward fewer farms and localized geographic intensification of the poultry industry. In Virginia, for example, which ranks 8th in broiler production and 4th in turkey production (USDA, 1999b), 75% of all poultry sales are generated from a four-county complex in the Shenandoah Valley (USDA, 1999a).

Intensification of the poultry industry has resulted in a large percentage of poultry farms being located in geographical areas that are grain deficient, resulting in a reliance on grain that is imported for poultry feed. This intensification of the industry has also resulted in the production of large quantities of poultry manure as well as other by-products from poultry processing plants, within relatively small geographical areas. For example, Virginia's largest poultry sector of the Shenandoah Valley, has a large deficit for corn and soybeans (Pelletier, 1999). Recent estimates by Pelletier (1999) revealed a 84% deficit for corn and a 95% deficit for soybean consumption by the animal industry in this region. In the Shenandoah Valley, poultry litter production is approximately 364,000 tons per year. If one accounts for the estimated 50,000 to 100,000 tons of litter that are utilized outside the valley that leaves approximately 265,000 to 315,000 tons of litter that must be properly managed (Pelletier, 1999). This quantity of litter corresponds to a range of 16.5 to 19.6 millions lbs. of nitrogen (N) and 16.7 to 19.8 million pounds of phosphate (P_2O_5).

Management options for litter in Virginia have included land application, using as livestock feed and alternative uses.

Throughout the U.S., most poultry manure is land applied as a source of nutrients for crop production and most of this valuable by-product will continue to be land applied in the future. However, poultry manure will need to be managed in such a way as to maximize its potential agronomic value while at the same time minimizing environmental contamination. In this review, I will briefly discuss the environmental concerns regarding land application of animal manure and some of the basic concepts that need to be considered in writing a comprehensive nutrient management plan for the utilization of poultry litter.

Potential Environmental Concerns:

Land application of nutrients in excess of crop needs can lead to a number of water quality problems. Recent reports have indicated that agricultural runoff and nutrients from livestock and poultry are the greatest pollutant in 60% of the rivers and streams identified by the USEPA as impaired (U.S. Senate Committee on Agriculture, Nutrition and Forestry, 1997). In addition, animal waste was reported to degrade nearly 2,000 bodies of water in 39 states and pollutants from concentrated animal feeding operations (CAFOs) impairs more miles of U.S. rivers than all other industry sources and municipal sewers combined (U.S. Senate Committee on Agriculture, Nutrition and Forestry, 1997). Recent fish kills in 1997 which were attributed to outbreaks of the toxic dinoflagellate *Pfiesteria piscicida* in coastal rivers of the Eastern U.S. may have been influenced by nutrient enrichment (Burkholder and Glasgow, 1997; USEPA, 1996), however, the direct cause of these outbreaks is unclear.

In the future, it will be important for poultry producers as well as users of poultry litter to be aware of the environmental concerns of the watershed where they are geographically located and whether their operation is within the watershed of impaired streams. For example, the poultry industry in the Mid-Atlantic states is located primarily within the Chesapeake Bay Watershed. Improving water quality of the Chesapeake Bay has been a major concern for a number of years. In 1983, a six-year study by the USEPA (1983) revealed that runoff from farmland is contributing to water quality decline in the Chesapeake Bay. It was estimated that non-point sources were contributing 67% of the nitrogen (N) and 39% of the phosphorus (P) that is entering the Bay. The report also estimated that agricultural crop land is contributing 60% of the N and 27% of the P that is entering the Chesapeake Bay. As a result, the 1983 Chesapeake Bay Agreement called for a 40% reduction in N and P loading in the Bay by the year 2000. In addition, the 1997 fish kills that were attributed to *Pfiesteria* resulted in heightened public attention on the management of poultry waste. This public concern was strong enough that Delaware, Maryland and Virginia have recently passed legislation which will regulate the land application of nutrients in poultry litter. Users of poultry litter in these states will need to become familiar with these new regulations.

Nutrients and Environmental Impact:

Nitrogen (N), phosphorus (P) and potassium (K) are the major plant nutrients in poultry litter (Table 1) that are normally managed. All three elements are key essential nutrients in most crop production systems, however, N and P are both potential major pollutants if applied in excessive amounts.

Nitrogen:

Nitrogen is an essential nutrient for both plants and animals. Poultry litter is an excellent source of N and when litter is land applied the N is taken up by the crop or it can be lost to the environment. Due to its dynamic nature, N can be lost from soils through surface runoff, by leaching and/or as a gas. Nitrogen in litter is bound as a combination of organic N forms, ammonium-N (NH_4-N) and nitrate-N (NO_3-N).

Nitrate-N (NO_3-N) is non-adsorbed by soils and is readily subject to leaching losses. Nitrate is considered as one of the most extensive sources of non-point source pollution in U.S. ground water supplies (Patrick et al., 1987). Groundwater contamination with NO_3-N is considered a serious human health risk. Consumption of NO_3-N by young infants (up to $\sim$ six months in age) can lead to Methemoglobinemia or "blue baby syndrome." To protect the public from this condition, the USEPA limit for NO_3-N in drinking water is 10 mg NO_3-N /L (USEPA, 1985).

Another concern with NO_3-N in the environment is its contribution to eutrophication of surface waters. Eutrophication is the increase in the fertility status of surface waters with limiting nutrients which leads to accelerated growth of algae and aquatic plants. Most fresh water ecosystems are P limited, but certain lakes and coastal estuarine systems are limited by N (NRC, 1978). When compared to drinking water standards much lower levels of NO_3-N can lead to eutrophication. Prolific algal growth can occur at NO_3-N levels of 1-2 mg L^{-1} when no other limitations exist (Walker & Branham, 1992).

Nitrogen can also be lost to the atmosphere as a gas. Surface applications of poultry litter can result in a loss of ammonium-N (NH_4-N) through ammonium volatilization, which is also considered to be a serious environmental impact. Volatilized NH_4-N will be re-deposited to soils where it can be taken up by plants or to surface waters where it can contribute to eutrophication. In a recent estimate in North Carolina, atmospheric deposition ranged from 10 to 50% of the external N loading in estuarine and coastal waters (Paerl and Fogel, 1994).

Phosphorus:

Phosphorus (P) is also essential for plant growth and most crop production systems are limited by the level of P in the soil. Phosphorus in soils is very insoluble and P is delivered to surface waters primarily through surface runoff. Possible exceptions could include soils with high water tables, deep sandy soils, and P-saturated soils that would increase the potential for phosphorus leaching (Sims et al., 1998). During runoff events, water moving across the soil surface can transport both soluble and particulate forms of P to surface waters (i.e., streams, rivers, lakes and oceans) which results in increased levels of bioavailable P. Most aquatic systems are P limited and eutrophication thresholds have been reported to range from 10 to 100 µg P/L (Correl, 1998; Mason, 1991; USEPA, 1986).

Losses of P in surface runoff will be affected by a number of site specific factors including hydrological conditions and farming practices (i.e., crop rotations and the application of manure and fertilizers) (Lennox et al., 1997). In recent years, a number of investigations have indicated that land application of P as animal manure can contribute to the concentration of P in surface waters and eutrophication (Williams et al., 1999). In particular, the increase in P concentrations in agricultural drainage water over time reflects the accumulation of P in soils, especially when P is applied in excess of crop P needs (Sharpley et al., 2000). Currently, the greatest concerns with respect to water quality impacts of P are in those areas with intensive agriculture where P imports have exceeded P exports in agricultural products. In particular, in those geographic areas having concentrated animal feeding operations, soils that have been tested commonly show very high levels of soil test P. Research has shown that the potential for P losses (dissolved and particulate P) increases as the P status of a soil increases. Recent findings have also suggested that the concentration of dissolved P in surface runoff increases with increasing levels of soil test P. Thus, the significant build-up of P in soils treated with poultry litter is a serious environmental concern in areas that are proximate to sensitive natural waters, such as the Chesapeake Bay.

Poultry Litter Management: An Example of N vs. P-Based Management

Historically in Virginia and other Mid-Atlantic States, poultry litter has been applied to agricultural land primarily for its nitrogen (N) value. Poultry litter is an excellent source of both N and P, and dry poultry litter contains similar amounts of N and P when expressed as P_2O_5 (Table 1). In a N-based scenario, poultry litter is applied at a rate that will supply the N needed by the crop to which it is applied (Fig. 1). When poultry litter is applied on the basis of its N value, considerable P is also applied to the soil regardless of whether a crop can use it (Fig. 1). An excess of P is applied, since plants take up and remove only about 10 to 25 percent as much P as N (Table 2), and an accumulation of P in soils treated with animal waste results because of a net surplus of P. In the past, it was thought that over applying P would not create an environmental problem, especially if soil erosion is controlled, since P is very immobile in soils. However, long term application of P in excess of crop P needs can result in a significant build up of soil P which may lead to the degradation of water quality.

With the passage of nutrient management legislation in several of the Mid-Atlantic States, the poultry industry will begin moving toward P-based nutrient management planning. The potential for generating excess litter and nutrients on some farms is illustrated in Table 3. This example assumes that poultry litter generated from a house with 200 animal units of poultry is applied to corn that is grown on a productive soil with an expected yield of 150 bushels/acre. A crop of 150 bushels of corn would be expected to remove 135 pounds N and 53 pounds P_2O_5/acre. If applied to supply the expected N needs of the corn, litter would be applied at a rate of 4.2 tons/acre and only 36 acres of crop land would be needed to utilize all of the litter. However, applying litter on a N basis results in a net excess of 215 pounds P_2O_5/acre above what the corn crop can remove. As a comparison, if litter is applied at a rate to supply the amount of P_2O_5 that is removed in the harvested grain, only 0.83 tons/acre are needed and a total of 181 acres of crop land would be needed to utilize all of the litter. In this P-based system, the corn crop would have a net deficit of 120 pounds N/acre which would need to be applied as commercial fertilizer and approximately 5 times more crop land would be needed as compared to the N-based system. If the producer is land limited, he will either need to look at ways of decreasing the amount of P coming into the farming operation (Fig. 2), or he will need to find ways of moving excess litter from the farm boundary.

Table 1. Typical values for the nutrient content of poultry litter sampled in Virginia (DCR, 1993).

Manure	Total Nutrient Content (pounds/ton)		
	Nitrogen	P_2O_5	K_2O
Dry Brolier Litter	63	62	29
Dry Turkey Litter	62	64	24

Table 2. Nutrient removal by selected Virginia crops (DCR, 1993)

Farm Product	Plant Part	Yield	Nutrient Removal In Harvested Product		
			N	P_2O_5	K_2O
Corn	Grain	150 bu	135	53	40
	Stover	4.5 tons	100	37	145
	TOTAL		235	90	185
Wheat	Grain	80 bu	100	45	49
	Straw	2.0 tons	34	9	113
	TOTAL		134	54	162
Alfalfa	Hay	4 tons	180	40	180
Coastal Bermudagrass	Hay	8 tons	60	20	60
Timothy	Hay	2.5 tons	60	25	95

Fig. 1. Nutrient removal by a 150 bu/acre corn crop as compared to applying poultry litter to meet the corn N needs and the P removal needs. Litter would be applied at rates of 4 and 1 ton/acre to meet the N and P needs, respectively.

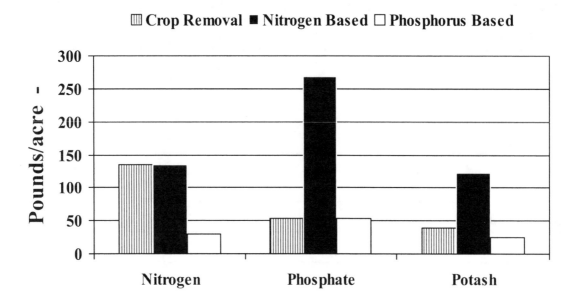

Table 3. An example of applying poultry litter on a N and a P_2O_5 basis to a high yielding corn crop. The available crop land has an expected yield of 150 bushels corn/acre and the operation is assumed to have 200 animal units of broilers (20000 birds) in one house.

Background Information
•   Farm produces 200 animal units of broilers or 20000 broilers •   Expected grain yield of 150 bu/acre •   The harvested corn grain will remove 135 pounds N and 53 pounds P_2O_5 •   The litter contains 62 pounds total N (15 pounds NH_4-N + 47 pounds organic N) and 64 pounds P_2O_5/ton. •   Litter is surface applied in the spring without incorporation. Thus 50% of the NH_4-N is lost due to volatilization and 60% of the organic N becomes available the year of application. Total available N = 36 pounds/ton. •   Nitrogen use efficiency = 1 pound available N/bushel of corn (Virginia Tech Recommendations)

Litter Data	Corn Data
Total litter = 150 tons/year Available N = 5400 pounds N/year Available P_2O_5 = 9600 pounds P_2O_5/year	Yield = 150 bushels/acre N Required = 150 available N/acre P Removal = 53 pounds P_2O_5/acre

Comparing the methods of Application	
N Basis	P_2O_5 Basis
Litter rate = 4.2 tons litter/acre 150 pounds available N applied/acre 268 pounds P_2O_5 applied/acre Corn N needs met 215 pounds P_2O_5/acre surplus Land Required = 36 acres	Litter Rate = 0.83 tons litter/acre 30 pounds available N applied/acre 53 pounds P_2O_5 applied/acre 120 pounds N deficit/acre Corn P_2O_5 removal needs met Land Required = 181 acres

Fig. 2. Movement of nutrients through various pathways in a poultry farm (NARES-132).

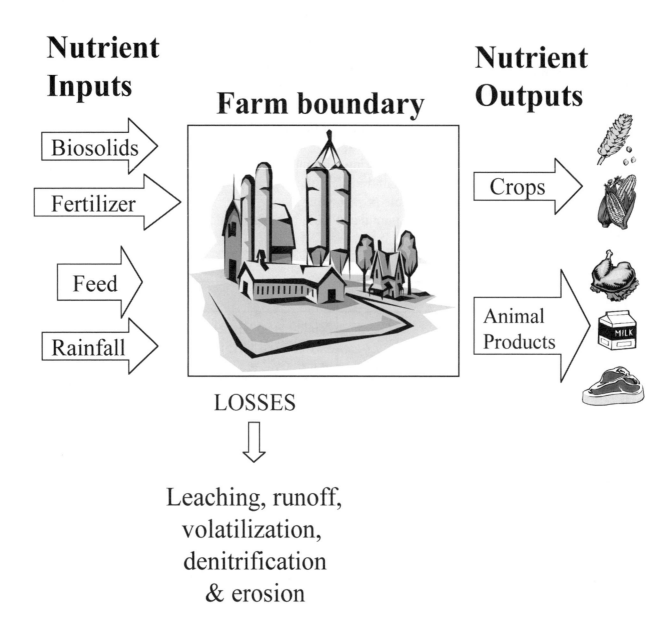

Nutrient Inputs

Biosolids

Fertilizer

Feed

Rainfall

Farm boundary

LOSSES

Leaching, runoff, volatilization, denitrification & erosion

Nutrient Outputs

Crops

Animal Products

Source: Adapted from *Poultry Waste Management Handbook,* NRAES–132 (1999)

Comprehensive Nutrient Management Plans:

Nutrient management planning is a tool that can be used to implement practices that permit efficient crop production and protect water quality from nutrient pollution (Nagle et al., 1997). A comprehensive nutrient management plan is a written, site-specific plan that addresses both crop production and water quality issues. The goal of any nutrient management plan should be to minimize the environmental impact of nutrient use while optimizing farm profits. A nutrient management plan is also a tool that a producer can use to maximize the utilization of on-farm resources and purchased inputs. Nutrient management plans should be site-specific and should be developed to reflect specific soil and crop production systems. Reducing N and P pollution of surface and groundwater has become a major goal in the U.S. and nutrient management planing will be an effective tool for ensuring that these nutrients are not over applied.

All comprehensive management plans should be developed site-specifically and strategies will not be the same for all farms (Beegle et al., 2000). In addition, each state may have various approaches, but the following general steps will be essential (Nagle et al., 1997):

1. Gather accurate soil information for each field or management unit and analyze representative soil samples from each management unit.
2. Determine crop yield potential for each field, based on the known productivity of the soils present coupled with the intended management practices.
3. Identify the plant nutrient needs to achieve this expected yield potential.
4. Determine the plant-available nutrients in manures or biosolids to be used, considering the type of organic material and method of application.
5. Estimate the nutrient contribution that can be expected from residual effects or carryover from previous fertilizer, manure or biosolids applications.
6. Include credit for N supplied to row crops following legumes.
7. Recommend application rates for manure and/or commercial fertilizers to supply the needed nutrients at the appropriate time for optimal crop production.

This step-by-step process should be followed for each field and production system that is covered by an individual nutrient management plan. It is extremely important for the plan writer to obtain as much current and site specific data as possible during the planning process.

Dealing with Environmentally Sensitive Areas:

The overall potential for N and P to migrate to surface and groundwater will be largely dependent on soil and site conditions. In nutrient management planning it will be important to recognize and delineate environmentally sensitive sites and conditions within a panning area. Soils and landscape features vary considerably within the Mid-Atlantic states, however, the following features and properties are conducive to nutrient losses from agricultural fields: 1) Soils with high leaching potentials. The leaching index for a soil can be obtained from the USDA-NRCS Field Office Technical Guide (USDA Soil Conservation Service, 1990); 2) Land with karst (sinkhole) drainage regimes. Sinkholes are landscape features commonly found in areas underlain by limestone bedrock. Producers need to be aware that sinkholes form a direct connection between surface water and groundwater; 3) Shallow soils over fractured bedrock; 4) Tile drained land. Tile outlets provide a direct connection to surface watersheds; 5) Excessively sloping lands; 6) Flood plains and other land near surface waters.

The list above is not inclusive, but it includes most of the major types of soil/landscape concerns in the Mid-Atlantic states. Appropriate setbacks or buffer areas should be established between sensitive areas and fields receiving nutrients, and intensive nutrient management practices should be employed to minimize nutrient losses. The Phosphorus Index (Lemunyon and Gilbert, 1993) that is being evaluated by most states in the Mid-Atlantic region would account for most of these sensitive areas.

Whole Farm Nutrient Balancing:

Selection of site-specific nutrient management strategies will require a knowledge of nutrient cycling and how nutrients flow through a farming operation (Fig. 2). It will be critical for a producer to know the quantities of nutrients that are entering, leaving and remaining on the farm. Nutrients are brought on the farm in the form of purchased inputs which include feed and fertilizer. Nitrogen can also be added by legumes and as precipitation. Nutrients are exported from the farm boundary in the form of farm products as constituents of meat, milk and crops. In farming operations that include concentrated animal feeding operations, more nutrients are typically brought onto the farm than are exported as farm products and these excess nutrients are recycled in the form of animal manure and crop residues (Fig. 2). In this unbalanced system, the probability of nutrient losses by erosion and leaching increases.

A goal of nutrient management planning should be to balance the nutrient flow for a given farm, realizing that a farm may be nutrient-deficit, nutrient-balanced or nutrient-surplus (Beegle et al., 2000). Nutrient-deficit farms are farms where nutrient imports are less than exports (Fig. 2) and are usually low intensity/low animal density operations. In this type of operation, manure production is not adequate for crop production needs which will require the additional input of nutrients as fertilizer or other sources. Nutrient-balanced operations are farms where nutrient imports are approximately equal to exports. In this scenario, manure can meet most if not all of the nutrients required for crop production. Nutrient-surplus operations are farms where nutrient inputs exceed exports. In this scenario, nutrients in manure exceed the needs for crop production and all of the manure cannot be safely applied on-farm. Thus, it will be important for a producer to recognize what type of operation they have and plan accordingly.

High density poultry farms are nutrient-surplus operations and proper management of excessive nutrients will be a key factor in nutrient management planning. In these types of operations a producer will need to know the amount of poultry litter that is produced annually, the nutrient content of the litter, and have access to adequate storage facilities for the litter. Producers should have access to recent manure analysis data to accurately determine the quantities of nutrients that are being generated as manure. These types of systems will be land limited and it will be important for the producer to use cropping systems that ensure a high uptake and removal of N and P. In addition, the producer will need to schedule manure and fertilizer applications to ensure optimum utilization of the applied nutrients by the crop. Producers will also need to find off-farm outlets for their excess manure. Successful management of excessive nutrients will depend on securing adequate land suitable for manure spreading on farms in the vicinity and on developing a manure transport system (Beegle et al., 2000). Alternatives to a lack of land may also include composing and the production of litter derived fertilizer products. The producer may also work to increase the feeding efficiency (i.e., adding the phytase enzyme to poultry rations) to reduce nutrients excreted by poultry and ultimately reduce the input of nutrients into the farm.

As noted earlier, manure is a good source of N, P and K. However, due to water quality concerns manure should be managed on the basis of N or P as the most limiting plant nutrients. Several states are moving toward P-based nutrient management planning for poultry litter and other nutrient sources. Some states are also looking at the development/implementation of a site-specific Phosphorus Index as a component of P-based nutrient management planning. The Phosphorus Index (PI) is a field-scale assessment tool for ranking the vulnerability of fields as sources of P loss in surface runoff (Lemunyon and Gilbert, 1993). Using the interactions between P source and transport processes that control P export, the PI could be used to identify and manage critical source areas within a farm that are most vulnerable to P losses.

For those growers using nutrient management plans based on N, it will be important to monitor soil test P levels and manage fields to prevent P levels from becoming too high. All growers should use appropriate conservation measures (i.e., cover crops, buffer strips, conservation tillage, etc.) to reduce soil erosion and runoff losses of P.

Summary:

Nutrient management planning for poultry litter will be an important tool for protecting water quality in the future. The plans should be site-specific and developed to maximize profits and minimize environmental impacts. Farmers should become informed about the environmental concerns regarding nutrient use, especially if their operation is proximate to sensitive natural waters. For poultry litter, nutrient management planing will continue to concentrate on N and P as the limiting nutrients, with some states moving to P-based planning. Nutrient management plans should be site-specific and will require farmers to become familiar with the quantities of nutrients that enter, exit and remain within the farm boundary.

References

Beegle, D.B, O.T. Carton, and J.S. Bailey. 2000. Nutrient management planning: Justification, theory, practice. J. Environ. Qual. 29:72-79.

Bagley, C.P., R.R. Evans, W.B. Burdine, Jr. 1996. Broiler litter as a fertilizer or livestock feed. J. Prod. Agric. 9:342-346.

Burkholder, J.A., and H.B. Glasgow, Jr. 1997. *Pfiesteria piscicida* and other Pfiesteria-dinoflagellates behaviors, impacts, and environmental controls. Limnol. Oceanogr. 42:1052-1075.

Correl, D.L. 1998. The role of phosphorus in the eutrophication of receiving waters: a review. J. Environ. Qual. 27:261-266.

Lemunyon, J.L., and R.G. Gilbert. 1993. The concept and need for a phosphorus assessment tool. J. Prod. Agric. 6:483-496.

Lennox, S.D., R.H. Foy, R.V. Smith, and C. Jordan. 1997. Estimating the contribution from agriculture to the phosphorus load in surface water. P. 55-75. *In* H. Tunney, O.T. Carton, P.C. Brookes, and A.E. Johnston (ed.) Phosphorus loss from soil to water. CAB Int. Press, Cambridge, UK.

Mason, C.F. 1991. Biology of freshwater pollution, 2nd ed. Wiley, New York.

Nagle, S., G. Evanylo, W.L. Daniels, D. Beegle, and V. Groover. Chesapeake Bay Region Nutrient Management Training Manual, F. Coale (ed.). Chesapeake Bay Program.

NRC. 1978. Nitrates: An environmental Assessment. National Research Council, National Academy of Sciences, Washington, DC. 435-84.

Paerl, H.W., and M.L. Fogel. 1994. Isotopic characterization of atmospheric nitrogen inputs as sources of enhanced primary production in coastal Atlantic Ocean waters. Mar. Biol. 119:635-645.

Patrick, R., E. Ford, and J. Quarles. 1987. Groundwater contamination in the United States. 2nd ed. University of Pennsylvania Press, Philadelphia, PA.

Pelletier, B.A. 1999. Virginia grain handling practices and corn for poultry litter exchange program. M.S. Thesis, Virginia Tech, Blacksburg, VA.

Sims, J.T., R.R. Simard, and B.C. Joern. 1998. P losses in agricultural drainage: Historical perspective and current research. J. Environ. Qual. 27: 277-293.

Sharpley, Andrew, Bob Foy, and Paul Withers. 2000. Practical and innovative measures for the control of agricultural phosphorus losses to water: An overview. J. Environ. Qual. 29:1-9.

U.S.D.A. Soil conservation Service. 1990. Field Office Technical Guide. Section II-iii-L, Water Quantity and Quality. Soil Ratings for nitrate and Soluble Nutrients. Publication. No. 120-411.

United States Department of Agriculture. 1999. 1997 Census of Agriculture. Virginia, State and County Data. Volume 1, Geographic Area Series. Part 46. AC97-A-46.

United States Department of Agriculture. 1999. Poultry Production and Value. Final Estimates 1994-97. Statistical Bulletin Number 958. http://usda.mannlib.cornell.edu/usda/reports/general/sb/b9580399.txt.

U.S. Environmental Protection Agency (USEPA). 1983. Chesapeake Bay program: Findings and recommendations. United States Environmental Protection Agency, Philadelphia, PA. 48p.

U.S. Environmental Protection Agency (USEPA). 1985. National primary drinking water regulations: Synthetic organic chemicals, inorganic chemicals, and microorganisms: proposed rule. Fed. Reg. 50:46935-47022.

U.S. Environmental Protection Agency (USEPA). 1996. Environmental indicators of water quality in the United States. EPA 841-R-96-002. USEPA, Office of Water (4503F), U.S. Gov. Print. Office, Washington, D.C.

U.S. Senate Committee on Agriculture, Nutrition and Forestry. 1997. Animal waste pollution in America: an emerging national problem. Environmental risks of livestock & poultry production. http://www.senate.gov/~agriculture/animalw.htm/. pp 1-23.

Virginia Department of Conservation and Recreation. 1993. Nutrient Management Handbook. Second Edition. Department of Conservation and Recreation, Division of soil and Water Conservation, Richmond, VA.

Walker, W.J., and B. Branham. 1992. Environmental impacts of turfgrass fertilization. P. 105-219. *In*: J. C. Balogh and W.J. Walker (eds.) Golf course management and construction. Lewis Publishers, Boco Raton, FL.

Williams, C.M., J.C. Barker, and J.T. Sims. 1999. Management and utilization of poultry wastes. Rev. Environ. Contam. Toxicol. 162:105-157.

Nutrient Management Plans — Swine

William J. Rogers
Nutrient Management/Environmental Division Manager
Brubaker Consulting Group
(formerly Brubaker Agronomic Consulting Service)
4340 Oregon Pike
Ephrata, Pennsylvania 17522

Biographies for most speakers are in alphabetical order after the last paper.

Introduction

Over the past fifteen years swine production has drastically changed. Many years ago swine were housed in outside open lots with very little confinement. Today most swine are completely confined. Since the industry has moved toward complete confinement, more manure is collected than with the older style open feed lots. With an increased awareness of environmental issues and continuing concerns about clean water, legislators have enacted nutrient management regulations on confined or concentrated animal agricultural operations.

Nutrient management plans have evolved over the last ten years. Previously most nutrient management planners attempted to only match manure applications to meet crop needs. At times, their primary concern was to use 100% of the manure produced on a farm at that farm. Planners now realize there is a value to swine manure. The swine industry is trying to encourage the perception that manure is not something that requires disposal but something that can be utilized. The application of swine manure is an excellent was to meet the agronomic needs of a crop.

The method of developing nutrient management plans has also evolved. Previously nutrient management plans dealt with meeting crop needs, most commonly by meeting the nitrogen needs of the crop. There were no formal plans developed dealing with manure handling, barnyards, or the areas that were to receive manure applications. Nutrient management planners are re-evaluating the development of nutrient management plans and their content. Their incentive is new regulations, concern over phosphorus, and continued concern about water quality. These new complete nutrient management plans are now referred to as Comprehensive Nutrient Management Plans (CNMP). At a minimum, a CNMP should address animal numbers, manure production, manure storage, manure application rates, manure application timing, manure handling in the barnyard, and soil conservation on the acres that are planned to receive manure applications.

Swine production has become more specialized. Each individual production phase or animal growth stage is conducted on separate sites. A few operations continue to complete the entire production on one site (farrow to finish). The standard production phases that will be reviewed are sow units, offsite nurseries, and finishing units. Each production phase encompasses unique issues and concerns that need to be evaluated during the development process of a CNMP. This report will discuss the above listed minimum requirements for each production phase; including potential problem areas and concerns that need to be analyzed while developing a CNMP for a swine operation.

Animal Numbers

Determining animal number for a swine operation is relatively simple. Most farms regularly track either daily or weekly animal inventories housed. Each type of unit will have different animal groups that are located on the operation. This variability is most common with the sow units and less common with offsite nurseries and finishing units.

Sow Units:
When working to develop a CNMP for a sow unit you need to determine the number of gestating sows, sows with piglets (farrowing sows), and boars. These three groups are fairly standard at all sow units. The only variation in these numbers will be the boar numbers. These numbers are slightly different in a unit using 100% artificial insemination than those units that are not using 100% artificial insemination.

Beside the standard animals groups (gestating/farrow sows and boars), there are numerous ways a sow unit may be operated. Other animal groups that could be located on a sow unit could be replacement gilts and nursery pigs. Again, if these animals are located on the unit they will be included in the inventory. The average weight of the replacement gilts should always be documented. Replacement gilt weight will vary from operation to operation, depending on whether the unit is raising their replacement gilts onsite or raised offsite and later brought onsite. In addition, replacement gilts may or may not be housed within the main buildings. If they are housed in a separate building they may have their own manure handling system that you will need deal with in the CNMP.

Nursery pigs do not have to be housed on the sow unit. Some sow units have the nursery onsite, some use offsite nurseries, and some send wean pigs directly to the finishing operations. During the information gathering process you will need to determine how each individual sow unit is handling the nurseries. As with the replacement gilts you will need to determine the average weight and manure handling system of the nurseries.

Offsite Nurseries:
Determining animal numbers in an offsite nursery is somewhat simple. There are two numbers that the operator will document - the total capacity of the unit and the number of animals that go through the unit in a given time period. For example, a unit may have a total capacity of 4,000 pigs, but have run 6 to 7 groups through per year; therefore the total animals could range from 24,000 to 28,000 per year. For development of a CNMP it is necessary to determine the total capacity of the unit or the number of animals that can be located on the site at any one given time. This number of animals is used to estimate manure production.

When determining animal numbers, determine the length of time they are located onsite, how many days the room is empty between groups, and how long it takes to fill and empty a room. Most nurseries are built with many rooms containing small groups of pigs in each room. The barn can be filled completely in one day from one source or continuously filled. Some units do not fill completely at one time; they rotate rooms to receive animals continuously from a variety of sources. Since there are times that the rooms are not in use there are no animals onsite producing manure. Determining these down days are needed when you begin to estimate manure production.

Finishing Units:
Determining animal numbers in a finishing unit is very similar to the offsite nurseries. There will be a total capacity of the unit and a total number of animals through a unit in a year. The number needed for the development of a CNMP is the total capacity at any one time. The operator should have animal numbers from past groups finished. For a new use the total capacity for which the building was designed for when developing the first CNMP for that operation. A producer may also be able to provide the number of animals placed in and sold out of the facility. There are mortalities during the cycle. You should use the average between the number of animals placed and the number of animals sold out of a unit. Using the farmer's records is the most accurate method to determine animal numbers. If these records are not available, use the designed capacity of the facility.

The main differences found in finishing units are the average weight of the animals that are housed in the unit. Some producers are placing pigs out of the nursery (feeder pigs) and others are placing animals directly weaned from the sow (wean pigs). A unit is either wean-to-finish (weanisher) or feeder-to-finish unit (finisher), determined by the age of the placed animals. This does not change the animal numbers placed in the unit. It does change the number of days that that unit is empty throughout the year and the average weigh to the animals in the unit. Weanisher units house two groups per year while finisher units house 2.5 to 3 groups per year. There is more down time with the finisher than with the weanisher. In addition, the weanisher houses animals around 12 pounds selling out at 240 to 260 pounds, while the finisher houses at 40 to 50 pounds and sells out at the same weight. The average weight in the weanisher is 131 pounds whereas the finisher has an average weight of 145 pounds.

Manure Production

Manure production within any unit is dependent on animal groups and numbers housed in the unit. The amount of material that needs to be used or land applied will vary depending on the additional dilutions that are added to the manure storage system. Dilutions come from rainfall, washdown water, and human waste. Many older facilities placed the shower and toilet water in the pit. This is not legal in most states. These dilutions have a direct impact on the nutrient content of the manure and the amount of material to be land applied, but not on the amount of manure produced.

The following sections discuss how to estimate manure production. These estimates only include the manure produced. This does not include any dilutions that should be added to determine the actual amount of material that will need to be distributed.

There are no standard book values for the amount of washdown water added to the manure storage. Some operators may know or can determine the amount of washdown water that is used. A rule of thumb is to use 1.6 gal/Animal Unit/day with a sow unit, 1.2 gal/Animal Unit/day at an offsite nursery, and 1.0 gal/Animal Unit/day at a finishing unit. Next the amount of rain/snow that falls on the manure storage facility should be added. This is easily calculated using the surface area of the manure storage facility, the average amount of excess rainfall, and conversion factors. The issue arises of which excess rainfall value should be used. Some people feel that when evaporation exceeds natural rainfall you have no excess rainfall for that month. Others feel that in those months you have a negative rainfall since water is being evaporated off the manure storage and there is a reduction in the volume of material stored. My approach is that if the manure storage facility does not contain a heavy crust, water can be evaporated off that storage and therefore there is a negative rainfall for that month. Rainfall calculations are determined using the following equation:

(surface area in acres) x (excess rainfall) x (27,154 gallons/acre-inch).

The manure production discussed below for each phase of production will need to have the appropriate rainfall and washwater dilutions added to determine the actual amount of material produced on that site.

Animal groups and the numbers of animals within each group were previously determined. Each animal type within the production cycle has a standard average weight. These weights are fairly standard throughout the swine industry. The average standard 'book' weight will vary depending on the state where you are working. Each individual state has a standard weight and book value to be used for manure production determination. The most accurate value to use would be actual records from the operation. If the operator has recorded the weight of animals entering and leaving his facility, use this average weight and not the standard book values. It is recommended that you use the minimum of one full year's worth of records for this determination. When you reference the manure production tables within a state's regulations or guidance documents, manure production is based on some manure production weight or volume per average animal weight or animal unit. This varies tremendously from state to state. Herein lies one of the problems with developing a CNMP for a new unit that has no production records. Which of these production guidelines is most correct? Also, most of these tables list varying animal types not specific animal unit or production unit.

Sow Units:

When developing a CNMP for a sow unit you need the standard animal groups (boars, gestating sows, and sows with piglets). There may or may not be replacement gilts and nursery pigs. Below is a table representing the estimated manure production from a 4,000-head sow unit using three state's estimated manure production tables. There are 3,600 gestating sows, 500 sows with piglets, 50 boars, and 125 replacement gilts. All these animals are assumed to be on the operation 365 days/year.

Estimated Manure Summary Table			
Manure Source	**Pennsylvania**	**Virginia**	**Maryland**
Sows w/piglets	1,091,058 gal.	902,400 gal.	617,580 gal.
Gestating Sows	1,702,944 gal.	2,102,400 gal.	1,715,558 gal.
Boars	65,536 gal.	32,850 gal.	21,325 gal.
Replacement Gilts	50,014 gal.	49,663 gal.	26,039 gal.
Total Manure	**2,909,552 gal**	**3,087,313 gal.**	**2,380,502 gal.**

There is variation in the different state's estimated manure production for this operation. This calculation is also slightly time-consuming, determining the estimated manure production for each individual animal group. It would be easier and possibly as accurate if one standard manure production value was developed for sow units with and without a nursery. The majority of the manure is from the sows (96% for PA, 97% for VA, and 98% for MD). The boars and replacement gilts have limited impact on the total manure production. As discussed later, nurseries do produce large amounts of manure and can have an impact on the total manure produced, therefore I propose different values for operation with and without nurseries.

Offsite Nurseries:

Estimated manure production for an offsite nursery is slightly easier than for a sow unit. There is only one animal group with which to work. Once the average capacity and number of days onsite are determined, the number of animal units can be easily determined by the following equation:

(Number of animals x average weight x days onsite ÷ 365 days ÷ 1000 lb./animal unit)

A quandary with estimating manure production for an offsite nursery is that Maryland is the only state that lists an estimated manure production for nursery pigs. Virginia and Pennsylvania do not list a separate listing only for nursery pigs. In the table below, finishing pig manure production value was used to estimate manure production in Virginia and Pennsylvania. This estimated manure production table is for a 4,000-head nursery unit.

Estimated Manure Summary Table			
Manure Source	Pennsylvania	Virginia	Maryland
Nursery Pigs	317,607 gal.	315,288 gal.	534,386 gal.

There is variation among the three states in the amount of manure they estimate to be produced by this unit. Remember that Maryland is the only state that has actual nursery pig estimated manure values.

Comparing these estimates to actual figures shows that Maryland's nursery pig values are correct and that using finisher pig values for nursery units would underestimate manure production. The manager at this 4,000-head operation maintains very accurate manure hauling and washwater records. The manure at this site is stored in an under house storage facility therefore there are no rainfall additions to the manure system. They produced a total of 542,100 gallons of manure in 1998 and 547,400 gallons in 1999. These values are consistent with Maryland's book value and demonstrate that there is a difference in the amount of manure produced by a nursery pig than that produced by a finishing pig. If your state does not provide separate nursery pig manure production values, it is recommended you use other values beside the finishing pig values to estimate manure production.

Nursery pigs produce a larger amount of manure per animal unit when compared to finishing pigs, sows, replacement gilts, and boars. If a sow unit has a nursery it would have a significant amount of additional manure produced due to the nurseries. This is the logic used to previously propose a separate manure production figure for sow units with and without nurseries.

Finishing Units:
Finishing units are closely linked to offsite nurseries in the procedure used to determine the estimated manure production. There are two standard types of units - weanishers and finishers. Each of these units will have a different number of empty days and average animal weights. If we evaluate the three states' estimated manure production for a standard 2000-head finishing barn, the estimated manure production is very similar.

Estimated Manure Summary Table			
Manure Source	Pennsylvania	Virginia	Maryland
Finisher	745,038 gal.	740,175 gal.	749,768 gal.
Weanisher	653,978 gal.	649,042 gal.	658,130 gal.

There is very little variation between states, but a significant difference between the finishers and the weanishers. The difference between finishers and weanishers is not quite as large as it appears. The manure production for the weanisher is determined by using a lower average weight, but the weanisher units contain pigs that normally would have been placed in a nursery unit. The previous discussion demonstrated that nursery pigs produce more manure per animal unit than finishing pig. The manure production for a weanisher is slightly lower than that of a finisher, but not as dramatic as presented in the table above. Data to defend this statement is currently being collected.

The difference in total material produced between a finishing and a weanisher will be greater. A finishing unit completes 2.5 to 3.0 cycles per year while a weanisher only completes 2 cycles per year. Therefore, there is less washwater added to a weanisher manure storage than added to a finisher manure storage.

Nutrient Analysis

Determining the amount of manure or total material produced at an individual unit is half the battle but not the most important part of a CNMP. In a CNMP we are attempting to use the material produced in the most efficient and environmentally safe way as possible. Therefore, the next step in the plan development is the attempt to estimate a nutrient value for the material produced. For an existing operation this can be as easy as collecting a manure sample from the manure storage. One problem with a manure sample is attempting to collect a manure sample that accurately represents that material stored in the manure storage facility. In an under house concrete storage or concrete tank that is agitated before the manure is removed a composite sampling tube can be used to collect a complete sample from top to bottom of the pit. When manure is stored in an earthen or HDPE-lined storage facility the only way to collect this sample is to float out onto the storage and collect a composite sample with a core style sampling devise. I do not recommend this approach. For these types of storages I recommend collecting a sample during the land application process. Secondly, one sample may not accurately represent the contents of the manure storage facility. It is best to have the operator collect a sample before or during each manure application. From a few years history of manure nutrient analysis you can begin to more accurately determine the average manure nutrient content.

For existing operations a single manure sample is a start for determining the nutrient content of the stored material. For new operations you need to attempt to estimate the nutrient content of the manure from tables and charts provided by each individual state. Below is a table containing estimated manure nutrient contents from under house stored manure by animal group type.

Animal Type	Pennsylvania			Virginia			Maryland		
	Lb./1000 gal.								
	N	P_2O_5	K_2O	N	P_2O_5	K_2O	N	P	K
Gestating Sows	25	10	17	--	--	--	25	10	17
Sows w/Piglets	40	13	13	--	--	--	29	15	23
Grower Pigs	52	23	18	--	--	--	52	22	18
Nursery Pigs	--	--	--	--	--	--	40	13	13
Swine Lagoon	--	--	--	10	5	6	--	--	--
Mixed Swine	--	--	--	41	27	30	--	--	--

These figures do include the added washwater dilutions, but do not include any additional rainfall dilutions. To determine an estimated nutrient content for a given operation the additional rainfall dilution will need to be accounted. A ratio between the manure/washwater and the manure/washwater/rainfall needs to be calculated to determine the dilution percentage for the standard book value nutrient content (this value should be less than 1). For nursery operations the nutrient content values provided by Maryland are very close to those that have actually been measured on operating farms in under house manure storages. This is not true for sow and finishing units. The values listed for finishing units are an average of all samples submitted to the state testing laboratories including finishers as well as weanishers. As presented earlier the finisher has more washdown water dilution than that of a weanisher. Is it significant enough to change the nutrient analysis? Most likely yes, but when developing a CNMP for a new operation you are only estimating the nutrient analysis, therefore this figure is accurate enough for a starting point. Once the operation is under production manure samples should be collected to validate the estimates made in the original plan. For sow operations you have to calculate the nutrient contributions from each of the individual animal groups to determine the overall manure nutrient content. Here again is a place that I would propose that we move away from the typical animal groups and move toward having book values for sow units with and without nurseries. The sows and the nurseries are the two major contributors to the manure. The boars and replacement gilts contributions are small and have little impact on the final nutrient content of the material stored.

Application Rates and Application Timings:

With older traditional nutrient management plans this section or part of the nutrient management plan was the body of the plan. In this section of a plan recommended manure applications are given to meet the nutrient needs of the planned crop. Historically these applications have been based on the nitrogen needs of the planned crop. In the near future, many manure applications may be limited by crop phosphorus needs. Since this is currently only regulated in Maryland the following discussion will be based on crop nitrogen needs.

In determining manure applications and application timing the manure nutrient content, nutrient needs of the planned crop, manure storage capacity, material production, and incorporation timing need to be accounted. Previously discussed where manure production and nutrient analysis. Therefore you already have determined the amount of manure produced and its nutrient value. In this section of the plan you are attempting to match manure applications to meet nutrient needs of the crop. The first determination is developing a realistic yield goal for each crop. Again, producer historical records are the best gauge for determining realistic yield goals. If these are not readily available each state has yield goal potential by soil type available for most crops. Understand that many of these tables are many years old and possibly outdated. In some crop groups there have been significant yield improvements over the last ten years. As a rule of thumb I use these table, the producer's yield estimates, and my professional judgement to determine a realistic yield goal.

Once you determine the yield goal you need to determine the nutrients required to meet that yield goal. From this you now know the amount of nutrients required for a given crop. The plan needs to make manure applications meeting that need. Since most plans are based on nitrogen you also need to account for residual nitrogen from past manure applications and legumes along with nitrogen availability based on the operators incorporation timing. Manure incorporation timing determines the nitrogen availability. Manure injected has a higher nitrogen availability than manure that is surface applied and incorporated a week later. The producer needs to be involved with these decisions since he will have to implement the plan. If a plan is developed with injected manure and the producer does not or can not complete this application he can not implement the plan.

In addition to meeting the crop needs manure application timings need to account for the manure storage capacity. Again, if a producer has only six months worth of storage capacity and the plan is developed stating all the manure is applied in the spring the operator will have a manure storage facility that is overflowing half the year. When developing a CNMP for an existing operation the manure storage capacity needs to be determined and included in the entire planning process. If you are developing a plan for a new operation the operator may allow you to assist him determine his manure storage capacity needs, and then the plan can be based off this decision.

If you are able to work with the operator to develop the manure storage capacity it is recommended to evaluate the crops to be produced, the manure application window for his area, and the equipment available for manure applications. If an operator is growing all corn, some planners could consult him to have one year's worth of manure storage. But, if there is only a short period of time that manure can be applied in the spring this may not be practical. This operation may only need six to eight months of storage and apply some manure in the fall with a cover crop and the rest in the spring. The manure application schedule needs to meet the nutrient needs of the crops along with the storage capacity and manure application timing ability on the facility.

Manure Storages

As with any animal species there are many different types of manure storage facilities that the industry standardly uses. In the following sections different types of manure storages will be discussed. This discussion will include the effects of the structure on manure nutrient values, items that need to be evaluated during the development of the CNMP, and some best management practices (BMPs) that should be included in the plan. With most new regulations and the federal CAFO/AFO strategy, CNMPs are beginning to evaluate existing manure storage facilities structural integrity. There are best management practices that should either be implemented with each different manure storage type.

In general for all manure storage facilities you should development of an emergency action plan for the facility. This plan states what has been done during construction to potentially lesson the chance of a manure spill, the maintenance items complete to retain the structural integrity, and what is to be completed if the structure fails. This plan would detail each of these items along with who should be contacted in case of an emergency. The development of an emergency action plan is a newer item that should be included in the developed of a CNMP. Along with a full emergency action plan, the planner should develop a one-page action list and emergency contact numbers for the catastrophic failure of a manure storage structure. When a storage facility fails or a tankers overturns most people are running around in high state of confusion. Developing this one page summary and then posting this action page and emergency list of numbers any employee can quickly determine the actions they need to take and who to contact.

In addition, each type of manure storage facility should be secured. This would include a fence for an earthen or HDPE-lined facility, a fence and lock for a concrete tank, and covers and locks for under house storages. Securing manure storages stops accidents from happening and keeps unwanted people out of manure storage facilities.

Earthen Manure Storage:
This type of facility is maybe the least environmentally favorable swine manure storage structure. If managed correctly they can safely store manure for many years. The management required for an earthen facility is the highest of all discussed facilities. There are many BMPs that must be added to maintain the structural integrity of these facilities. Many CAFO regulations are requiring some sort of groundwater monitoring system. The earthen structure can not have a leak detection system added so the only ground water monitoring that can be used is monitoring wells, adding to the initial capital cost of the facility. If these structure fail it is harder to cleanup from the failure.

Earthen manure storages collect a large amount of dilution rainfall water in the stored manure. This dilution water must be accounted for in the nutrient content of the manure, the total material that is produced, and the storage capacity needs of the operation. New CAFO regulations require 2 feet of freeboard space be maintained for these facilities. Some older state standards allowed 12 to 18 inches of freeboard. The freeboard is to allow for the facility to collect the rainfall from a 25-year/24 hour rainfall event. This is normally 6 to 8 inches in the Mid-Atlantic States. The additional space is required to maintain the structural ability of the earthen berm to hold back the force exerted on the berm by of the manure stored behind it.

During operation there are many BMPs that are required with the management of these facilities. With the development of the CNMP these BMPs should be listed in the plan and discussed with the operator. Many of these BMPs are not followed on current animal operations. The major concerns are the maintenance of freeboard space, maintaining the vegetative cover reduce soil erosion on the inside and outside of the berms, removing burrowing animals from the berms when found, and removing woody brush and trees. During the development of a CNMP if items are in need of repair they should be listed in the plan along with any ongoing maintenance BMPs.

An additional management practice to be reviewed and discussed is the removal of the stored manure. Will the operator only remove the liquid every year leaving the solids to build up? If so the CNMP needs to address the nutrient content of only the liquid. A discussion on when solids should be removed and how to remove these solids should be added to the plan.

The removal of solids will also require an update to the CNMP. If the operator plans to agitate this facility and remove both solids and liquid on a regular basis, the plan should discuss the BMPs required when agitating an earthen manure storage facility to maintain the integrity of the earthen liner.

HDPE-Lined Manure Storages
These facilities have a thick plastic liner installed inside an earthen storage area reducing the potential for nutrient leaching into ground water. Many of these facilities also contain a ground water drainage system that should be sampled and used as a leak detection system. HDPE-lined facilities reduce ground water contamination potential over the earthen lined facilities but can still overflow the berms if not regularly emptied. Therefore an emergency action plan should be developed for the manure storage facility.

The major items to evaluate on these facilities are the same as for the earthen storages. The one difference is it is more difficult to rupture these liners when agitating the manure, although it is still possible. These facilities also collect rainfall and snow diluting the manure and requiring the 2-foot freeboard. One BMP not needed with earthen liners that should be added when using an HDPE-lined facility is an escape mechanism or route from the facility. You can walk out of most earthen lined facilities. Once you place the HDPE-liner inside the earthen berm they are very hard to walk out of. If someone was to fall in they need to have some mechanical assistance to climbing out of the structure. This can be as simple as a nylon rope attached to a post located several places around the facility.

Outside Concrete Tanks
A few newer swine facilities use concrete tanks. These are not widespread since they are costly to construct and have limited manure storage capacity when compared to earthen or HDPE-lined facility. With the use of these facilities there is again a lesser possibility for catastrophic failure and leaching of nutrients into ground water. Most concrete tanks contain a foundation drain that can also be sampled and used as a leak detection system. Concrete tanks normally are designed with less freeboard than earthen and HDPE lined facilities. They must have at least the freeboard to contain a 25-year/24 hour return interval storm. Some state CAFO regulations are still requiring a two-foot freeboard on these facilities.

During the operation of these facilities there is less chance that they will be damaged during unloading of manure. One item to evaluate is the unloading method. Many older tanks are designed with gravity unloading systems. If the valve on the tank leaks or ruptures the tank will quickly empty. If the operator plans to continue to use the gravity unloading valve a containment area should be added that could contain the manure stored within the tank.

Since concrete tanks can be built deeper and taller than earthen and HDPE facilities they normally have less surface area to collect rainfall dilutions. This increases the nutrient content of the manure and reduces the volume of material to be handled. Concrete tanks are used more regularly with other animal species than with swine.

Under House Concrete Tanks

Over the past few years this form of manure storage has become more prevalent with finishing units. The construction of under house manure storage systems works easily with newly constructed finishing units. It is comparable to adding a basement under a house. The designs of these limit the possibility of manure leaking from the facility into ground water. Under the concrete manure storage is a foundation drain that if sampled can act as a leak detection system. The leak detection system can be sampled routinely to monitor the facility and also can be used if a leak is detected to contain the movement of nutrients outside the storage into groundwater. These facilities also lower the possibilities of a catastrophic failure since they are build mostly underground.

Manure stored in these facilities does not have the extra rainfall dilutions to be accounted for in the amount of manure that needs to be land applied and the nutrient content of the manure. Manure stored under house is the easiest swine manure to export to neighboring farms since it has fewer dilutions there is a higher nutrient content in the manure. One drawback to manure stored in these facilities is the gases that are emitted just below the animals. Ventilation needs to be increased in these facilities drawing these gases out of the building. This does add to the operator's management responsibilities. Lastly, there is an additional public perception positive for under house manure storages, 'out of sight out of mind'. Due to the positives of this type of storage there are a few sow units that are beginning to use this form of manure storage.

Barnyard and Soil Conservation

There are no traditional barnyards that need to be evaluated for manure and stormwater management. The barnyards need to be evaluated for stormwater management. Most animals are housed in large buildings. This can cause some stormwater management problems. Around new swine units stormwater management, if not addressed, can cause soil erosion and ponded water. Recommendations should be made to efficiently move the water away from the buildings while maintaining adequate vegetation. I like to see facilities well graded and seeded with a 2-foot gravel strip around the edge of the buildings.

Soil conservation in the land application areas should be a priority with the application of swine manure. Swine manure is the thinnest of most manure types. Therefore it has the highest potential for runoff from the land application areas; soil conservation should be a strong concern when developing a CNMP for a swine operation. Each field or farm that is to receive manure applications should have a fully implemented soil conservation plan. Manure application that are greater than 8,000 gallons per acre should be split applied with a specified number of dry days between manure applications. When manure is surface applied with no incorporation on steep slopes recommendations should be made to start at the top of the hill so that if manure begins to run down the hill it can be noticed with the next load of manure. Some land application areas may have terraces with pipe outlets, the areas need to be buffered and monitored during land application so that no manure enters the piped outlet. Most piped outlets drain directly to the closest creek or stream. Therefore, once manure enters these outlets it has a direct path to surface water. Soil conservation on the land application areas can also be the first line of defense against soil phosphorus losses.

Conclusions

Development of a swine comprehensive nutrient management plan includes many different items. Plans should be developed including the following items:

- Animal numbers for each animal group
- The average weight of each of these animal groups
- Manure production
- Dilutions (Washwater and Rainfall) additions to the manure
- The manure storage facility and associated management problems
- Required best management practices around the barnyard and the manure storage
- Field evaluations of the areas that will be receiving manure applications

The last item that should be included in a CNMP is a list of the records that are needed to implement the plan. The operator should be informed of the records that he needs to maintain. A system should be developed so that they can easily implement the plan and collect the required data. A CNMP should be a continuing evolving plan that needs to be evaluated annually, incorporating the records that have been collected.

Nutrient Management Plans — Dairy

David DeGolyer
Managing Consultant
Western New York Crop Management Association, Inc.

Biographies for most speakers are in alphabetical order after the last paper.

Introduction

One of the biggest challenges in the dairy industry is complying with current environmental standards. Developing a comprehensive nutrient management plan (CNMP) to meet these demands involves the whole farm operation. The CNMP planner needs to become part of the dairy operation team in order to address all the farm's environmental concerns while keeping a strong focus on its economic viability.

Developing a CNMP plan takes time and implementing the plan can be costly. The cooperation of many individuals, though time consuming, is essential to the successful creation and implementation of the farm plan. For example, a farmer, working with the farm's experts, will focus on developing a business plan while the nutritionist prescribes ways to increase forage consumption and limit the amount of imported nutrients, especially phosphorus. A Certified Crop Advisor (CCA) will utilize the field nutrients indicated on soil tests and prescribe cost effective and environmentally responsible rates of nutrients to be applied through farm waste and chemical fertilizers. NRCS and Soil and Water District personnel will design conservation plans that keep soil loss to a minimum and the certified engineer will design structures that can contain the waste during a 25 year storm event. The goal of the planner, working with these outside experts and farm employees, is to help the grower design a business strategy that addresses water quality issues and improves the farm's profitability.

Main Components in the Plan

In designing a Dairy CNMP there are three main components:

- Farmstead Plan
- Waste Utilization Plan
- Records

Farmstead Plan

The farmstead component is divided into four sections. The information the planner gains from the first three; the executive summary, a five year business plan, and evaluation of the farmstead's present environmental conditions, is used to develop the final section, a plan to bring the farm into compliance with the current standards. There is no one-size-fits- all method for reaching farm compliance. Every aspect of each farm must be considered individually.

The first section of a CNMP is the the executive summary describing the farm setting, soils, total animal units, herd average, crop acres, and the average production for each commodity grown over the past three years. It also identifies water sheds and the sensitivity concerns for each water body.

The water shed drives the whole plan. The distance from the local water body and the sensitivity within that water body need to be considered. Information concerning the watershed priority nutrient(s) of concern may be obtained from the county water quality committee and may include sediments, biochemical oxygen demand (BOD), specific nutrients such as nitrogen or phosphorus, and pathogens. As well as addressing watershed concerns, the planner determines realistic nutrient application rates based on the past three years' average crop yields and assesses the concentration of animal units based on available acreage.

The second component, the business plan, defines the goals and objectives for the farm over the next five years. It helps the owner to formulate a course of action and develop budgets. If the farmer intends to expand, then a plan is needed to define where the new barn site will be and limit the possibility of adversely affecting the environment. It is also necessary to determine if there is a large enough land base for spreading manure or if solutions for exporting the excess nutrients must be considered.

The third component, determining environmental compliance, includes assessing the farmstead's ability to withstand a 25 year/24 hour storm event. This is more challenging for dairy operations than with other livestock. Wastes from barnyards, forage bunks, and parlors need to be either collected or treated using a proper filter system. Clean water should be diverted from barnyards and bunks through gutters, tile lines, or gravel drip trenches in order to keep it clean.

A barnyard is designated as any open area for cattle where there is no supporting vegetated areas to feed upon. An overgrazed pasture with limited vegetation can be classified as a barnyard area. With these exercise lots, the bottom line is, during a heavy rain event, is there enough area to filter the contaminated water before it enters a water body? Possible solutions for problem areas include reducing the size of the barnyard area and increasing the filter field, moving the exercise lot to another location to provide enough grass filter strips between the lot and the water course, or increasing the pasture size to reduce over grazing. The planner will work with the farmer to find the solution that best fits the operation.

Leachate from the bunk typically is treated using two systems. One is a low flow collection that gathers high concentrate silage juices and diverts them to a underground tank or sends them to the manure storage. The other system is the high flow system which takes over during times of heavy precipitation. A percentage of the leachate that is diluted with the clean water will continue over the low flow collection to the filter system. This system must be able to handle a 25 year/ 24 hour storm event.

A problem peculiar to dairy operations is dealing with parlor waste. The BOD from the milk house is too high to be stored in a leachate field. The most practical way to store parlor waste is in a manure storage, although another possible solution is to filter it through a grass strip.

After evaluating the farmstead, the planner will make recommendations to bring the farm up to compliance. These will also take into consideration future plans outlined by the farm operator. At this point in the process, an engineer may be employed to design any necessary systems.

WASTE UTILIZATION PLAN

The purpose of the waste utilization plan is to measure whether or not the farm has a large enough land base to accommodate the waste produced, determine if storage needs to be constructed or expanded, and designate the proper time to spread manure. This is determined based on two major aspects :

1. Accounting for the total amount of nutrients to be spread or exported:
   - Manure produced
   - Uncollected Manure - Days in pasture
   - Types of Manure - Liquid, semi solid, solid
   - Bedding types and amounts
   - Milk house waste if collected
   - Other liquids, including rainfall
   - Manure sample

2. Define the Farm Land Base
   - Design farm field maps with acres and field ID
   - Crop rotations designating highly erodible fields
   - Soil names with drainage
   - Soil test crop fields (all fields sampled within the past 3 yrs)
   - Hydrologicly sensitive areas (HSA)
   - Slope length and gradient
   - Flooding Frequency
   - Winter Access
   - Neighbor relations

449

The first step in the plan is to determine the total manure produced from each storage system. The total amount of manure produced should be estimated based on the milk production of cows and the weight of calves and heifers (see Table 1). Bedding will add to manure and must be calculated into the equation. If it is a liquid system, sawdust is approximately 85 percent void while sand is 35 percent void, therefore a straight addition calculation is ineffective. Annual precipitation and evaporation must be considered as well when dealing with an outside manure storage. Other liquids from barnyard runoff and silage leachate should be estimated and calculated into the total waste.

The formula for calculating the total waste to be spread is:

Total Waste = Manure Produced + Bedding + Parlor Waste + Other liquids (silage Leacate, rainfall, etc.) - Uncollected Manure (days in pasture or barnyard)

Table 1. Manure Production Per Cow Based on Milk Production

Milk Production Rate Lbs cow /Day Lbs cow/Yr	Manure Gal/Day	Manure Gal/Yr	Manure Tons/Yr
50 lbs/Day 15250 lbs/Yr	15.1	5,511	22
55 lbs/Day 16775 lbs/Yr	16.2	5,913	23.6
60 lbs/Day 18300 lbs/Yr	17.2	6,278	25.1
65 lbs/Day 19825 lbs/Yr	18.3	6,679	26.7
70 lbs/Day 21350 lbs/Yr	19.3	7,044	28.2
75 lbs/Day 22875 lbs/Yr	20.1	7,336	29.3
80 lbs/Day 24400 lbs/Yr	20.8	7,592	30.4
85 lbs/Day 25925 lbs/Yr	21.5	7,848	31.4
Dry Cows	9.7	3,540	14.2
Heifers 1 Animal Unit/ 1000 pounds	10.6	3,878	15.5

Each collection system will need to be identified as liquid (gallons), solid, or semi solid (tons). For example, a typical farm may have pack manure from the dry cows (tons), lactating cow manure from a liquid storage (gallons), and heifer manure that is spread daily (semi solid, tons), each from completely different storages, as well as calf manure removed from hutches (tons). Each manure system should have a sample analysis taken annually that measures organic nitrogen, ammonia nitrogen, phosphate, potash, and dry matter. From this, the total N, P, and K is calculated. The next step is to decide proper application sites, rates, and timing.

The average rate of manure to be applied per acre is calculated by dividing the total waste by the available acreage. Crop nutrient needs are based on rotation, soil sample results, and realistic historical yield. By combining the total nutrients and crop needs, a quantitative analysis can be determined. This analysis will help the planner and grower determine if export of nutrients will be needed (see Example 1).

Example 1   Example of a (Simple) Waste Application Plan

Facts of the Example Farm
- The herd includes 200 Cows (20 are dry) and 100 heifers. The average milk production is 80 lbs and the average heifer weight is 800 pounds.
- The watershed nutrient of concern is nitrogen.
- All the manure is collected in a manure storage. This earthen lagoon measures 50 by 150 ft.
- 1000 yards of sawdust are used per year.
- The parlor and wash use 3 gallons of water per cow per day.
- There are 250 acres of corn, 50 acres of alfalfa/ grass new seedings and 150 acres of alfalfa/ grass. The rotation is 4 yrs. hay and 4 yrs. corn.
- No additional water is coming into the system.
- The average corn silage yield is 20 tons/acre and alfalfa/grass averages 4.5 tons of DM.
- All the soils are well drained with good yield potential.
- Manure sample   26 lbs of total N
                   14 lbs of organic N
                   12 Ammonia N
                   12 lbs Phosphate
                   25 lbs Potash

A. Total Estimated Waste (see Table 1)

Number of lactating Cows ------> 180 cows (averaging 80 lbs) * 7592 (gal/yr)	1366560
Dry Cows---> 20 cows * 3540 (gal/yr)	70800
Number of Heifers---->((100 heifers * 800 lbs) / 1000 (lbs/AU)) * 3878 (gal/yr)	310240
Parlor Waste ---> 3 gals * 180 cows *365 days	197100
Bedding ((1000 yards * 27 ft/cu ft. * 12.5 lbs per cu ft)/7.5 v. gal.) *85 % voids	38250
Precipitation (35 inches)- evaporation (5 inches) /12 * 50 * 150 / 7.5 v. Gal	2500

Total Estimated Waste   1,985,450 gals
Average Gallons per acre   4,412 gals

B. Calculation of Manure Rate:

Plan:

- All farm fields within the rotation will receive manure.
- Nitrogen needs are 150 lbs of N, based on 20 ton corn silage yields
- Nitrogen from alfalfa/grass sod (very little legume left in sod before being plowed): 1st yr - 100 lbs, 2nd yr - 40 lbs, 3rd yr - 15 lbs
- Manure will be incorporated within 1 day on all corn fields except fall applications and fields coming from sod.
- Organic N released per season: 1st yr. - 35%, 2nd yr. -12%, and 3rd yr. - 5%.

Calculation of Application Rate for Various Production Years of Corn

Production Yr.	1 st Yr	2nd Yr	3rd Yr	4th Yr
Nitrogen Needs	150	150	150	150
Sod Residual	100	50	20	0
N Starter/ Fertilizer	20	20	20	20
Manure Residual N	0	10	16	17
Balance of N Required	30	90	94	113
Maximum Rate (Gallons) Per Acre	6,122	*7,086	*7,402	**12000

* manure incorporated after one day
**2 applications of manure, 5000 gallons applied in fall and 7000 incorporated in spring.

C. Manure Recommendations based on production year and N balance

Crop	Prod. Year	Acres	Recommended Rate	Total Manure
Corn	1	60	6,000	360,000
Corn	2	65	7,000*	455,000
Corn	3	65	7,000*	455,000
Corn	4	60	12,000**	720,000

Total 1,990,000

- Manure for first year and 4th year (5000 gallons) corn fields will be applied in the fall.
- Corn after corn fields receive 7,000 gallons applied and incorporated in the spring.
- Manure applied to fourth year corn does not exceed nitrogen needs, however it is being overloaded with phosphorous based on the current year crop uptake. Based on soil sample results, this should provide adequate phosphorus for four years of alfalfa/grass mix. The key to the plan is to have the farm apply manure to every acre in the rotation while keeping phosphorous relatively within balance.
- Historical yield and soil sample information should be considered in determining realistic yield goals for each field.

One issue that is not addressed by the Example Waste Application Plan, but requires attention from the planner is the nutrients being fed to the cow. The total mixed ration that is fed to the cows can certainly impact the amount of nutrients that are excreted. Lowering the amount of dietary phosphorous from .5 to .38 provides a significant change in waste nutrients being applied to the soils. With this change, a cow will secrete approximately 90 fewer pounds of phosphate per year. Increasing the forage consumption and importing less grain is the key to nutrient cycling, cow health, and increased profits. A reputable nutritionist whose focus is on the farm's bottom line and environmental concerns is crucial to the development and implementation of a nutrient management plan.

Once the amount and composition of the manure is determined, the timing for application must be addressed. Each field must be assessed for the risk of runoff and leaching. Runoff is rated along 4 different risk levels in each of 6 categories; drainage, areas of concentrated flow, slope gradient, slope length, winter access, and neighbor relations (see Table 2). Areas of concentrated flow or hydrologically sensitive areas (HSA) are the most evident environmental problems associated with land application of manure. Fields that are HSA are prone to nutrient runoff that could affect a nearby water body. These sites should receive manure with caution. HSAs are seasonal, only hydrologically active during certain times of the year.

Hydrologically sensitive areas are grouped into 3 main categories:

1. Runoff areas including saturated and open lots or other compacted surfaces prone to concentrated flows to water bodies- Major pollutants from runoff areas include pathogens and soluble P and N.

2. Erodible areas that are prone to being washed- Steep slopes with no cover are likely to have concentrated flow. Pollutants from erosion areas include P and sediments.

3. Groundwater recharge areas near springs and wells with well drained soils- Pollutant threats come mainly from N and pathogens.

Table 2.   Estimated Risk to Minimize Impact on Surface Water Quality

Field Characteristic	Risk Level 1	Risk Level 2	Risk Level 3	Risk Level 4
A. Slope gradient Annual crops Perennial Crops	0-5 % 0-8%	6-10 % 9-15 %	10 + % 15 + %	Not applicable
B. Slope length	0-300 ft.	300-500 ft.	500 + feet	Not applicable
C. Flooding frequency	None or rare	occasional	frequent	Not applicable
D. Drainage Class	Well drained	Moderately well drained	Somewhat poorly	Not applicable
E. Areas of concentrated flow	no	no	yes	Not Applicable
F. Winter Access	Unlimited	Sometimes limited	Usually limited	Not Applicable
G. Closeness to neighbors	No problem	No Problem	No Problem	Problem

Best months for spreading based on Risk Level:

Risk Level 1: Year- round

Risk Level 2: Primary - April to December
                      Secondary - January to December
                      (If not enough Risk Level 1 fields available.)

Risk Level 3: Mid-April to October

Risk Level 4: Restricted - no spreading

If all factors are determined to be in risk category 1 and the slope is greater than 500 ft., the field is classified in risk category 3 and no manure should be applied during the winter season. If necessary, parts of fields can be broken down through the revised soil loss equation and application timing can be determined for each subdivision according to the 6 factors.

The second factor in determining the timing of manure application is the leaching index and is divided into three categories; low, intermediate, and high. Fields that have high leaching, especially if there is a well nearby, require strict management practices. For example, manure should not be applied during the early fall, unless a cover crop is established to take up the free nitrogen. If additional nitrogen is needed, side dress nitrogen instead of preplant nitrogen should be applied.

Another management concern is pathogens, especially from calf manure, that can pose serious health risks. Separating calf manure from that of other livestock and land applying it to minimal risk fields (fields the likelihood of runoff to streams and other surface waters is low) greatly reduces the risk of water contamination. Another possible control method is to land apply calf manure during times when the chance of runoff is minimal and avoid spreading when the ground is frozen or saturated. Typically, the total amount of manure from calves is minimal compared to the rest of the livestock, yet the danger it can pose warrants extra precautions.

After assessing the risk levels and the leaching index of each field, the planner can determine the size of storage recommended based on cow numbers and plans for future expansion. If necessary, the farmer can expand existing storage or build new facilities that meet current environmental regulations.

Records

In order to show compliance with the written plan, it is crucial that the farmer keeps complete and accurate records. Three areas of primary importance are manure application, yield information, and operation and maintenance of the farmstead systems.

It is imperative that precise manure application records are maintained. These would include the number of loads of manure applied, the rate of application, and the date that it was spread. To insure accuracy, farm staff must be trained in proper calibration of the manure spreading equipment because rates may vary within a given field, as well as between fields. It may also be wise to have farm workers initial the record of work that they completed to encourage accountability.

Accurate yield information is important in developing nutrient management recommendations for each individual field. Although soil sample results play a big part in recommended fertilizer application, it is imperative to know previous yield results. More nutrients can be applied to fields that have shown consistently high yields.

After the farmstead has reached compliance with the environmental standards, it is important to keep all the systems functioning effectively. This may include cleaning gutters, mowing around lagoons, cleaning silage low-flow collection systems, mowing filter strips and diversion ditches, and maintaining fences. These maintenance steps should be completed as frequently as necessary and dates should be recorded with each task accomplished.

The more comprehensive and accurate the farm's records, the easier it is to prove compliance with the comprehensive nutrient management plan. These records also allow the farmer to see which fields might not have received the recommended manure rate and can therefore supplement with fertilizer.

Conclusion

A Comprehensive Nutrient Management plan provides a beneficial situation for both the environment and the farm. The farmer can save on fertilizer costs by utilizing his manure resources through proper timing and accurate application rates. A business plan that addresses environmental issues allows him to stay competitive in a global market. When proper steps are taken to insure that nutrients remain in their designated places, the growing crops benefit and the risk of water contamination is minimal.

References

Cornell Cooperative Extension Publication, Cornell Field Crops and Soils Handbook, Second Edition, 1987.

Natural Resource , Agriculture, and Engineering Services. Earthen Manure Storage Design Consideration, April 1999.

Stecman, S.M., Rossiter, C, McDonough, P, and Wade, S.et al. Animal Agriculture and the Environment. Proceedings at Animal Agriculture and the Environment North American Conference, December 1996.

United States Department of Agriculture Soil Conservation Service, Agricultural Waste Management Field Handbook.

Van Hourn, H.H., et al. Dairy Manure Management: *Strategies for Recycling Nutrients to Recover Fertilizer Value and Avoid Environmental Pollution*. Circular 1016. Institute of Food and Agriculture Sciences, University of Florida, Gainesville, December 1991.

Nutrient Management Plans: The Professional Assistance Needed and the Cost

Richard F. Wildman, President
Agricultural Consulting Services, Inc. (ACS)
1634 Monroe Avenue, Rochester, New York 14618
(716) 473-1100 • RFW@ACSOFFICE.COM

Biographies for most speakers are in alphabetical order after the last paper.

Plan Development

No matter how good a nutrient management plan may be, it is worthless if never adopted. Take a minute to think about nutrient management planning from a farmer's perspective. For the farmer, nutrient management is just one of many components making up the overall operation of a farm. As with any other business, a farm must successfully balance the needs and requirements of many, varied aspects of the business to insure both short- and long-term profitability.

The manner in which a farm manages manure and nutrients can have a profound impact on profitability. Hauling distances, loading rates, supplemental nutrients needed, application timing and methods vary widely in cost from farm to farm. As a result, a given farm could be at a competitive disadvantage based upon costs associated with manure handling. In some instances, herd size--and thus gross income—may be limited by the land-base requirements of the nutrient management plan.

The perception of farms and farmers as responsible environmental stewards is a critical public relations element. If a farm is perceived as a poor environmental manager, it can expect criticism from the community in which it operates. Examples of criticism include such things as opposition at zoning permit hearings, odor complaints, spill and nuisance complaints, and even possible lawsuits. At the very least, these types of criticism will divert the attention of management. In the worst case, they could result in large, costly legal judgments or fines.

Therefore, a nutrient management plan should also address environmental management issues on the farm in a manner that is credible to both environmental regulators and the public at large. A plan that is considered vague or lax will not ultimately serve the best interests of the farm.

It is well known that the number of farms has been steadily declining, and there is no indication this trend will change. The bottom line? Farms with higher-than-average-costs of production tend to be less profitable, and less profitable farms tend to go out of business. The challenge facing farm managers today is how to develop and implement methods for nutrient management that will be credible with the public while still allowing for potential expansion and controlling costs.

Farms in the Northeast can vary dramatically in the environmental risks related to their actual locations. One farm may operate at a site that requires extensive engineering and limitations on manure application to avoid environmental contamination, while a nearby farm may be located on a site that presents very few constraints.

There is little question that well-managed farms understand the serious impact environmental issues can have on business. It is essential that a planner understands that profitability is always going to be a core element of any decision a farmer makes. If a plan does not lead to maintaining or improving profitability, it will not be relevant to the farm manager.

When defining profitability, one must also recognize that decisions made by a manager do not all necessarily lead to an immediate return. Some decisions may result in costs that will eventually be returned through risk reduction. Other decisions could lead to improved public relations, a better public image, and greater public support. Nonetheless, the desire of a farm manager to be a good environmental steward will not allow for unsound economic decisions.

If the focus of a planner is limited to environmental and engineering issues, then the planning process will miss the opportunity to balance operating efficiency and the manager's business goals. The result is likely to be a plan that does not optimize the profitability of the business —and a plan that is unlikely be fully adopted.

Planners should understand and be highly trained in the complex factors facing farm mangers when dealing with design and operating processes which impact environmental quality. A solid nutrient management plan will typically require the involvement and contributions of several professionals, each one contributing an area of specialization. The focus of all planning team members should be to address operating processes in a manner that can improve a farm's profitability while meeting environmental goals *at the same time*. And if profitability is the ultimate goal, the plan is likely to be adopted.

Individual operations will vary in their needs for service by planners. Some managers may choose to oversee the planning process themselves and only utilize specialists as needed. More typically, however, farm managers choose to delegate the planning process to outside contractors who can both coordinate the planning process and maintain accountability for the development of a viable plan.

Planning Costs

There is no standard cost for plan development, and costs can vary widely based upon the complexity of each situation. Some plans can be completed for well under $1,000 while others may require closer to $50,000 for full implementation. Regardless of the fees charged, however, the planning process is governed by the same realities of business and economics faced by farms and farmers. Ultimately, only planners that are cost-effective for their clients *and* profitable business operations themselves will survive in the marketplace.

No matter how a planner charges for services--by the acre, per cow, flat fee, or hourly— much of a planner's costs will be directly related to the amount of time spent serving clients. Although professional service firms vary in the efficiency of their operations, surveys indicate that a multiplier of three is typical in determining fees. In other words, a client can expect the fee to be two-and-a-half to three times the direct salary costs of the consultant. This means a team of planners can actually be more cost efficient than a single planner. With a team approach, less complex tasks can be handled by lower-compensated staff, thus leaving only the more difficult and complex functions to be completed by staff members who are the most highly trained and, therefore, more highly compensated.

Conclusions

When assessing the cost of plan development, it is important to recognize that many aspects of a plan should be intrinsic to the farm's general operations. It is unrealistic to separate the cost of developing environmentally-responsible strategies from the cost of developing a profitable strategy overall. For many consultants, the process of providing nutrient management plans to meet CAFO (concentrated animal feeding operations) requirements will not add any significant cost to the services they are already providing their clients.

Obviously if a farm is not currently doing any planning, then the costs of planning will be an increase. However, if a farm has not been doing any systematic planning, it can expect operating efficiencies—and, thereby, profitability--to increase as well.

Professional consultants providing services to farm businesses need to be compensated for their work. At times, the nature of services provided require specialized technical or scientific expertise from someone a farmer may consider to be relatively highly-paid. While it is appropriate for any farm manager to carefully consider all costs of doing business (including the use of a professional planner), it is very unlikely that the expense of hiring a good planner will significantly impact a farm's profitability.

In fact, any significant costs related to nutrient management planning are not associated with the cost of doing the planning but rather with the cost of *implementing* the plan. In the final analysis, planning costs should be carefully evaluated against the potential gains or losses from operating efficiencies that result from developing and carrying out a good plan.

Speaker Biographies
(presented in alphabetical order)

Charles W. Abdalla

Charles W. Abdalla is an associate professor and extension specialist in the Department of Agricultural Economics and Rural Sociology at Penn State University, University Park.

Dr. Abdalla's research and extension programs address economic and policy issues related to natural resources and the environment. His recent research addressed the impacts of industrialization of the food and agricultural system on rural communities and environmental quality. His extension programs have focused on water quality and quantity; land use change at the rural-urban interface; and off-site impacts of animal agriculture. He is a member of a task force on animal confinement policies and co-chairs a task force on land use issues sponsored by the Farm Foundation. He is a member of the national planning team for the Animal Waste Initiative – Promoting Environmental Stewardship of the land grant system.

Dr. Abdalla received a B.S. in environmental resource management from Penn State University and an M.A. in economics, M.S. in agricultural economics, and Ph.D. in agricultural economics from Michigan State University.

Roselina Angel

Roselina Angel, a native of Colombia, obtained a master's in monogastric nutrition and a Ph.D. in poultry nutrition from Iowa State University. After finishing her Ph.D., she worked for Purina Mills, Inc., doing research into nutrient requirements of exotic animals as well as with poultry.

Roselina joined the faculty of the Animal and Avian Sciences Department at the University of Maryland in April 1998. Her research focus at Maryland is on feed management of broilers to minimize phosphorus excretion.

Obie D. Ashford

Obie is a native of Mississippi, where he grew up on a diversified farm consisting of cotton, hogs, and dairy. He received his B.S. degree in agriculture from Alcorn State University, Lorman, Mississippi, in 1965.

His career in the agriculture field began in 1965 upon graduation from college. During his career in the agriculture field, Obie has been involved in various aspects of the field activities such as working as a state resource conservationist, area conservationist, district conservationist, and soil conservationist at various locations in New Jersey, Georgia, Pennsylvania, and Maine between 1965 and 1993. In 1993 he became the director of the Northeast Regional Technical Center in Chester, Pennsylvania. His career entailed providing technical support for the thirteen eastern states, from West Virginia to Maine. Moreover, in 1995 he became the team leader of oversight and evaluation for NRCS's East Regional Office in Beltsville, Maryland, in which one of his major duties was to evaluate state operations. His term as team leader ended in 1998, when he became the national leader for the Animal Husbandry and Clean Water Programs Division, Natural Resources Conservation Service, in Beltsville, Maryland, where he currently resides.

John C. Becker

John Becker is a professor of agricultural economics on the faculty of Penn State University, University Park. He is a 1969 graduate of LaSalle University with a B.A. degree in economics and a 1972 graduate of the Dickinson School of Law with a J.D. degree. He is a member of the Pennsylvania Bar and is of Counsel to the Camp Hill, Pennsylvania, law firm of Zeigler and Zimmerman, P.C. His teaching programs and publications include bulletins, circulars, and independent study courses that focus on legal issues such as environmental law and regulation, estate tax, estate transfer, land owner liability, real estate tax assessment, and employer-employee issues. He is Director of Research at the Agricultural Law Research and Education Center of the Dickinson School of Law at Penn State University and an adjunct faculty member.

Mr. Becker is a member of the Centre County, Pennsylvania, and American Bar Associations and the American Agricultural Law Association. He is a past chairperson of the Agricultural Law Committee of the Pennsylvania Bar and has served as a director of the American Agricultural Law Association and vice-chair of the Agricultural Law Committee of the General Practice Section of the American Bar Association. In 1994 he was elected a fellow of the American Bar Foundation.

Mr. Becker has organized and presented several continuing education programs for the Pennsylvania Bar Institute, Dickinson School of Law, the American Agricultural Law Association, and the American Association for the Advancement of Science. His published legal research appears in *Drake Law Review, Dickinson Law Review, Indiana Law Review, William Mitchell Law Review, the Drake Journal of Agricultural Law, The Journal of Soil and Water Conservation,* and the *Pennsylvania CPA Journal.*

Mr. Becker is a retired member of the Pennsylvania Army National Guard, having served as Command Judge Advocate of the 28th Infantry Division (Mech.) holding the rank of Colonel in the Judge Advocate General Corps.

Gregory D. Binford

Greg is a native of Crawfordsville, Indiana, where he was raised on a 1,200-acre corn-soybean farm. He received his B.S. (1986) degree in agronomy from Clemson University, and his M.S. (1988) and Ph.D. (1991) degrees from Iowa State University in soil science.

His research career began as a graduate student at Iowa State University, where his work focused on the development of diagnostic tools for improving nitrogen (N) management in corn. His research also looked at differences in N response between continuous corn and corn after soybean. Nitrogen-15 tracers were also used to monitor N uptake by the corn crop and losses from the soil profile.

Following completion of his graduate studies at Iowa State, he joined the faculty at the University of Nebraska-Lincoln as a Research and Extension Soil Fertility Specialist at the Panhandle Research and Extension Center in Scottsbluff. His research at Nebraska included: N response of corn following sugar beets, N response and development of N diagnostic tools in sugar beets, N & P response in sunflowers, chloride requirements of winter wheat, N response of winter wheat, iron-chlorosis in turf, N response and inoculation response in dry edible beans, and micronutrient responses in sugar beets.

He moved to the East Coast in 1995 to join Pioneer Hi-Bred International, Inc., as a regional field sales agronomist. In this position, his primary responsibility was to train sales staff on product characteristics and general agronomics. His regional area included the Delmarva Peninsula, New Jersey, and Eastern Pennsylvania.

In the summer of 1999, Greg joined the faculty at the University of Delaware in the Department of Plant and Soil Science. His current position is Assistant Professor of Soil and Water Quality, and he will be working in both research and extension. His main areas of interest include: the development and refinement of diagnostic tools for monitoring nutrient status of crops, water quality/nutrient management issues, manure management, and crop response to N & P.

Eldridge R. Collins, Jr.

Eldridge Collins is a native of Florida. He earned B.S. and Ph.D. degrees in agricultural engineering at Auburn University, with an interlude between degrees for four years of service in the U. S. Marine Corps, where he was discharged with the rank of Captain in 1966. In 1971 he joined Virginia Polytechnic Institute and State University, where he is Professor and Extension Agricultural Engineer in the Biological Systems Engineering Department and serves as Extension Project Leader in the department. He has 29 years of extension and research experience at Virginia Tech in the areas of livestock waste management, point and nonpoint source pollution from agriculture, and livestock and poultry facilities and environmental control. His work has included on-farm consultation with producers, work with the agricultural service sector, applied research, and cooperative work with other state and federal agencies. He routinely is involved in conducting grower meetings and other extension programs related to good practices for handling and utilizing agricultural wastes and protecting water quality. He played a key role in the 1980s to help establish farm nutrient management in Virginia. Dr. Collins is the author or coauthor of over 155 reports, papers, or articles in his field, and has worked on development projects overseas and done private consulting in his field.

David DeGolyer

David is a native of Castile, New York, where he spent his early years on a dairy farm. He received his B.S. degree in agronomy from Cornell University in 1986 and became a certified crop advisor (CCA) in 1994 and a certified CAFO planner in 2000.

David completed an internship at Cornell Cooperative Extension in Cattaraugus and Chautauqua Counties in the summer of 1985. In the summer of 1986, he worked for Cornell Cooperative Extension as an integrated pest management scout.

David has been employed as a consultant for the Western New York Crop Management Association (WNY CMA) since 1987 and has served as the managing consultant since 1995. WNY CMA is a grower-owned cooperative that serves nine counties in western New York and encompasses over 90,000 acres. The service provides full crop consulting services, integrated pest management scouting, soil sampling, and CAFO planning. In his position as manager, David oversees four crop consultants and designs services for the members.

Anthony J. Esser

Anthony Esser is the National Program Manager for EQIP for the USDA-Natural Resources Conservation Service in Washington, D.C. Mr. Esser (Tony) is a 27-year career employee of NRCS and has been a conservation planner, district conservationist, RC&D project coordinator, liaison to the NYS Department of Environmental Conservation, and NRCS water quality coordinator and EQIP coordinator in Syracuse, New York. Additionally, Tony has worked as a soil erosion control specialist for the New Jersey State Soil and Water Conservation Committee and as a natural scientist with the New Jersey Pinelands Commission. Mr. Esser received his B.S. in agricultural science from the University of Rhode Island and M.S. in agricultural engineering from Rutgers University.

Gregory K. Evanylo

Greg is a native of New Jersey (Exit 10), where he spent his formative years raising hell. He earned a B.A. in biology from the University of Connecticut in 1974, an M.S. in plant and soil sciences from the University of Massachusetts in 1978, and Ph.D. in agronomy from the University of Georgia in 1982.

Following a one-year stint as a postdoctoral researcher at the University of Kentucky, in 1984 Greg joined the staff at the Virginia Truck and Ornamentals Research Station (now the Eastern Shore Agricultural Research and Education Center) as a research soil scientist, where he conducted research on vegetable and agronomic crop soil fertility and nitrogen management for water quality.

Greg joined the Department of Crop and Soil Environmental Sciences at Virginia Tech as an assistant professor with extension and research responsibilities in nutrient management and sustainable agriculture in 1989. He was promoted to associate professor in 1992, at which time he assumed additional responsibilities in waste management. Since 1992, Greg's extension and research programs have been designed to investigate and promote sustainable soil management through the composting and utilization of yard wastes, biosolids, manures, and industrial wastes.

Greg organized Virginia's state composting association — the Virginia Recycling Association's Organics Recycling and Composting Committee — in 1995 and is presently chair of its Technical Standards, Research and Methods Subcommittee. He is a member of the Composting in the Southeast Conference Planning Committee and chair of the Year 2000 Conference that will be held in Charlottesville, Virginia. Greg has served on numerous state regulatory technical committees for nutrient management, biosolids use, and composting. He is a member of the Soil Science Society of America, Council for Agricultural Science and Technology, Soil and Water Conservation Society, U.S. Composting Council, Virginia Recycling Association, and Water Environment Federation.

Daniel G. Galeone

Dan is a native of Bucks County, Pennsylvania, where he was born in 1962. He received a B.A. from LaSalle University in biology (1984) and an M.S. from Penn State University in ecology (1989), with an emphasis on forest hydrology, under the watchful eye of Dr. David R. DeWalle.

Galeone began his career with the federal government back in the mid-1980s when he was employed by the U.S. Forest Service in University Park, Pennsylvania, as a temporary field technician/statistical analyst. He began working for the U.S. Geological Survey in Pennsylvania as a hydrologist in March 1991. While with the USGS, Dan has worked on numerous projects pertaining to wellhead protection, agricultural-related water-quality issues, and wastewater-effluent land application.

William J. Gburek

William J. (Bil) Gburek is a hydrologist with the Pasture Systems and Watershed Management Research Laboratory, USDA-ARS, at University Park, Pennsylvania.

He was born and raised in Buffalo, New York, and received B.S. and M.S. degrees in civil engineering from the State University of New York at Buffalo. He escaped to central Pennsylvania in 1967, where he began his career with the Agricultural Research Service as a research hydrologist, a position he maintains to this day. While with ARS, he received his Ph.D. in civil engineering from the Pennsylvania State University.

His research career has been generally focused on understanding and quantifying the interactions between hydrology and water quality in the watershed context. Current research projects are in the general areas of groundwater recharge, subsurface flow and nutrient transport in fractured aquifers, and hydrology of the near-stream environment as related to storm runoff production and phosphorous loss from the watershed. He is currently participating in two interagency efforts: a joint ARS-NRCS effort to unify the transport factors within the Phosphorus Index, and an ARS-USGS effort on groundwater age-dating to quantify expected lag time between nitrate management at the land surface and effects at the watershed outlet.

Janet K. Goodwin

Janet Goodwin is Environmental Scientist with the Environmental Protection Agency. She has been with EPA for twenty years, specifically with the effluent guidelines program. She has worked on a number of different regulations, ranging from metal-forming regulations to chemical and pesticide manufacturing regulations. She has worked on the Feedlots regulation since late 1997.

Robert E. Graves

Robert E. Graves is a professor of agricultural engineering in the Agricultural and Biological Engineering Department at The Pennsylvania State University. He is recognized internationally for his work in dairy production systems and manure handling. At Penn State, he is responsible for developing educational programs and materials on farm buildings, manure handling, and composting. Bob is a native of New York State and holds degrees from Cornell University and the University of Massachusetts. He came to Penn State in 1982 from Massachusetts, where he was manager and part owner of a 350-cow, 500-acre, dairy crop farm. Between 1972 and 1977, he was extension agricultural engineer at the University of Wisconsin.

Bob has observed and worked with farmers in North America, Europe, Asia, and Africa. He is involved in regional, national, and international agriculture engineering activities and has authored over 200 articles, bulletins, and handbooks. He has received numerous honors for his work in farm buildings and facilities. In 1993, he was awarded the Henry Giese, Structures and Environment Award from the ASAE. In 1994, he was honored by the Pennsylvania Dairymen's Association with their Extension Award.

Allen F. Harper

A native of south-central Virginia, Allen Harper received his B.S. and M.S. degrees in animal science from Virginia Tech in 1979 and 1982, respectively. His master's degree research focused on the effects of feed additives and pen space allowance for growing pigs. From 1982 to 1989, Allen served as Agricultural Extension Agent in Suffolk, Virginia, where he was responsible for livestock production educational programming. He returned to graduate school in 1989, and in 1992 received the Ph.D. degree in nonruminant nutrition at Virginia Tech. His Ph.D. graduate work investigated the importance of supplemental folic acid in the diet of breeding gilts and sows.

Since 1992, Allen has served as assistant professor and extension animal scientist with Virginia Tech, where he is responsible for swine production and management extension programming throughout Virginia. He is also involved in applied swine research at the Virginia Tech Tidewater AREC Swine Unit located at Suffolk, Virginia. In 1998 he was promoted to his current rank of associate professor and extension animal scientist for swine programs.

Allen's current extension and research programs focus on nutrition and management for improved swine performance and environmental protection, proper use of antimicrobial feed additives and improved reproductive efficiency in sows and boars. He also serves as secretary and educational advisor to the Virginia Pork Industry Association and as superintendent of the Virginia State Fair Junior Market Hog Show.

Donald Hilborn

••

Donald Hilborn, P. Eng., is a by-product management specialist with the Ontario Ministry of Agriculture, Food, and Rural Affairs (OMAFRA). He is very involved with all aspects of manure management in Ontario. He has conducted courses on pollution control, concrete manure storages, custom manure application, and nutrient management.

During the last two years, a major portion of his work has been focused on linking manure system with cropping processes. He has been in the forefront with the ministry's effort to develop user-friendly yet environmentally acceptable nutrient management information. Donald and a team of other OMAFRA staff have just completed the 2000 version of the Ontario Nutrient Management Software. Already this software has been used as a key tool in the development of many comprehensive nutrient management plans for Ontario farmers.

William E. Jokela

••

Bill Jokela is an associate professor and extension soils specialist in the Plant and Soil Science Department at the University of Vermont, where he has done extension and research work in soils since 1985. Bill grew up on a farm in Minnesota and received a B.A. degree in biology from Carleton College in 1969. He was a crop and livestock farmer before returning to school for his M.S. and Ph.D. degrees in soil science at the University of Minnesota. Between graduate degrees he worked for three years as an extension specialist in soil fertility.

In his current position Bill carries out a statewide extension program in soils, including the soil testing and nutrient recommendations program and recent development of a phosphorus index for use in nutrient management planning. He has conducted research on fertility of corn and hay forage crops, conservation tillage practices, soil testing, and manure management. Current research projects include manure application techniques to reduce ammonia and other losses and evaluation of water quality effects of BMPs such as vegetated buffer strips and cover crops.

Kenneth B. Kephart

Ken is a native of Huntingdon County, Pennsylvania. He received his B.S., M.S., and Ph.D. degrees from The Pennsylvania State University. Ken served as extension livestock specialist at the University of Delaware for two years before returning to Penn State in 1985 to lead the swine extension program.

Ken is currently Associate Professor of Animal Science in the Department of Dairy and Animal Science at Penn State. He currently teaches a course in swine management and serves as extension swine specialist for Pennsylvania. His extension activities include a close interaction with the packing industry in Pennsylvania in the development of grade and weight programs. In recent years, Ken's extension programs have focused on environmental issues. In 1997, he worked closely with the National Pork Producers Council in the development of the Pollution Prevention Strategies Module to complement NPPC's Environmental Assurance Program. Ken's extension programs have also involved him in applied research, as he has worked in the area of nutrition and ventilation. His current research projects involve the evaluation of marketing practices and the development of odor-reduction technologies.

Peter J. A. Kleinman

Peter is a soil scientist with USDA's Agricultural Research Service (ARS) in State College, Pennsylvania. As part of his responsibilities, he is charged with conducting research for the National Phosphorus Project, providing a scientific foundation for the Phosphorus Index as well as development of integrated nutrient management strategies.

Peter received his M.S. from the Department of Natural Resources at Cornell University, carrying out research on nutrient cycling under slash-and-burn agriculture in West Kalimantan, Indonesia (Borneo). As part of his research, he examined the critical role of fallow duration in conserving soil phosphorus, a primary limit to agronomic productivity in this system.

Turning his focus to domestic agronomic issues, Peter obtained his Ph.D. in pedology from Cornell's Department of Soil, Crop, and Atmospheric Sciences. His dissertation research focused on the cycling of phosphorus in agricultural soils of the Catskills, as part of activities related to the New York City Watershed. Peter's research examined environmental indicators of soil phosphorus, as well as the dynamics of phosphorus in manure-amended soils.

Les E. Lanyon

....

Dr. Lanyon is a professor of soil fertility in research and extension. A native of southwestern Pennsylvania, he received his agronomy degrees from Iowa State University (B.S., 1970) and The Ohio State University (M.S., 1975, and Ph.D., 1977). He has been at Penn State since 1977, working on the soil fertility of forage crops, utilization of animal wastes in crop production systems, soil and water quality, and nutrient management.

He has researched the interactions of nutrient supply, plant varieties, and soil conditions with yield, mineral content, and disease development of common forage crops in Pennsylvania. He has evaluated the application of manure to cropland at disposal rates, the spatial and temporal dynamics of nutrients on crop/livestock farms in Pennsylvania, and animal manure applications as an integrated part of the nutrient supply in long-term crop sequences. He is currently studying the implications of soil management for soil quality. The variety of small plot and large scale studies have been complemented by computer simulation studies of nutrient management for dairy farm structure and performance, the role of farm information management in nutrient management, and the consequences of social perspectives on resource rights for farm performance.

Dr. Lanyon's extension program focuses on environmental quality impacts of crop production and animal agriculture. He highlights the potential contribution of different management levels to farm management and the role of various stakeholders in the implementation or outcomes of the farm management process. He is the state contact person for the Pennsylvania Farm*A*Syst program.

April B. Leytem

••

April Leytem is a postdoctoral research associate in the Department of Plant and Soil Sciences at the University of Delaware, Newark. She earned her Ph.D. in soil science at North Carolina State University in 1999.

Dr. Leytem has been working on the Phosphorus Site Index for the state of Delaware since May 1999. The Phosphorus Site Index is a tool, which is being developed to help identify critical source areas of P loss from agricultural sites.

J. J. Meisinger

••

John (Jack) is a native of Naperville, Illinois, where he was introduced to agriculture on a 320-acre mixed livestock farm with 30 cows, 50 steers, and 6 farrowing sows. He received his B.Sc. from Iowa State University in 1967, majoring in agronomy, and his Ph.D. from Cornell University in soil science.

His research career began in 1967 with USDA-ARS in Beltsville, Maryland, where he studied methods to estimate soil nitrogen mineralization, methods to estimate N fixation, and the use of N-15 in agricultural research. He has remained at the Beltsville Agriculture Research Center with ARS and has expanded his research into the effects of no-tillage on soil N cycle processes, use of soil N tests, and use of cover crops to protect water quality.

His most recent research activity related to animal agriculture is the evaluation of the pre-sidedress soil N test (PSNT) in Maryland, the use of field scale N budgets to estimate areas at risk for nitrate loss, and ammonia volatilization from manures. He is currently involved with a large interdisciplinary dairy manure project and a poultry phosphorus project studying methods to conserve nutrients in manure and thereby minimize losses to the environment.

Greg L. Mullins

Greg is a native of southwest Virginia, where he grew up on the family's part-time hillside farm. He received a B.S. degree in agriculture from Berea College in 1979, an M.S. degree in agronomy from Virginia Tech in 1981, and a Ph.D. from Purdue University in 1985.

After receiving his degree from Purdue, Greg joined the faculty in the Agronomy and Soils Department of Auburn University in May 1985. Greg spent approximately 14 years on the faculty at Auburn in a research/teaching position in the areas of nutrient management and soil chemistry. His research program concentrated primarily on the chemistry and fertility of phosphorus and potassium in relation to Alabama field crops. Greg's program also involved the land application of industrial and animal wastes. He was active in undergraduate student advising and taught undergraduate courses in introductory soils and a graduate level course in advanced plant nutrition.

Greg is a nutrient management specialist in the Department of Crop and Soil Environmental Sciences at Virginia Tech. He came to Virginia Tech on April 1, 1999. In his new position, Greg has extension and research responsibilities in the areas of nutrient management and water quality.

Roberta Parry

Roberta is a senior agriculture policy analyst with the Office of Policy and Reinvention at the U.S. Environmental Protection Agency in Washington, D.C. During her ten years with EPA she has worked on a variety of legislative, regulatory, and scientific agriculture issues, mainly dealing with nutrients and water quality. Her current focus is on controlling pollution from livestock operations. She has a Master's degree in public administration from the John F. Kennedy School of Government at Harvard University.

John B. Peters

● ●

John is a native of central Wisconsin, where he grew up on a small farm. He received his B.S. degree in soil science and resource management from the University of Wisconsin–Stevens Point in 1976 and an M.S. degree in soil science from the University of Wisconsin–Madison in 1978.

His career with the University of Wisconsin, which began in 1978, has included several years in West Africa involved in several studies to monitor the changing status of the soil fertility conditions in this sub-Saharan region.

John has been the director of the University of Wisconsin Soil and Forage Analysis Laboratory since 1984. His duties include research and extension programming working with soil fertility and waste disposal issues, particularly as they relate to soil test calibration, crop production, and nutrient management strategies. The development and promotion of diagnostic testing services is also one of his primary interests. This includes soil, livestock wastes, and feed and forage testing, as well as many other specialty tests. Currently, he is the chair of a regional committee working on the development of a manure-testing manual. The goal of this project is to standardize manure sampling, testing, and reporting across the NCR-13, NEC-67, and SERA-6 soil testing working groups. He is also responsible for the management of the Wisconsin FSA soil testing laboratory certification program.

For seven years prior to his current position, John was the assistant superintendent of the University of Wisconsin Agricultural Research Station at Marshfield.

William J. Rogers

Bill is a native of the Washington, D.C., area. He learned his love for agriculture on his grandparents' farm in eastern North Carolina. He received his B.S. degree in agricultural engineering technology and his M.S. in soil science from North Carolina State University in 1989 and 1997, respectively.

After graduating with his engineering degree Bill began to work at N.C. State as an agricultural research technician in the Soil Science Department. This work, in the area of soil conservation and fertility, led him to begin work on his master's degree. After finishing his master's degree he began working for Brubaker Agronomic Consulting Service LLC in August 1996. Currently, Bill is the manager of the nutrient management and environmental divisions within Brubaker Consulting Group (formerly Brubaker Agronomic Consulting Service). He continues to work with all animal species, developing comprehensive nutrient management plans and CAFO permits. Along with the nutrient management work, Bill continues to consult with growers in all areas of environmental issues dealing with agriculture.

Andrew N. Sharpley

Andrew N. Sharpley is a soil scientist at the USDA-ARS Pasture Systems and Watershed Management Research Laboratory, University Park, Pennsylvania, and adjunct professor of agronomy at The Pennsylvania State University. He was born in Manchester, England, and in 1987 became a U.S. citizen. He received degrees from the University of North Wales, United Kingdom, in 1973 and Massey University, New Zealand, in 1977.

Sharpley's research has investigated the cycling of phosphorus in soil-plant-water systems in relation to soil productivity and water quality and includes the management of animal manures, fertilizers, and crop residues. Most recently he has developed decision-making tools for field staff to identify sensitive areas of the landscape and to target management alternatives and remedial measures to reduce the risk of P loss from farms. Overall, he has focused on achieving results that are both economically beneficial to farmers and environmentally sound to the general public. He is a fellow of the American Society of Agronomy and Soil Science Society of America and the recipient of the ASA Environmental Quality Research Award, the Soil Science Applied Research Award, and the USDA-ARS Scientist of the Year Award. He is an editor of the *Journal of Environmental Quality* and *Nutrient Cycling in Agroecosystems*.

Daniel R. Shelton

Dan is a native of Chicago, Illinois. After one year at Northern Illinois University, he transferred to Florida State University (pre-Bobby Bowden) where he received a B.S. in biology. After an abbreviated stint in the Peace Corps, he received an M.S. in plant pathology and Ph.D. in microbial ecology from Michigan State University. He subsequently spent two years at the University of California, Riverside, as a postdoc.

Dan joined the Agricultural Research Service, Beltsville, Maryland, in 1985. He spent the first twelve years of his career isolating and characterizing xenobiotic-degrading microorganisms, elucidating metabolic pathways, developing/implementing bioremediation technologies, assessing tillage effects on herbicide fate and efficacy, and examining the impact of cover crops on nitrogen availability and biotransformations. Three years ago he was redirected into pathogen research. His current research mission includes (i) developing functional relationships between pathogen transport and soil/hydrological parameters; (ii) determining rates and extent of pathogen transport/dispersal and survivability in pasture, crop, and vegetable production systems; and (iii) evaluating best management practices to minimize transport and dissemination of pathogens from manures to waters and fresh produce.

Dan matriculated from the George Mason University School of Law in 1993 and is a (non-practicing) member of the Virginia State Bar. He has also taught graduate courses in Environmental Microbiology and Bioremediation in the Department of Environmental Engineering, Johns Hopkins University.

Dan currently lives in Falls Church, Virginia, with his wife (Kathrine) and two collie dogs (Sheba and Hermes).

J. Thomas Sims

• •

J. Thomas (Tom) Sims is a professor of soil and environmental chemistry in the Department of Plant and Soil Sciences at the University of Delaware, Newark. Dr. Sims is also Director of the Delaware Water Resources Center and the University of Delaware Soil Testing Program. He received his B.S. degree in agronomy (1976) and his M.S. degree in soil fertility (1979) from the University of Georgia. He earned his Ph.D. degree in soil chemistry (1982) at Michigan State University. He is a fellow of the Soil Science Society of America and the American Society of Agronomy and recipient of the Outstanding Research Award from the Northeast Branch of the American Society of Agronomy and Soil Science Society of America.

Dr. Sims teaches courses in environmental soil management and advanced soil fertility (undergraduate and graduate) and conducts an active research program directed towards many of the environmental issues faced by agriculture in the rapidly urbanizing northeastern United States. His research has focused on the development of nitrogen and phosphorus management programs that maximize crop yields while minimizing the environmental impact of fertilizers and animal manures on ground and surface waters. Other research has evaluated the potential use of sludge composts, coal ash, and other industrial by-products as soil amendments. Again, the goal has been to develop environmentally sound management programs for these materials, based on their reactions in the soil and effects on plant growth and water quality. In his role as director of the University of Delaware Soil Testing Program, he has developed and evaluated soil tests for environmental purposes such as soil nitrate testing, environmental soil tests and field rating systems for phosphorus, and soil testing strategies for heavy metals in waste-amended soils.

Richard A. Smith

. . . .

Richard A. Smith grew up in Baltimore, Maryland, and attended high school at the Baltimore Polytechnic Institute. He received his B.S. and M.S. degrees in biology from the University of Richmond in 1967 and 1969. He worked for three years as a research associate in the Department of Geography and Environmental Engineering of the Johns Hopkins University and received his Ph.D. in environmental engineering from Johns Hopkins in 1975.

Dr. Smith's research career with the U.S. Geological Survey spans nearly 25 years and includes investigations into many aspects of water quality and related topics in ecology, public health, statistics, and modeling. His early years at the USGS were devoted to development of an ecological risk assessment model for the Federal Government's offshore oil leasing program. More recently, he has been a codeveloper of the SPARROW water quality model currently in wide use by the USGS.

For more than a decade he has written and spoken widely on the subject of national and regional water quality conditions. His peer-reviewed publications have appeared in numerous journals including *Science, Nature, Environment, Tropical Medicine and Hygiene, Ecological Modeling, Biogeochemistry,* and *Water Resources Research.*

Susan M. Stehman

Dr. Stehman grew up in southeastern Pennsylvania. She received her B.S. degree in biology from Penn State in 1977 and her M.S. in animal science with an emphasis in microbiology from University of Massachusetts, Amherst, in 1981. After completing her degree in veterinary medicine at the University of Pennsylvania in 1985, she spent four years as a food animal clinician, specializing in dairy medicine, on staff at the Ambulatory Clinic at the College of Veterinary Medicine at Cornell University. Dr. Stehman then spent two years in private practice in a nine-person predominately dairy practice in northern New York.

Dr. Stehman joined the Cornell Diagnostic Laboratory as a field extension veterinarian in 1991. She is involved in coordinating disease investigations and developing preventive ruminant herd health programs for veterinarians and producers in New York and surrounding states. Dr. Stehman has been actively involved in development the New York Cattle Health Assurance Program (NYSCHAP). NYSCHAP is an integrated cattle disease prevention program which utilizes a farm's team of animal health advisors to develop a farm-specific herd health and quality assurance plan to address infections of environmental, animal health, public health, and food safety concern. Along with Dr. Chris Rossiter, Dr. Stehman is actively involved with Johne's disease prevention and control and in field research focusing on diseases with public health and food safety implications, including E. coli O157:H7 and salmonellosis.

T. P. Tylutki

Tom is a native of Amsterdam, New York. He came to Cornell as an undergraduate, where he received his B.S. in 1990 and his M.S. in animal nutrition in 1994.

In September 1993, he began working as an extension associate with Danny Fox and Alice Pell in the Animal Science Department at Cornell. His work focused on the impact of the animal on nutrient management within the New York City Watershed. He also worked with a group of faculty and staff members on a comprehensive study involving two farms in Cayuga County, New York. This work represents the beginning steps in integrated nutrient management.

In November 1995, Tom became the extension dairy specialist on the Cortland, Chemung, Tompkins, Tioga, Schuyler County Dairy and Field Crops Team with Cornell Cooperative Extension. He focused on the animal's impact on nutrient management, forage quality, calf rearing, commodity trading strategy, and computers in agriculture.

In May 1999, Tom returned to the Department of Animal Science at Cornell University as a research support specialist. He is focusing on the integration of animal and crop production in order to develop integrated nutrient management plans. A portion of his time has been spent developing tutorials for dairy cattle nutrition that will be used with the Cornell Net Carbohydrate and Protein System version 4.0. Tom has been using these tutorials in a classroom setting with senior dairy students.

Tom is currently working on his Ph.D. through the employee degree program at Cornell University. He is focusing on developing and implementing quality control protocols in the feeding system for commercial dairy operations.

Harold M. van Es

Harold Mathijs van Es is an associate professor of soil and water management in the Department of Crop and Soil Sciences at Cornell University. He received a Ph.D. degree in soil physics from North Carolina State University in 1988, an M.S. degree in soil management from Iowa State University in 1984, and a Kandidaats degree in physical geography in 1981 from the University of Amsterdam, Netherlands, his native country.

Harold's current job responsibilities are in extension, research, and teaching. Research activities focus on hydrology, chemical movement, tillage, soil compaction, precision agriculture and soil statistics. Extension activities include the education of extension specialists, farmers, and other professionals on sustainable management of soil and water resources. Teaching activities include an undergraduate course in soil and water management and a graduate course in spatial and temporal statistics. Harold currently serves as chair of Division S-6 (Soil and Water Management) of the Soil Science Society of America and also chairs the Cornell Environmental Outreach Council. He served as an associate editor of the *Journal of Environmental Quality,* guest editor of the journal *Geoderma,* and a member of the editorial committee for *Methods of Soil Analysis, Part 1: Physical Properties,* Soil Science Society of America.

Richard F. Wildman

• •

Richard F. Wildman is founder and president of Agricultural Consulting Services, Inc. (ACS), an independent crop consultant company headquartered in Rochester, New York. He is a certified professional crop consultant—independent, CAFO (concentrated animal feeding operations) planner, and CCA (certified crop advisor). He holds a Bachelor of Science degree from Colorado State University and is a graduate of the Executive Program for Agricultural Producers (Texas A & M).

Mr. Wildman and his company work with many of the most successful farms in the Northeast, providing independent, unbiased counsel and state-of-the-art technical information. By developing a close working relationship with a client's management team, ACS becomes an actively involved partner in monitoring and resolving operational issues and needs.

Mr. Wildman's personal involvement with public policy issues of importance to the agricultural industry includes being an integral part of the high-profile Southview Farms case in 1993, which became a key factor in formulating new standards for concentrated animal feeding operations.

A nationally recognized expert in his field, Mr. Wildman regularly speaks at regional, state, and national conferences. He has worked with committees at Cornell University and served as a resource for state agencies such as the Department of Environmental Conservation and Department of Agriculture and Markets. He has also advised the governor's office on bills related to competitive and economic issues facing the farming community in New York State.

Mr. Wildman currently serves on the Northeast Region Certified Crop Advisor (CCA) Board, the New York State Agricultural/Environmental Steering and Management Committees, and the board of the New York Crop Research Facility. He chairs the Cooperative Extension Vegetable Advisory Committee for several New York counties and is a past officer of the National Alliance of Crop Consultants. He is also a member of the National Association of Independent Crop Consultants and the Association of Agricultural Production Executives.

Peter Wright

•••

Pete grew up on a dairy farm in central New York. He earned a B.S. and MEng. degree from the agricultural engineering program at Cornell University in 1977 and 1978. The major subject areas for both degrees were in soil and water engineering and environmental engineering.

Since joining the Cornell Cooperative Extension system in 1994, Pete has been working with producers, extension agents, agribusinesses, and others to improve the efficiency and effectiveness of animal waste handling systems, both to reduce costs and to limit the negative effects on the environment. This includes performing applied research on farms to determine the functioning of different production and treatment systems as well as producing technical information for the agricultural community.

Prior to 1994, he was an agricultural engineer for the Soil Conservation Service. While there, he analyzed, designed, and installed many different types of agricultural waste handling and treatment systems in New York and Virginia over a sixteen-year period.

Conference
Sponsors

(presented in alphabetical order)

Agricultural Consulting Services, Inc. (ACS)

....

ACS, Agricultural Consulting Services, Inc., is an independent, fee-based nutrient management and environmental engineering firm serving agriculture throughout the Northeast.

Agricultural Consulting Services, Inc.
1634 Monroe Avenue
Rochester, New York 14618

Phone, toll-free: (877) 310-1100
Phone: (716) 473-1100
Fax: (716) 473-1765
E-mail: ACS@ACSOFFICE.COM
Web site: WWW.ACSOFFICE.COM

Bion Technologies, Inc.

....

Bion Technologies, Inc., designs and operates advanced waste and wastewater treatment systems for the livestock industry, providing proven solutions to the nutrient and odor management problems associated with large dairy and hog farms.

Bion Technologies, Inc.
8899 Main St.
Williamsville, New York 14221

Phone: (800) 769-BION (2466)
Fax: (716) 633-4966
Web site: WWW.BIONTECH.COM

BioSun Systems

BioSun's patented, bacterial process eliminates odor and reduces solid waste for livestock producers. This process can save up to 50% on disposal costs, protects the environment, and helps producers and their neighbors coexist.

BioSun Systems
5775 Wayzata Blvd., Suite 700
Minneapolis, MN 55416

Phone: (612) 525-2251
Fax: (612) 417-0729

Brubaker Consulting Group
(formerly Brubaker Agronomic Consulting Service)

Brubaker Consulting is a leader in agricultural consulting in the United States. We provide full service crop consulting, environmental assessments, nutrient management services, several research farms, biosolids management, turf management, engineering and land planning services.

Brubaker Consulting Group
4340 Oregon Pike
Ephrata, PA 17522

Phone: (717) 859-3276
Fax: (717) 859-3416
E-mail: BILLR@BRUBAKERAG.COM
Web site: WWW.BRUBAKERAG.COM

Potash & Phosphate Institute (PPI)

. . . .

The Potash & Phosphate Institute (PPI) develops and promotes scientific information that is agronomically sound, economically advantageous, and environmentally responsible in advancing the worldwide use of phosphorus and potassium in crop production systems.

Potash & Phosphate Institute (PPI)
655 Engineering Drive, Suite 110
Norcross, GA 30092-2837

Phone: (770) 447-0335
Fax: (770) 448-0439
E-mail: PPI@PPI-FAR.ORG
Web site: HTTP://WWW.PPI-FAR.COM

Purina Mills, Inc.

. . . .

Purina Mills is "America's Leader in Animal Nutrition" and is committed to delivering the future of animal nutrition, today. Purina is the largest supplier of dairy nutrition in the United States and continues to expand in the Northeast.

Purina Mills, Inc.
475 St. John's Church Road
Camp Hill, PA 17011

Phone: (800) 518-6458
Web site: HTTP://DAIRY.PURINA-MILLS.COM

Suggested Readings

Suggested Readings

(Ordering information on page 503)

Agrichemical Handling

Fertilizer and Manure Application Equipment

NRAES–57 • 22 pages • This publication discusses types of fertilizer and manure nutrient values and provides guidance on equipment selection. Procedures for calibrating fertilizer and manure application equipment are reviewed. The publication includes over 30 illustrations, six tables, a plan for a fertilizer storage shed, and a glossary of terms. (1994)

Livestock and Poultry

Beef Housing and Equipment Handbook

MWPS–6 • 136 pages • Agricultural engineering recommendations for beef producers are summarized in this complete housing guide. Essential components for an efficient operation, such as building design, operation size, and equipment, are discussed. Drawings, tables, and discussions to help improve, expand, and modernize an operation are included. Topics covered include: cow-calf, cattle handling, and cattle feeding facilities; feed storage, processing, and handling; water and waterers; manure management; farmstead planning; building construction, materials, ventilation, and insulation; fences; gates; and utilities. (1986)

Sheep Housing and Equipment Handbook

MWPS–3 • 90 pages • This handbook presents valuable information for planning an efficient sheep system based on operation size, housing system choice, building needs and location, feeding methods and location, environmental controls, and manure handling. Sections include: planning sheep facilities, barns, barn and lot layouts, manure management, feed storage and handling, treating and handling facilities, equipment for raising orphan lambs, utilities, and construction materials. (1994)

Swine Housing and Equipment Handbook

MWPS--8 • 112 pages • Whether building a new facility or redesigning an old one, this is a handbook you should not be without. It is a complete guide to swine building design, operation, and equipment. A detailed environmental control section covering mechanical and natural ventilation, manure pit ventilation, cooling, and insulation is featured. Chapters cover site selection; remodeling; scheduling; handling; grain-feed centers; manure handling; utilities; and building design factors, types, and sizing. (1983)

Suggested Readings

(Ordering information on page 503)

Dairy

Calves, Heifers, and Dairy Profitability: Facilities, Nutrition, and Health

NRAES–74 • 378 pages • This proceedings is from the Calves, Heifers, and Dairy Profitability National Conference, which was held in January 1996. The publication focuses on raising high-producing replacement heifers at a minimum cost. Included in the proceedings are 35 papers divided into nine categories: heifer growth and development, calf and heifer housing, labor, contract raising, pasture management, calf nutrition, heifer nutrition, reproduction, and calf and heifer health. (1996)

Dairy Feeding Systems: Management, Components, and Nutrients

NRAES–116 • 402 pages • Dairy feeding systems affect milk production, labor requirements, capital investments, cropping systems, nutrient levels, and overall profitability. This is the proceedings from the Dairy Feeding Systems: Management, Components, and Nutrients Conference, which was held December 1998 in Camp Hill, Pennsylvania. Included are 31 papers divided into nine categories: feeding systems, feed storage facilities, feed inventory management, feed delivery management, feed consumption area, monitoring and managing feed costs by monitoring and managing intake, feed quality control, feeding system economics, and herd nutrition and cropping management. The book will be a valuable resource for producers and farm managers; producer advisors and consultants; extension and university educators; nutritionists; crop specialists; feed, seed, and equipment sales representatives; nutrient managers and agronomists; veterinarians; agricultural engineers; facility designers; policy makers; lenders; and the agricultural media. (1998)

Dairy Freestall Housing and Equipment

MWPS–7 • 136 pages • This, the sixth edition of this book from the MidWest Plan Service, presents freestall dairy facility designs and equipment planning. Discussion is based on total herd management by production groups, management by age groups, and replacement animal housing. Some of the new, expanded, and revised discussions in this book include: designing and maintaining the freestall area, dry cow housing, milking parlor environment, foot baths, safety passes, personnel passes, designing raised ridge caps, summer ventilation management, designing commodity storages, and designing silos. Chapters include: total dairy facility, replacement housing, milking herd facilities, milking centers, special handling and treatment areas, building environment, manure and wastewater management, feeding facilities, and utilities. With the new book layout, information is easier to find. (1997, 6th edition)

Suggested Readings

(Ordering information on page 503)

• •

Dairy Housing and Equipment Systems: Managing and Planning for Profitability

NRAES–129 • 456 pages • This is the proceedings from the Dairy Housing and Equipment Systems: Managing and Planning for Profitability Conference held February 1–3, 2000 in Camp Hill, Pennsylvania. The proceedings presents and documents guidelines for managing existing housing systems and planning new systems to improve profitability, reduce labor requirements, and improve cow comfort. Included are 36 papers divided into eight categories: cow comfort, decisions, and management; planning new facilities; system management; environmental control for cow comfort; freestall design and management; facilities management and health; designing and managing the feed area; and "special" cow needs and management. The educational program is designed for a diverse group of agricultural and industry professionals, including dairy farm owners and managers, producer advisors and consultants, builders and facility designers, extension and university educators and researchers, veterinarians, lenders, state and federal regulatory staff, equipment suppliers, agricultural engineers, dairy scientists, and agricultural economists. (2000)

Dairy Reference Manual

NRAES–63 • 293 pages • Faculty and staff of The Pennsylvania State University have put together the third edition of the *Dairy Reference Manual*, a compendium of information on all facets of dairying—from youngstock to nutrition to housing. This wire-bound manual will be invaluable to extension educators, farm planners, consultants, engineers, veterinarians, manufacturers, and salespeople. Topics covered include farm management, dairy housing, handling and behavior, dairy nutrition, reproduction, milking equipment, and more. Much of the information is included in the manual's 240 tables, 88 illustrations, and three appendixes. (1995)

Environmental Factors to Consider When Expanding Dairies

NRAES–95 • 44 pages • While a multitude of resources exist to help producers project net farm income from a proposed expansion, a concise resource that examines potential impacts on the environment is difficult to find…until now. This publication presents environmental factors that producers and their advisors should examine as part of expansion planning. Chapters discuss land and water considerations, nutrient management, odors, common concerns associated with expansion, concentrated animal feeding operations (CAFOs), and benefits of whole-farm planning. A producer needs to consider much more than the information in this publication when planning an expansion, but expanding without careful consideration of the potential environmental effects can be a business disaster. (1999)

Guideline for Milking Center Wastewater

NRAES–115 • 34 pages • This publication helps producers and their advisors plan and assess milking center wastewater reduction and treatment systems. Topics covered include wastewater characteristics and estimating the amount of wastewater produced; source control of waste-

Suggested Readings

(Ordering information on page 503)

• •

water; the milking center drainage system, including codes and regulations, components, and drainage systems; and treatment alternatives, including liquid manure system, short-term storage and land application with manure spreader, settling tanks, grass filter, aerobic lagoon, organic filter bed, septic system, constructed wetlands, stone-filled treatment trench, spray irrigation, lime flocculator treatment, and aerated septic system. Safety and health concerns are also summarized. Published in cooperation with the Dairy Practices Council. Includes 13 illustrations and 8 tables. (1998)

Guideline for Planning Dairy Freestall Barns

NRAES–76 • 52 pages • This publication covers the many aspects of planning a dairy freestall barn, from the various components of a freestall housing system to the design and construction of the barn and stalls. Topics covered include construction materials, lighting, wiring, management and maintenance of the stalls, ventilation for both insulated and uninsulated barns, manure and liquid waste handling, and regulatory considerations. Thirty-seven illustrations are included in the guideline along with several freestall housing plans. (1995)

Planning Dairy Stall Barns

NRAES–37 • 22 pages • Dairy stall barn construction is outlined in this guide. Site selection, construction details, barn arrangements, ventilation systems, and electrical service are described. Thirteen figures and three tables supplement the text. A detailed four-page construction plan for a single-story sloping tie stall dairy barn is included as well. (1988)

Manure and Waste Management

Animal Agriculture and the Environment: Nutrients, Pathogens, and Community Relations

NRAES–96 • 386 pages • This proceedings from the Animal Agriculture and the Environment Conference, held in December 1996 in Rochester, New York, includes 33 papers that discuss the following topics: the fate of pathogens and nutrients, protecting the environment, land application, nutrient management plans, odor management, feeding management to reduce nutrients in manure, considerations in public policy, and cost to the farmer. The proceedings will be of interest to dairy, poultry, and livestock producers and their advisors; community officials and their consultants; regulatory agencies; cooperative extension and university educators; consultants; rural landowners; soil and water conservation district staff; federal government staff; and watershed managers. (1996)

Suggested Readings

(Ordering information on page 503)

. . . .

Concrete Manure Storages Handbook

MWPS–36 • 72 pages • Now the science and art of designing concrete storages are combined in one handbook. This reference book provides design and construction procedures for rectangular and circular storages based on current codes, standards, specifications, and engineering practices. The handbook is intended primarily for engineers, designers, and builders who are familiar with material and functional design requirements. Natural Resources Conservation Service personnel will find this book a valuable resource. It can also be useful to owners and users of the facilities. (1994)

Dairy Manure Management

NRAES–31 • 285 pages • Thirty papers from the 1989 Dairy Manure Management International Symposium are included in this proceedings. Agricultural engineers, animal scientists, agronomists, soil conservationists, industry representatives, and extension specialists contributed papers about dairy manure and the environment, manure utilization and processing, and handling and storage. The economics of manure management and its effects on the environment are addressed as well. A two-part paper discusses production and utilization of biogas and anaerobic digestion on dairy farms. Results from experimental farms using various methods of manure processing are also presented. (1989)

Earthen Manure Storage Design Considerations

NRAES–109 • 100 pages • Earthen manure storages are becoming more common for economic, environmental, and management reasons, but there is a lack of information about safe, environmentally sound, practical designs. This book was written to meet the needs of producers, engineers, and design professionals who are seeking information about designing, constructing, and managing earthen storages. It covers environmental policies (both existing and pending legislation); design standards and planning documents (such as nutrient management and waste management plans); manure characteristics; storage planning (determining size and location, loading and unloading methods, on-site soils investigations, and regulations); storage design (stability and drainage issues, types of liners, and safety); construction (quality assurance, earthwork, topsoil placement, seeding, and documentation); management (maintaining the structure, clearing drains, and manure management); and liability. A lengthy appendix provides guidelines and calculations for soil liners; other appendixes provide pump information, cost estimate information, and addresses for helpful organizations. Includes 26 illustrations and 14 tables. (1999)

Suggested Readings

(Ordering information on page 503)

• •

Evaluation of Anaerobic Digestion Options for Groups of Dairy Farms in Upstate New York

ABEN 97 • 180 pages • Anaerobic digestion can reduce odor, nutrient runoff, and emissions of greenhouse gases (methane and carbon dioxide). This publication summarizes the results of a one-year study of anaerobic digestion options in York, New York. The goal of the study was to determine the technical and economic feasibility of converting dairy wastes to useful products in a centralized facility serving the York community — an area including approximately 100 dairy farms that maintain more than 30,000 animals within a 20-mile radius. This publication includes an 11-page executive summary and seven chapters on the following topics: dairies, water pollution, and anaerobic digestion; the dairy manure resource and energy; dairy waste management survey results; transportation of manure to centralized digesters; anaerobic digester analysis; economic feasibility; and discussion and conclusions. Included are 19 tables, 47 figures, and four appendixes. (1998)

Guideline for Dairy Manure Management from Barn to Storage

NRAES–108 • 36 pages • When a producer is considering a new or improved dairy manure management system, he or she must predict the costs, risks, savings, and operating changes that may occur. This publication concisely reviews information essential to the planning process for dairy farmers and their advisors. Published in cooperation with the Dairy Practices Council, this guideline covers the following topics: planning the development or improvement of a manure handling system, getting technical information and assistance, and meeting regulations; manure characteristics and production; alternatives for manure management; options for transferring manure from barn to storage; and manure storage types and storage management. Fourteen illustrations and 14 tables are included. (1998)

Liquid Manure Application Systems Design Manual

NRAES–89 • 168 pages • This is the most up-to-date, complete book available about designing liquid manure application systems. The abundantly illustrated manual covers the following topics: characteristics and testing of liquid manure; evaluating sites (Is your site suitable for application? What regulations exist? How do soil type, land slope, tillage practice, and crop cover limit the system?); liquid manure from the barn to the field (What is the best way to move manure? What types of pumps can be used? Where will manure be stored?); field application (Which methods of application are feasible — traveling gun, center pivot, drag hose, tanker, boom sprayer? Should you surface apply, spray, inject, or incorporate?); management (How do you avoid runoff and leaching? What volume and rate of application should be used? What can be done to reduce odors?); and applying the design procedure. Also includes 69 illustrations, 20 tables, work sheets, many example problems, and a list of manufacturers. (1998)

Suggested Readings

(Ordering information on page 503)

• •

Liquid Manure Application Systems: Design, Management, and Environmental Assessment

NRAES–79 • 220 pages • This is the proceedings from the Liquid Manure Application Systems Conference that was held in December 1994. It includes 26 papers and is divided into five categories: livestock manure systems for the 21st century, design of liquid manure systems, planning environmentally compatible systems, custom application, and managing for economic and environmental sustainability. (1994)

Livestock Waste Facilities Handbook

MWPS–18 • 112 pages • Recommendations, federal regulations, and design procedures for most manure handling and management alternatives for livestock are discussed in this handbook, including scrape systems, gravity drain gutters, gravity flow channels, infiltration areas, and waste transfer to storage. (1993)

Nutrient Management: Crop Production and Water Quality

NRAES–101 • 44 pages • This full-color publication is divided into two sections: "Basic Principles" and "Field Management." "Basic Principles" focuses on nutrient pathways and their behavior. "Field Management" centers on management guidelines that promote efficient distribution of nutrients to reduce fertilizer costs and the potential for loss. Two workbooks supplement this book; they are sold separately. (1997)

Poultry Waste Management Handbook

NRAES–132 • 72 pages • Waste management has been a concern in poultry operations for many years. Problems with proper storage, handling, management, and utilization of byproducts of production have come to the forefront in planning, establishing, and operating poultry farms. In addition, growers have become sensitive to the potential for nuisance litigation should their farms generate odors, insects and vermin, or runoff that offends neighbors. This publication covers all aspects of solid, semi-solid, and liquid poultry waste management, including: manure production and characteristics, environmental regulations and hazards, poultry housing design and waste management, manure storage systems, waste treatment (including composting, anaerobic/facultative lagoons, anaerobic digestion, and incineration), nutrient management, application equipment, dead bird management, and alternative uses for manure. Forty-two illustrations and 14 tables are included. (1999)

Suggested Readings

(Ordering information on page 503)

• •

Composting

Composting for Municipalities: Planning and Design Considerations

NRAES–94 • 126 pages • How can municipal composting benefit a community? How much of the municipal waste stream can be composted? How much does it cost to start a composting facility? Written for municipal planners, policy makers and regulators, facility operators, and consultants and designers, this book will explain everything from planning and siting a facility to making compost and marketing the finished product. Seven chapters are included. The chapter on planning discusses assembling a planning team, conducting a market survey, and identifying costs. The feedstocks chapter centers on raw materials most common in municipal operations and reviews current sorting and separation technology. The final chapter, "Planning for Long-Term Success," explains management strategies that will help ensure a useful and lasting facility that has a good relationship with the community. Appendixes include a sample market survey, sample contract documents, characteristics of common raw materials, and a compost pad area calculation. Also included are 41 illustrations, 15 tables, sample calculations, and a glossary. (1998)

Field Guide to On-Farm Composting

NRAES–114 • 128 pages • This book was developed to assist in day-to-day compost system management. It is spiral bound, is printed on heavy paper, and has a laminated cover for durability. Chapter tabs make finding information a snap. Topics discussed in the book include: operations and equipment; raw materials and recipe making; process control and evaluation; site considerations, environmental management, and safety; composting livestock and poultry mortalities; and compost utilization on the farm. Highlights of the guide include an equipment identification table, diagrams showing windrow formation and shapes, examples and equations for recipe making and compost use estimation, a troubleshooting guide, and 24 full-color photos. This book is intended as a companion to the highly successful *On-Farm Composting Handbook*, NRAES–54, which is described below. (1999)

On-Farm Composting Handbook

NRAES–54 • 186 pages • A perennial favorite among NRAES customers (we've sold 20,000 copies since 1992), the *On-Farm Composting Handbook* contains everything you ever wanted to know about on-farm composting — why to compost (the benefits and drawbacks), what to compost (raw materials), how to compost (the methods), and what to do if something goes wrong (management). The ten meticulously organized chapters also discuss site and environmental considerations, using compost, and marketing compost. Highlighting the text are 55 figures, 32 tables, and sample calculations for determining a recipe and sizing a compost pad. This book is so informative and comprehensive, it is used as a college textbook. (1992)

Suggested Readings, Ordering Information

Sludge Composting and Utilization: A Design and Operating Manual

NJ–1 • 315 pages • Published in 1982, this technical, research-based publication focuses on primary sludge. The book examines sludge composting using as a model a facility that serves the city of Camden, New Jersey. It is one of the first books to provide a comprehensive treatment of municipal sludge composting and a must for any professional who desires a complete library on composting. The manual focuses specifically on a static pile, forced aeration composting system, but much of the information is useful for other methods of composting as well. Twelve chapters cover all aspects of sludge composting — from the basics (What is composting? What is sludge?) to the finer points (How does one ensure a sufficient airflow rate in a sludge compost pile? In what temperature range does decomposition occur most rapidly?). Four chapters focus on the crucial parameter of airflow. Highlighting the text are 74 figures and 65 tables. (1982)

Ordering Information

Publications listed on pages 495–503 can be ordered from NRAES. Before ordering, contact NRAES for current prices and for exact shipping and handling charges, or ask for a free copy of our publications catalog.

NRAES (Natural Resource, Agriculture, and Engineering Service)
Cooperative Extension
152 Riley-Robb Hall
Ithaca, New York 14853-5701

Phone: (607) 255-7654
Fax: (607) 254-8770
E-mail: NRAES@CORNELL.EDU
Web site: WWW.NRAES.ORG

See the inside back cover for more information about NRAES, including a list of NRAES member universities.

Conference Notes